Rat Hepatic Neoplasia

The MIT Press
Cambridge, Massachusetts, and
London, England

[Workshop on Rat Liver Neoplasia, Woods Hole, Mass., 1974.

Rat Hepatic Neoplasia

edited by

Paul M. Newberne

W. H. Butler

This book was set in IBM Century by Eastern Composition, Inc., printed and bound by Halliday Lithograph Corporation, in the United States of America

Library of Congress Cataloging in Publication Data
Main entry under title:

Rat hepatic neoplasia.

Includes index.
1. Liver—Cancer—Congresses. 2. Carcinogenesis—Congresses. 3. Pathology, Cellular—Congresses. 4. Rats—Diseases—Congresses. 5. Pathology, Experimental—Congresses. I. Newberne, Paul Medford, 1920- II. Butler, W. H.
RC280.L5R37 599'.3233 78-6220
ISBN 0-262-14029-2

Contents

Participants

R. W. Baldwin
Cancer Research Campaign Laboratories, University of Nottingham, Nottingham, England

P. Bannasch
Abteilung für Cytopathologie, Institut für Experimentelle Pathologie am Deutschen Krebsforschungszentrum, Heidelburg, Germany

Frederick F. Becker
Department of Anatomic and Experimental Pathology, M. D. Anderson Hospital, Tumor Institute, 6723 Bertner, Houston, Texas 77030

W. H. Butler
Imperial Chemical Industries, Ltd., Macclesfield, Cheshire, England

James R. Gillette
Chemical Pharmacology Laboratory, National Heart and Lung Institute, Bethesda, Maryland 20014

Glenys Jones
Shell Research Laboratories, Sittingbourne, Kent, England

D. L. Knook
Institute for Experimental Gerontology TNO, Rijswijk, (ZH), The Netherlands

H. L. Leffert
Cell Biology Laboratory, The Salk Institute for Biological Studies, P. O. Box 1809, San Diego, California 92112

P. M. Newberne
Department of Nutrition and Food Science, Massachusetts Institute of Technology, Cambridge, Massachusetts 02139

Adrianne E. Rogers
Department of Nutrition and Food Science, Massachusetts Institute of Technology, Cambridge, Massachusetts 02139

E. K. Weisburger
Carcinogen Metabolism and Toxicology Branch, National Cancer Institute, Bethesda, Maryland 20014

Observers

Gil Cloyd
Proctor and Gamble, Cincinnati, Ohio

Paul Estes
Pfizer, Inc., New York, New York

William Field
Schering Corp., Lafayette, New Jersey

L. S. Goyings
Pathology and Toxicology, Agricultural Division, The Upjohn Co., Kalamazoo, Michigan

Harold Grice
Toxicology Division, Food Research Laboratory, Tunney's Pasture, Ottawa, Ontario, Canada

Paul Harris
Department of Pathology, Eli Lilly Co., Greenfield, Indiana

W. A. Kelly
Mead Johnson Research Center, Evansville, Indiana

C. H. Keysser
The Squibb Institute for Medical Research, Princeton, New Jersey

V. A. Kirkland
Shell Chemical Co., San Ramon, California

Gert Laqueur
Laboratory of Experimental Pathology, National Institute of Arthritis and Metabolic Diseases, National Institutes of Health, Bethesda, Maryland

Göran Magnusson
Astra Pharmaceuticals AB, Toxicology Laboratories, Södertälje, Sweden

Frank McConnell
Ortho Research Foundation, Raritan, New Jersey

Leonard Merkow
Experimental Pathology and Electron Microscopy, Singer Memorial Research Institute, Allegheny General Hospital, Pittsburgh, Pennsylvania

J. W. Newberne
Drug Safety and Metabolism, Merrell-National Laboratories, Cincinnati, Ohio

F. W. Sigler
Norwich Pharmacal Co., Norwich, New York

Robert Squires
Tumor Pathology Laboratory, National Cancer Institute, National Institutes of Health, Bethesda, Maryland

Charles Tate
Merck, Sharp and Dohme Research Laboratories, West Point, Pennsylvania

W. Yoon
Merck, Sharp and Dohme Research Laboratories, West Point, Pennsylvania

Preface

The increasing use of the laboratory rat in basic biological research and in safety evaluations of drugs and food additives has taxed our capacity to interpret the lesions on which far-reaching decisions are based. So it seemed desirable to bring together the work of a group of scientists with divergent opinions and experience and to attempt to establish the state of the art relative to understanding liver lesions in the rat, as they occur spontaneously or as a result of exposure to chemicals. As a result, a small group representing the international community of pathologists convened for a Workshop on Rat Liver Neoplasia at Woods Hole, Massachusetts, in October 1974.

Because there was a considerable lag between the time the workshop was held and the date of this publication, most chapters have been updated. The basic nature of the subject matter makes the data no less valuable now than when the workshop was held, and the updated chapters place the information in this volume in the forefront of our understanding of the fundamentals of rat liver neoplasia.

The participants and core of subjects were the choice of the editors, but opinions expressed were those of individuals, not of any organization. Each of the organizations providing financial support to the workshop was invited to send one observer.

Paul M. Newberne
Cambridge, Massachusetts

Acknowledgments

The editors would like to thank the following organizations for the financial support that made possible the workshop from which this volume derives: William S. Merrell Co.; Shell Chemical Co.; Norwich Pharmical Co.; G. D. Searle Co.; Upjohn Co.; Hoffmann-La Roche, Inc.; Pfizer, Inc.; Eli Lilly and Co.; Schering Corp.; E. R. Squibb and Sons, Inc.; Astra Pharmaceutical Products, Inc.; Ortho Research Foundation; Merck Institute for Therapeutic Research; Vick Chemical Co.; Ayerst Co.; Meade Johnson Co.; Proctor and Gamble Co.; McNeil Laboratories; Ciba-Geigy Corp.; Hoechst Pharmaceuticals, Inc.; Wyeth Laboratories, Inc.; Lakeside Laboratories, Inc.; Smith, Kline and French Co.; and A. H. Robins.

We also wish to thank the participants who prepared the working papers in this volume and the discussants who provided fresh perspectives on various issues.

We are indebted to Louise Kittredge and Rosemary Burklin for editorial assistance and aid in preparation of the manuscript.

Rat Hepatic Neoplasia

1 Introduction

P. M. Newberne
Department of Nutrition and Food Science
Massachusetts Institute of Technology
Cambridge, Massachusetts 02139

W. H. Butler
Imperial Chemical Industries, Ltd.
Macclesfield, Cheshire
England

Much of the experimental work in carcinogenesis has been performed with rat hepatocarcinoma as the model. The histologic development of tumors, the influence of carcinogen dosage and environmental factors on tumor induction, the many aspects of enzyme, nucleic acid, or other biochemistry, and the immunologically active factors produced by tumors or by the host in responding to them, have all been studied in hepatocarcinomas of the rat, probably more extensively than in any other single tumor.

Despite the wide experience of many investigators with rat hepatocarcinoma, there is no general agreement on the histologic criteria for evaluating the biologic significance of the neoplasm or for the lesions that precede it, nor is there agreement as to the relationship between "premalignant" and malignant lesions. There was extensive discussion of this question at this workshop and also at a subsequent meeting at the National Cancer Institute (*Cancer Res.* 35:3214, 1975). Participants in the NCI discussion did not come to agreement on the definition of premalignant changes appearing prior to a well-defined nodule. They decided, however, to designate such nodules as "neoplastic," as opposed to hepatocarcinoma or "hyperplastic" nodules. This decision on terminology is arbitrary and perhaps was unwise, since the biological behavior of such nodules is unknown. In that workshop several histologic changes thought by some participants to represent "premalignant," though not necessarily irreversible, changes were illustrated and discussed. The need for better methods to define and follow premalignant cytologic changes was evident.

The importance of such definition lies, of course, in its applicability to prediction of the carcinogenicity or noncarcinogenicity of compounds evaluated early in the testing period or given in doses perhaps inadequate to induce tumors. However, the use of histologic descriptions to describe lesions that are neither definitely malignant nor proved to be irrevocably premalignant introduces the problem of false positive results and is not adequately supported by data at present. During the course of the workshop it became clear that there is a critical need to understand more about the histology of the liver from embryo to senescence. The changes in histochemistry and metabolism that affect histology can be then associated with the morphology of the liver in many different exposure situations.

In this workshop the following questions were considered:

1. What are the normal histologic and biochemical parameters of growth and aging in the rat liver?
2. What factors control cell division in the rat liver?
3. What is the range of the normal, premalignant, and malignant appearance of cells found in rat liver and, in particular, of the hepatocyte?
4. What are the distinctions between premalignant and degenerative transformations of the hepatocyte?
5. How strong is the correlation between premalignant changes in the liver and the ultimate development of carcinoma?
6. What influence has diet on chemical carcinogenesis in rat liver?
7. What are the enzymatic or chemical correlates for chemical carcinogens, their metabolites, and the induction of hepatocarcinoma?
8. What are the immunological characteristics of liver tumors and what is their relationship to chemical carcinogenesis?

The speakers and discussants covered these and related topics.

In brief, the major points can be summarized as follows: Development of the rat liver through gestation and the first six weeks of life is marked by changing relationships in number and volume of the different cell types that make up the liver. Rate of development, particularly of polyploidy, is strongly dependent on the strain of rat, as are life span and aging changes, including growth and "spontaneous tumors." Metabolic and organelle changes characteristic of aging can be demonstrated in isolated cell preparations. The results do not always agree with results obtained using cell fractions, probably because of mixing of components of the different cells that make up the liver and disruption of membrane structures that may be functionally important.

Detailed studies of factors contributing to control of growth and cell division of hepatocytes in vitro have demonstrated a major role for several hormones, the amino acid arginine, and serum factors not yet defined. There is significant inhibition of cell division by very-low-density lipoproteins (VLDL); the drop in serum content of VLDL after partial hepatectomy may be a signal for initiation of regeneration. A controlling or regulatory role for alpha-fetoprotein (α_1 F) is not indicated by available data, but secretion of α_1 F follows treatment with hepatic carcinogens in doses far below those that induce either cell division or tumors. It may be an indicator of interaction between surgical or chemical stimuli and the cells' genetic material.

In several sessions devoted to the histology and significance of changes induced in the liver by chemical carcinogens, the diversity of changes and the evidence relating them to tumor development were discussed. The sequential development of hyperplastic hepatocytes with abnormal histologic and histochemical appearance, of even more deranged nodules, and ultimately of malignant tumors was amply demonstrated, but the evolution of one stage to the next and the establishment of irreversibility were issues felt not resolvable at present. Furthermore, the variability of changes observed even between cells within one abnormal focus in a carcinogen-treated liver or between apparently similar foci induced by one or different carcinogens does not permit definition of structural, enzymatic, chromosomal, or other patterns characteristic of malignant or premalignant change. Examples of tumor induction preceded by no detectable premalignant alterations were described.

Requirement for increased cell division in initiation of hepatic carcinogenesis and enhancement of chemical carcinogenesis by partial hepatectomy were felt to be variable and perhaps dependent on the carcinogen and the dose used. The importance of using doses that induce neither hepatocyte necrosis nor cirrhosis was discussed. Distinction between degenerative and premalignant changes may be particularly difficult, and this problem elicited discussion of the significance of foci of glycogen-containing cells, vacuolated cells containing material other than glycogen, and cells arranged in acinar patterns which may be either true glandular formation or the residual cells from cords undergoing central necrosis. Induction of hepatocellular glycogenosis is characteristic of several carcinogens and has been related to both reduction of glycose-6-phosphatase and proliferation of smooth endoplasmic reticulum. Its significance was debated because it occurs in areas with decreased rather than increased cell division and can be demonstrated in cells which appear to be degenerative. Progression to hepatocarcinoma is, in any case, marked by glycogen depletion.

Morphologic classification of the tumors themselves may depend on evaluation at the electron as well as the light microscopic level to distinguish between the hepatocellular, biliary, or vascular elements of a poorly differentiated tumor. As with premalignant lesions, degenerative changes may obscure the true origin of a tumor (as is the case when necrosis of a trabecular carcinoma gives it a glandular appearance). The majority of tumors induced in the rat liver are hepatocellular and have more uniform characteristics at the electron microscope level than at the light microscopic level.

The influence of diet as well as of strain and sex of rat on tumor induction was described and discussed by several participants. Commercial diets often protect against chemical carcinogenesis when compared to semisynthetic diets. Diets deficient in lipotropic agents, protein, or certain vitamins tend to enhance chemical carcinogenesis or alter tumor distribution. Clearly, diet must be evaluated in both scientific studies and commercial tests of chemicals for carcinogenic activity.

Dietary and other environmental influences may affect carcinogen metabolism and binding to sensitive sites or trapping agents such as gluthathione in the target cell. Diet or other treatment may alter the quantity of trapping agent, thereby altering the amount of carcinogen available for binding at other sites. In vitro systems that demonstrate metabolism and covalent binding may be useful in predicting pharmacologic or toxic activity of compounds, but their usefulness in predicting carcinogenic activity has not yet been adequately demonstrated. Cell lines derived from hepatocytes transformed in vitro were described, and the possible significance of studies of their metabolism of chemicals was discussed.

Extensive studies of the identification and isolation of tumor or oncofetal antigens and of antibody response to them in carcinogen-treated rats were presented. Antigen-antibody complexes in tumor-bearing rats and induction of cytotoxic immunity to tumors were demonstrated and related to cell surface antigens.

It is evident from this summary and from the following papers and discussions that many questions arose which will require further study, questions not only common to cell studies of cancer but also specifically related to hepatocellular carcinoma in the rat and ranging from criteria for diagnosis to basic concepts of genetic control of metabolism and immunology. This volume presents certain aspects of the state of knowledge of hepatocellular carcinoma in the rat; together with the volume on mouse liver neoplasia, it may serve as a basis for research aimed at establishing uniformity in safety evaluation and carcinogenesis studies using rodents.

2 Embryology and Aging of the Rat Liver

D. L. Knook
C. F. Hollander
Institute for Experimental Gerontology TNO
Rijswijk (ZH)
The Netherlands

2.1 Introduction

There are many reports in the literature concerning aging phenomena of rat liver cells and their subcellular organelles. Fewer studies have been devoted to the embryonic development of this organ. Furthermore, most biochemical and microscopic studies on the adult rat liver deal exclusively with the function and structure of the parenchymal cells. Consequently, there are two underdeveloped areas in our knowledge of the developing rat liver: first, the embryonic growth of this organ and, second, the function and activities of cell types other than parenchymal cells. The purpose of this paper is to analyze and describe the roles of the different cell types in the rat liver during development and aging, although emphasis will still be on the parenchymal cells. Since little data is available on the prenatal stage, the main emphasis will be on the structural and biochemical changes occurring in the liver during the postnatal and adult phases of the rat's life.

2.2 Embryonic Development

The embryonic development of the rat can be divided into three arbitrary periods each of about seven days (Nicholas, 1949). The first of these periods—the first seven days of pregnancy—includes fertilization, segmentation, migration, and implantation. The second period of seven days involves the formation of embryonic membranes, the differentiation and organization of the germ layers, and the formation of the placenta. The final embryonic differentiation takes place during the third period, which is completed at parturition.

The formation of the liver begins in the second period; the earliest recognizable hepatic structure can be found sometime during the tenth day of development (Henneberg, 1937). The hepatic anlage arises as an endodermal diverticulum from the foregut. The primary hepatic bud possesses a thick epithelial layer, which shows a ventral growth. On day 11, epithelial cords are present whose strands come into contact with veins in the transverse septum to create a spongework of parenchyma in which spaces are occupied by a network of sinuses. The two component tissues of the septum, the endoderm and mesoderm, interact; the endoderm differentiates into hepatic cords of parenchymal cells, and the mesoderm gives rise to the connective tissue and the vascular endothelium (Croisille and Le Douarin, 1965). During days 12 and 13 the parenchymal spongework is further developed, resulting in a rapid expansion of the total liver. During this period the liver develops to become the main site of blood cell production (Metcalf and Moore, 1971). Hemopoiesis starts on the thirteenth gestational day (Dadoune, 1963), and numerous hemopoietic cells are found by day 15 (fig. 2.1) (Franke and Goetze, 1963; Greengard et al., 1972).

2.3 Cellular Composition

The rat liver is often considered to be a homogeneous organ, and many studies of its function and biochemical properties have not taken into consideration the distribution and contribution of the various discrete cell populations. Liver tissue is not homogeneous. During fetal and postnatal life the liver contains the following cell types:

1. hepatocytes or parenchymal cells
2. sinusoidal lining cells (mainly endothelial, Kupffer, and fat-storing cells)
3. hemopoietic cells
4. bile duct cells
5. connective tissue cells
6. blood vessel wall cells

Bile duct, connective tissue, and blood vessel wall cells represent minor components in the cellular population and account for less than 7 percent of the total number of liver cells in the young adult rat (Daoust and Cantero, 1959; Fabrikant, 1968).

The absolute and relative numbers of parenchymal, sinusoidal lining, and hemopoietic cells change during development and aging. The significance of these changes in the numerical proportions of the cell types on the interpretation of functional and biochemical data is obvious.

2.4 The Distribution of Cell Types as a Function of Prenatal and Postnatal Age

2.4.1 Parenchymal Cells

Hepatocytes, or parenchymal cells, represent the major cell type in the liver of the young, three-month-old rat. They constitute about 85 percent of the total liver volume (Striebich et al., 1953; Weibel et al., 1969; Greengard et al., 1972) but only about 60–65 percent of the total number of cells (Abercrombie and Harkness, 1951; Stowell, 1952;

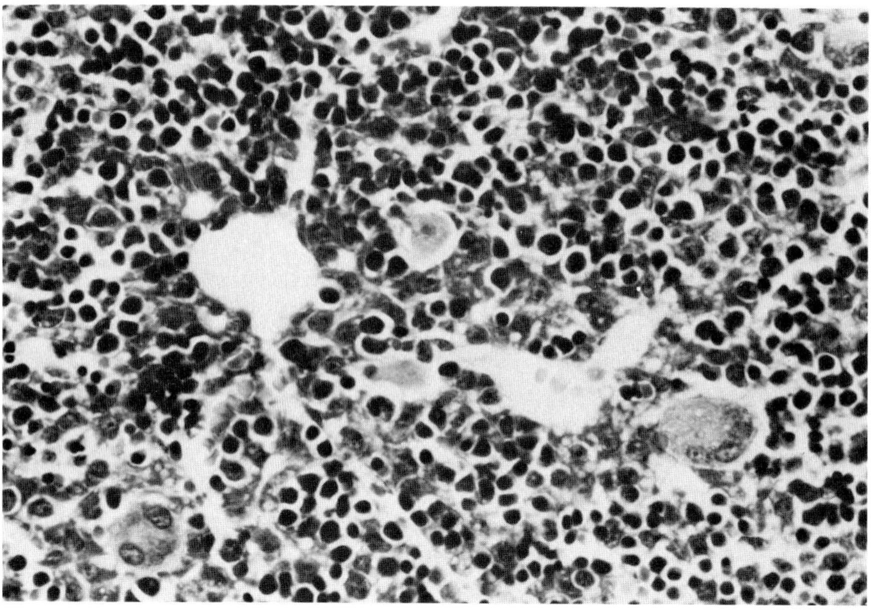

2.1 Liver of a WAG/Rij rat, day 15 post coitum, showing active hemopoiesis (HPS, × 370).

Sibatani and Fukuda, 1953; Grant and Rees, 1957; Daoust and Cantero, 1959; Weibel et al., 1969; Greengard et al., 1972). The variations in the values presented by various authors are due to a number of factors, such as the age of the animals investigated (in most cases only their weights are mentioned), differences in diet and strain of animal, and the technique employed (Daoust and Cantero, 1959). However, all of these values were obtained in morphometric studies in which the number of parenchymal *nuclei* is counted but the term parenchymal *cells* is used. Since we found that at the age of three months about 35 percent of all parenchymal cells are binucleate (Van Bezooijen et al., 1974), the terms parenchymal nuclei and parenchymal cells are far from equivalent. This observation leads to the conclusion that, in a count corrected for the presence of binucleate cells, little more than half of the cells in the young adult liver are parenchymal cells.

The relative number of parenchymal cells changes during fetal and adult life. Exact data are scarce, because only a few morphometric studies have been devoted to the developing rat liver. In the most extensive study, covering the fifteenth day of gestation to the hundredth postnatal day, a number of morphological parameters were determined (Greengard et al., 1972). Between the fifteenth and eighteenth days of gestation, parenchymal cells account for half of the liver volume, but their number amounts to about one-third of the total number of cells. From gestational day 18 to day 21 the number of parenchymal cells rises sharply so that 85 percent of the liver mass consists of parenchymal cells at term. Due to a parallel increase in the volume of the parenchymal cells, they still account for only 43 percent of all cells by number. The volume of liver occupied by parenchymal cells stabilizes during postnatal life, so that on the hundredth postnatal day these cells make up 85–90 percent of the total liver volume. At this time, the parenchymal cells, according to a count uncorrected for binucleate cells, represent 63 percent of the total number of cells (Greengard et al., 1972). This value declines to 56 percent at the end of the first year of life (Daoust and Cantero, 1959).

2.4.2 Sinusoidal Lining Cells

The nonparenchymal cells associated with the sinusoids have always been described as Kupffer cells in the light microscopic literature. These cells have been called macrophages, histocytes, and phagocytes with or without adjectives describing their position (hepatic, fixed, sinusoidal, littoral), shape (stellate, sternzellen), or function (phagocytic). However, besides the Kupffer cells (fig. 2.2), a number of other cell types, such as endothelial cells (fig. 2.3), fat-storing cells, pit cells, and monocytes, are present in the hepatic sinusoidal area (Wisse, 1974). The endothelial and Kupffer cells are often considered to be different functional expression of one cell type, but it has been recently demonstrated that they are different types of cells (Wisse, 1972 and 1974). In

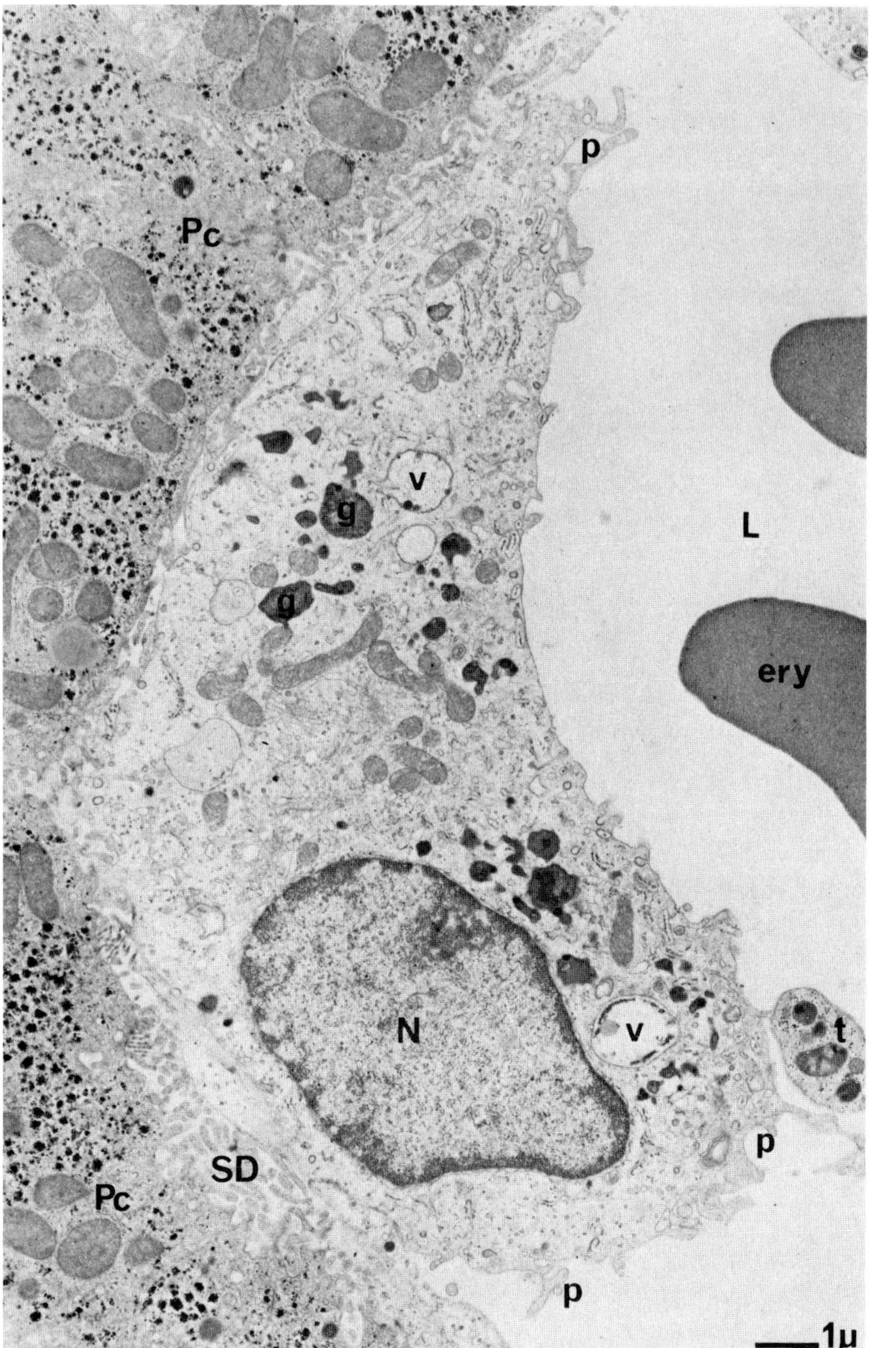

2.2 Kupffer cell of rat liver after perfusion fixation. As well as nucleus (N), a number of granules (G) or vacuoles (v) of varying shape, density, and diameter can be discerned. The surface of the cell bears pseudopodlike processes (p). L = sinusoidal lumen; ery = red blood cells; t = thrombocyte, SD = space of Disse; Pc = parenchymal cell. Bar = 1 μ.

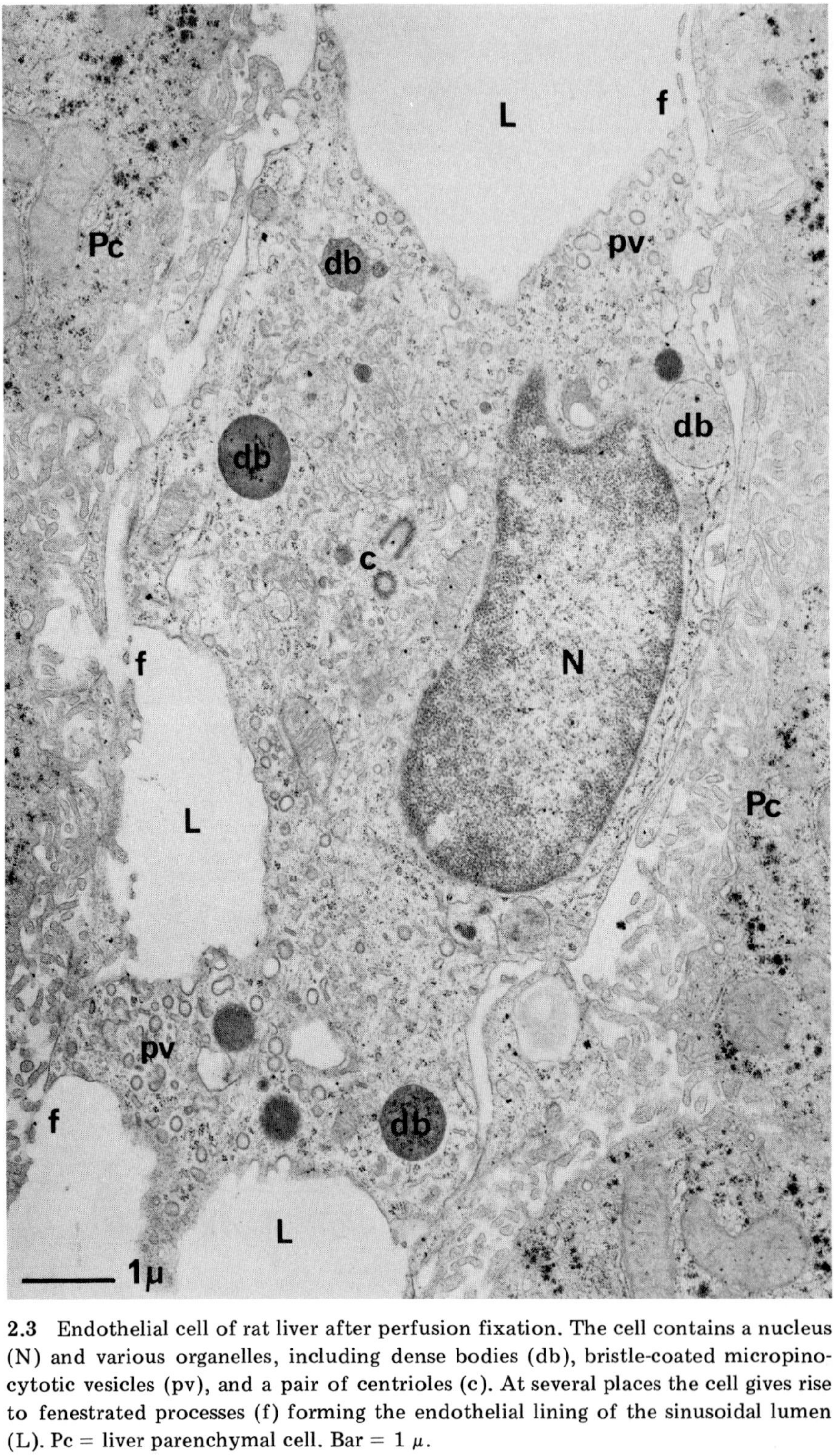

2.3 Endothelial cell of rat liver after perfusion fixation. The cell contains a nucleus (N) and various organelles, including dense bodies (db), bristle-coated micropinocytotic vesicles (pv), and a pair of centrioles (c). At several places the cell gives rise to fenestrated processes (f) forming the endothelial lining of the sinusoidal lumen (L). Pc = liver parenchymal cell. Bar = 1 μ.

the few morphometric studies on the contribution of nonparenchymal cells to the development of the rat liver, the poor resolution of the light microscope has made it impossible to distinguish between the different types of cells in the sinusoidal area.

During fetal and postnatal life, sinusoidal lining cells never constitute more than 3 percent of the total liver volume. The number of sinusoidal cells in the embryonic liver increases between the eighteenth day of gestation and term to 11 percent of the total number of cells and continues to increase during postnatal life (Greengard et al., 1972). The volume fraction occupied by sinusoidal cells does not change, and the rise in the relative number of Kupffer cells with age is due to both an increase in volume and a relative decrease in number of parenchymal cells (Daoust and Cantero, 1959; Van Bezooijen et al., 1974). About one-third of all liver cells in a three-month-old rat are sinusoidal lining cells (Daoust and Cantero, 1959; Fabrikant, 1968; Greengard et al., 1972). This proportion varies only slightly during the first year of the rat's life (Daoust and Cantero, 1959).

2.4.3 Hemopoietic Cells

The liver is the main hemopoietic organ in the embryo. During the gestational age of fifteen to twenty days, about one-third of the liver volume is hemopoietic cells (fig. 2.1); they account for more than half of all liver cells at this stage of development. A rapid loss of hemopoietic tissue occurs some hours before birth, but in the newborn rat part of the liver volume can be identified as hemopoietic tissue (fig. 2.4). During postnatal differentiation, the hemopoietic tissue volume further decreases, and the bone marrow becomes the major site of hemopoiesis. Only traces of hemopoietic tissue may be present in the liver of a three-month-old rat (Greengard et al., 1972).

2.5 Morphological Differentiation of the Parenchymal Cell in the Early Postnatal Period

In addition to alterations in the populations of the various classes of liver cells, morphological changes occur in the cells during the development of the liver. Electron microscopic studies show the early differentiation of the fetal hepatocytes (Spycher, 1967; Herzfeld et al., 1973). In the early postnatal period, a number of structural changes takes place in the cytoplasm of the parenchymal cell, but its morphological differentiation is not complete until the fortieth day after birth. Some of these morphological changes, illustrated in figure 2.5, concern glycogen storage, the lysosomes, and the mitochondria. The large glycogen pool is rapidly mobilized after birth and is nearly depleted within the first twelve hours of life. It has been suggested that lysosomal enzymes are, at least in some measure, involved in the glycogen catabolism (Phillips et al., 1967).

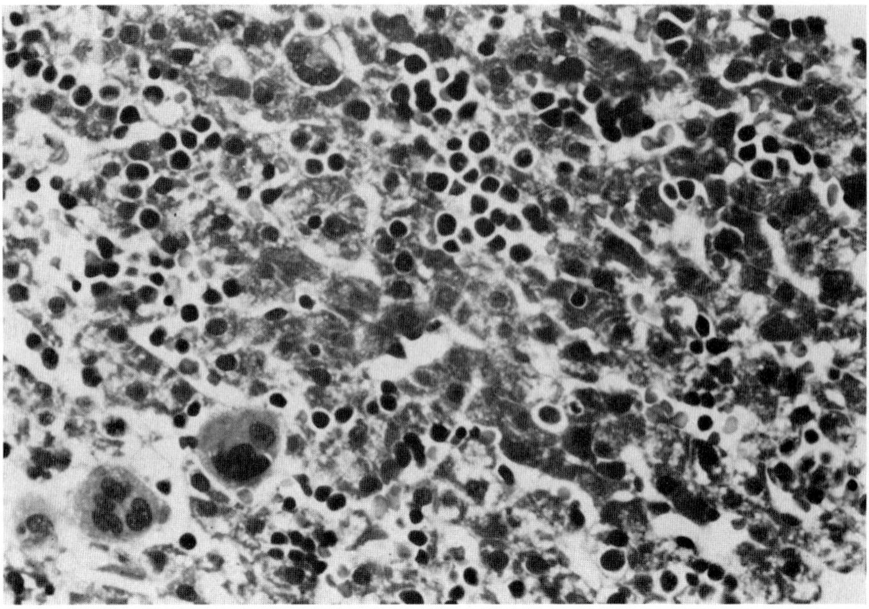

2.4 Liver of a 20-hour-old WAG/Rij rat, still showing numerous foci of hemopoiesis (HPS, × 370).

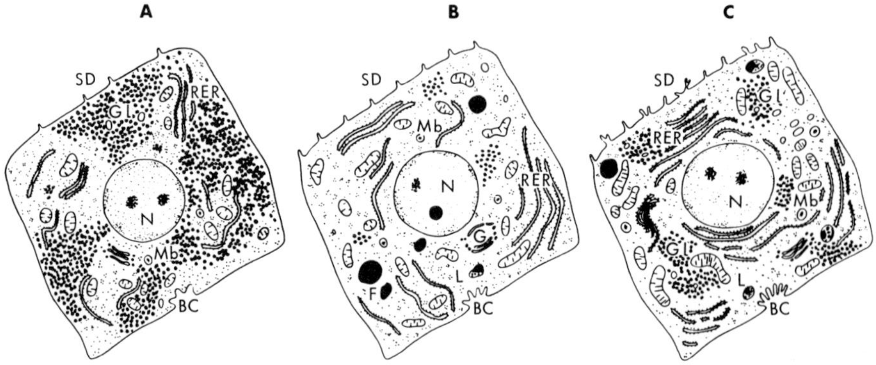

2.5 Schematic ultrastructure of liver parenchymal cells in (a) 30-minute-old rat, (b) 2-day-old rat, and (c) 29-day-old rat. *BC* = bile capillary; *F* = fat droplets; *G* = Golgi complex; *GL* = glycogen; *L* = lysosomes; *Mb* = microbodies; *N* = nucleus; *RER* = rough endoplasmatic reticulum; *SD* = space of Disse. (Modification of figure in Dvořák et al., 1967.)

It is well known that the heterogeneous group of lysosomal structures vary widely with the physiological and pathological state of the cell. Considerable variations also occur during early postnatal life. Only a few small round or elliptical lysosomes are present during the first two days after birth (Dvořák and Mazanec, 1967; Phillips et al., 1967). Afterward, the lysosomal structures enlarge and increase in number (Dvořák and Mazanec, 1967). The heterogeneity of structure and the cytoplasmic volume occupied by lysosomal structures increase during the life of the rat (Knook et al., 1975).

At term, the number of mitochondria and the fraction of cytoplasmic volume that they occupy have reached about half of their adult values (Herzfeld et al., 1973). There is a great increase in the volume occupied by mitochondria during the first two weeks after birth, and the adult value is reached at the end of the fourth postnatal week. Mitochondria of fetal and neonatal parenchymal cells are larger and contain less protein per unit volume than do those of adult cells (Franke and Goetze, 1963; Herzfeld et al., 1973). As shown in figure 2.5, the distribution of mitochondria in the cytoplasm is not constant during the early postnatal period. During the first day after birth the mitochondria are concentrated in small groups. A more adult distribution is observed during the following postnatal days. From this description of the morphological changes occurring after birth, it may be concluded that the morphological differentiation of the parenchymal cells is completed during the first four to six weeks after birth.

2.6 Polyploidy of Parenchymal Cells

After about the fortieth postnatal day, the ultrastructure of the parenchymal cell remains constant. The main changes (which began at birth) occurring after this period involve the cellular volume and the nuclear ploidy, two interrelated quantities. In the rat, the parenchymal cell population is predominantly diploid at birth and predominantly polyploid later in life. Six cytogenetic types of cells can be found in the adult liver, viz., mononuclear diploid, tetraploid, and octaploid cells and their binuclear equivalents. It is often stated that the shift to the polyploid condition can be considered to be a typical aging phenomenon. However, a gradual shift toward increasing polyploidy, accompanied by the appearance of binucleated cells, begins shortly after birth (Alfert and Geschwind, 1958; Nadal and Zajdela, 1966). Unfortunately, the literature presented no data on the distribution of mononuclear and binuclear polyploid cells during the whole life span of the rat, and thus no conclusions on shifts in polyploidy later in life could be made. Therefore, the distribution of the various nuclear classes of hepatocytes was investigated at our institute in two strains of rats from two weeks to thirty months of age. These studies showed that the most important changes in the ploidy of parenchymal cells in the WAG/Rij rat occur within the first three months of life (fig. 2.6); only minor changes were found afterward (Van Bezooijen et al., 1974). The age at which the main changes occur is notably different for the RU rats, which show the major shift from a diploid to a largely tetraploid parenchymal cell population between three and twelve months of age (De Leeuw-Israel, 1971). These results demonstrate that the most important shift to polyploid cells takes place relatively early in the rat's life. Rat and man are comparable in this respect, since the predominant shift to polyploidy occurs before puberty in man (Swartz, 1956).

2.7 Anatomy of the Rat Liver

The liver of the rat is divided into four parts: the median lobe with a deep fissure for the hepatic ligament; the right lobe, which is partially divided into anterior and posterior portions; the undivided large left lobe; and a small caudal or Spigelian lobe, which lies deep and fits around the esophagus. There are some differences between the livers of rats and mice. The rat differs from the mouse in having no gall bladder or cystic duct; ducts from the various liver lobes unite to form the bile duct.

2.8 How Old Is an Old Rat?

To establish what a young, adult, or old rat is, life span studies were done on the males and females of two rat strains. The survival curves are given in figure 2.7. The females of WAG/Rij and BN/Bi rats and the BN/Bi males show similar survival curves. The 50 percent survival time of WAG/Rij and BN/Bi females is 30 and 30.5 months, respectively, and the maximum life span is 47 and 54 months. Male BN/Bi rats have a mean life span of 28.5 months and a maximum life span of 44 months. Male WAG/Rij rats die earlier; they have a mean life span of 21.5 months and a maximum life span of 33 months. The last observation can be explained by the high incidence of pituitary adenomas in WAG/Rij males (Hollander, 1976). The neoplastic lesions observed in both strains have been described in detail elsewhere (Boorman and Hollander, 1972; Boorman and Hollander, 1973). Because of the pathological findings, female rats were used in our studies. Rats aged 30 months and older were considered to be old, as judged from the rapid increase in the mortality rate and from the average life span. Rats of 12 months of age were considered adults by the criteria of average life span and the occurrence of only minor changes in the mortality rate. Rats of 3 months of age were considered to be young.

2.9 Isolated Liver Cells as an In Vitro System for Development and Aging Studies

The reduction in the functional reserve capacity and other age-related changes in the liver of the rat (De Leeuw-Israel, 1971) are reflected in

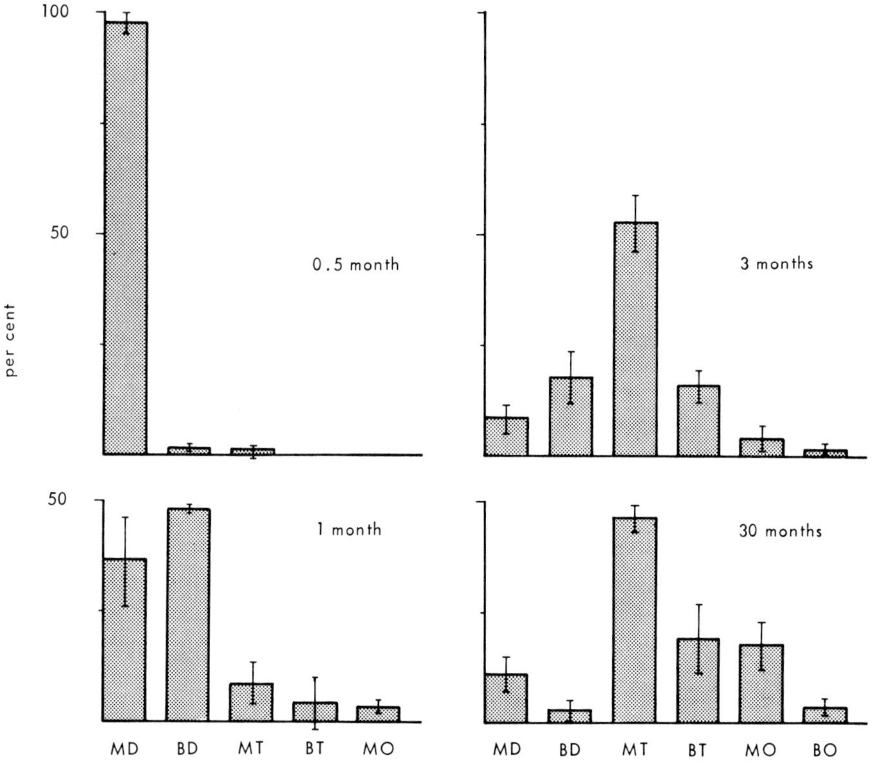

2.6 Percentage distribution of nuclear classes in parenchymal cells isolated from female WAG/Rij rats. Four animals were used per age group and 200 nuclei measured per animal. MD = mononuclear diploid; BD = binuclear diploid; MT = mononuclear tetraploid; BT = binuclear tetraploid; MO = mononuclear octaploid; BO = binuclear octaploid. (Van Bezooijen et al., 1974, *Mech. Age Dev.* 3:105–117.)

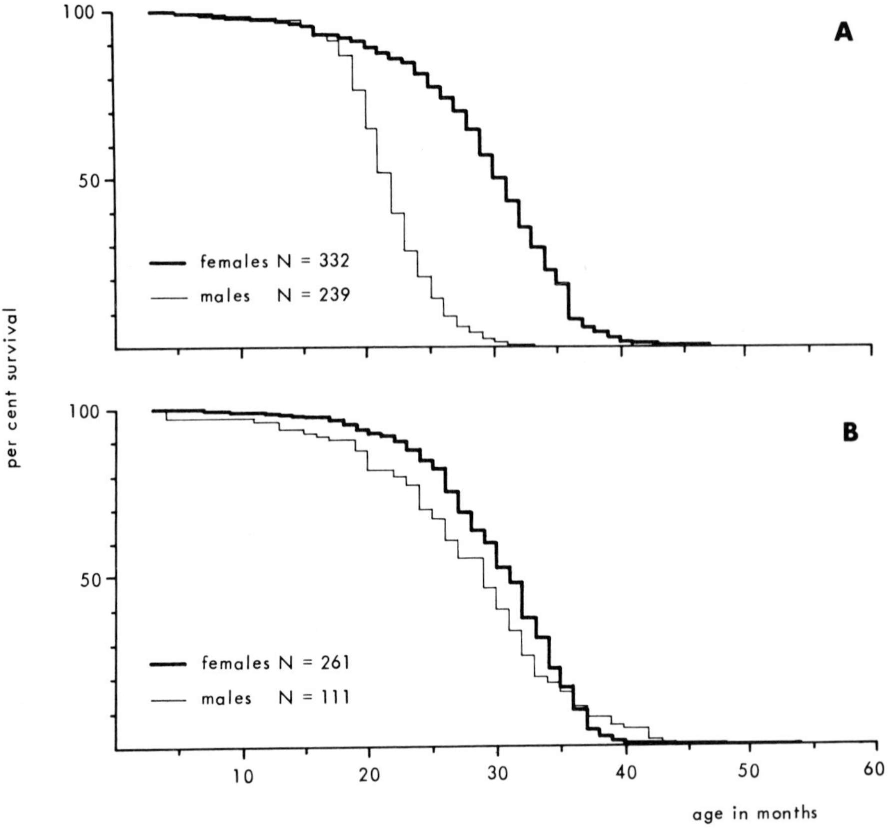

2.7 Survival curves of (a) WAG/Rij and (b) BN/Bi rats. The animals either died or were killed after they had become moribund.

metabolic changes at the cellular level (Knook, 1973). Until recently, nearly all studies on the metabolic activities of the developing and aging rat liver were performed with cell-free homogenates. These methods are performed with a mixture of extracts from all of the different cell populations of the liver. Since each cell type has its own enzyme composition, age-related changes in the activities of various liver enzymes in homogenates may result from shifts in the populations of the various cell types (Finch, 1972; Wilson, 1973). Even in the supposed absence of changes in cell populations, alterations in enzyme activities can be caused by differences in the age-related enzyme patterns within distinct liver cell classes. These difficulties could be overcome by a model system that allows the study of age-related changes in specific classes of liver cells. Suspensions of isolated, viable cells of the individual types present in the liver are such model systems.

Methods have been developed to separate the hepatocytes, or parenchymal cells, and the so-called nonparenchymal cells (which are mainly sinusoidal lining cells) from livers of rats aged two weeks to thirty-six months (Van Bezooijen et al., 1974; Knook et al., 1975). An enzymatic method consisting of collagenase and hyaluronidase digestion was employed for the isolation of parenchymal cells (Van Bezooijen et al., 1974). The final parenchymal cell suspension contained no liver cells of other types. The isolated hepatocytes met three criteria for viability: (1) they had a well-preserved ultrastructure, including a continuous plasma membrane and a regular distribution of the cytoplasmic organelles (fig. 2.8); (2) the nonpermeability of the plasma membrane could be demonstrated by adding trypan blue to the cell suspensions (85–90 percent of the isolated cells excluded this vital dye); and (3) the oxygen consumption of the isolated hepatocytes, when recalculated per gram of wet weight, was comparable to the oxygen consumption per gram wet weight of the perfused liver. (Van Bezooijen et al., 1974).

The nonparenchymal cells were also isolated by an enzymatic method. After perfusion of the liver in situ with a pronase-containing buffer system, small liver fragments were incubated in the same solution (Knook et al., 1975). The final nonparenchymal cell suspension consisted mainly of sinusoidal lining cells. Approximately half of the cells were endothelial, and one-fourth were Kupffer cells. The ultrastructure of the isolated Kupffer cells (fig. 2.9) and endothelial cells (fig. 2.10) was well preserved, and all cells excluded trypan blue (Knook et al., 1975). Fat-storing cells (fig. 2.11), lymphocytes, and some other blood cells were also present; parenchymal cells were absent.

2.10 Effect of Age on Endogenous Respiration of Parenchymal and Nonparenchymal Cells

The endogenous respiration rate of parenchymal and nonparenchymal cells isolated from female BN/Bi rats was measured with a Clark elec-

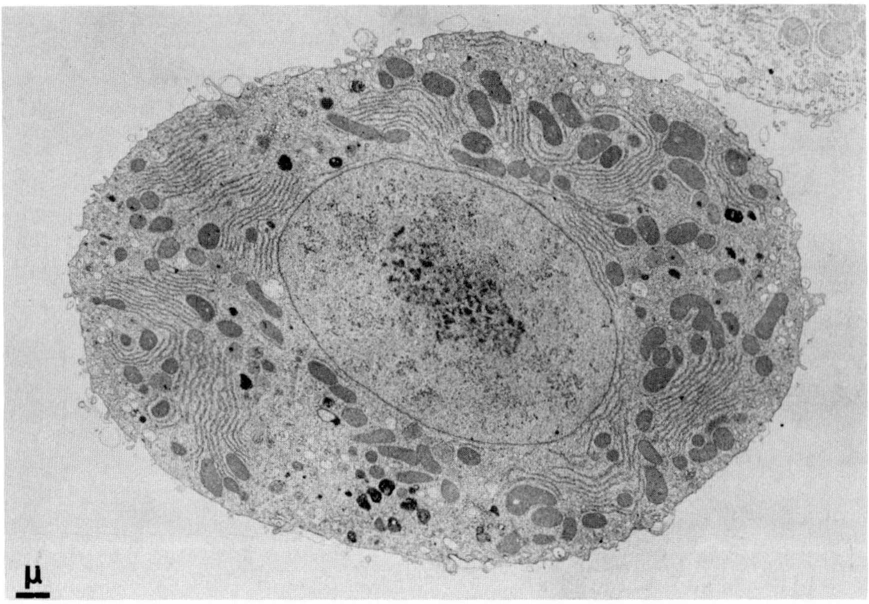

2.8 Isolated parenchymal liver cell from a 30-month-old WAG/Rij rat (× 5020).

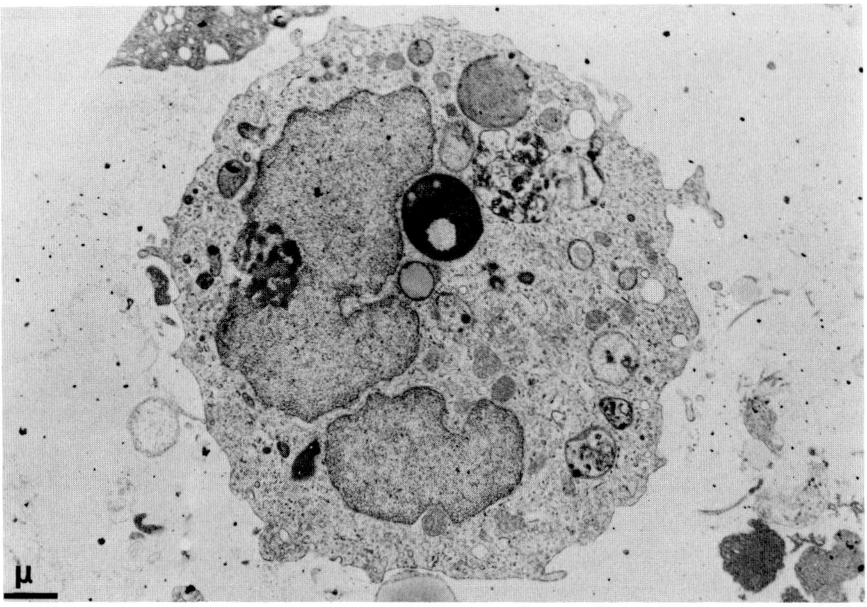

2.9 Isolated Kupffer cell from a 35-month-old BN/Bi rat with heterogeneous lysosomal structures (× 7450).

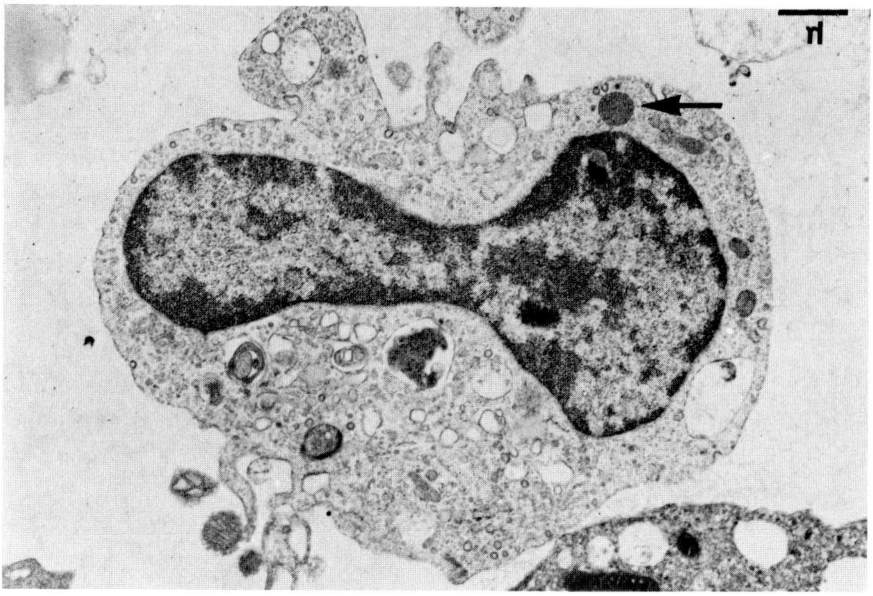

2.10 Endothelial cell isolated from the liver of a 35-month-old BN/Bi rat. The arrow indicates a lysosome (× 9500).

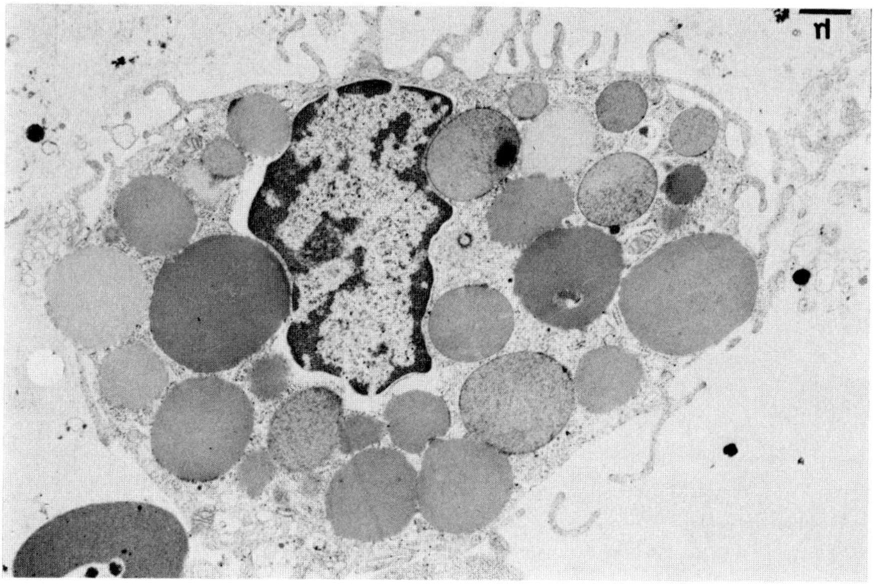

2.11 Fat-storing cell isolated from the liver of a 35-month-old BN/Bi rat (× 7050).

Table 2.1 Effect of age on endogenous respiration of liver parenchymal cells isolated from BN/Bi rats.[a]

Age (months)	Protein (mg/10^6 cells)	Endogenous respiration natoms/min/10^6 cells	Endogenous respiration natoms/min/mg protein
1	1.22 ± 0.15 (3)	24.6 ± 3.2 (3)	20.2 ± 0.2 (3)
3	2.15 ± 0.12 (13)	34.9 ± 2.1 (13)	16.4 ± 1.4 (12)
12	3.11 ± 0.15 (3)	41.6 ± 3.4 (3)	13.3 ± 0.6 (3)
30–35	3.03 ± 0.30 (13)	40.9 ± 6.7 (5)	13.6 ± 3.9 (5)

[a]Mean values ± standard error with the number of cell preparations in parentheses.

trode. The results obtained with parenchymal cells from rats aged one, three, twelve, and thirty to thirty-five months are summarized in table 2.1. In no case was the addition of succinate observed to stimulate respiration. This finding indicated the absence of cell remnants, particularly mitochondria, in the cell preparations (Van Bezooijen et al., 1974). The oxygen consumption per million cells was higher for cells isolated from adult and old rats. When expressed on a protein basis, the respiration of parenchymal cells decreases with age. These results are associated with an increase in the protein content per million cells with age, reflecting an increase in average cell size, which in turn results from a shift to more polyploid cells. Thus the larger and older cells show a lower specific respiratory activity than do the smaller and younger ones. Preliminary results also indicate a decreased respiratory rate in nonparenchymal cells with age.

A decrease in respiration and associated production of ATP through the coupled phosphorylation has been suggested to account, at least partly, for the decline in tolerance of older animals to physical stress. Experiments with liver slices or cell-free preparations have yielded conflicting results (Chen et al., 1972). Work with mitochondria isolated from adult and senescent rat livers showed no age-related decline in some parameters of mitochondrial function, such as the efficiency of oxidative phosphorylation and the respiratory control ratio (Gold et al., 1968; Bulos et al., 1972). In studies with isolated mitochondria, factors such as a selection of mitochondria during isolation and their original location in parenchymal or nonparenchymal cells cannot be controlled. Unfortunately, the extent to which the results obtained with liver mitochondria from old rats apply to the respiration phenomena of intact cells depends on those factors. The availability of viable parenchymal cells with an intact endogenous respiration overcomes the problems by offering the possibility of performing experiments on age-related alterations in the respiration and phosphorylation of mitochondria in situ.

2.11 Lysomal Enzyme Changes in Parenchymal and Nonparenchymal Cells

There are indications that lysosomes play a role in cellular aging processes in the rat liver. One is the increase with age in the percentage of cytoplasmic volume of parenchymal cells occupied by lysosomal structures (Knook et al., 1975). (table 2.2). Lysosomal enzyme activities in the liver have also been supposed to increase with age, although recent reviews (Finch, 1972; Wilson, 1973) included various reports of increased, decreased, and even unchanged lysosomal enzyme activity as a function of age. Those conflicting results were obtained with cell-free extracts or subcellular fractions from the total liver. The differences in the measured total activity, therefore, may have resulted from large variations in the contribution of the lysosomes from the different cell populations. In view of this possibility, we and our colleagues felt the study of lysosomal enzyme changes in distinct isolated cell classes would produce firmer results.

Lysosomal enzyme activities were determined in suspensions of isolated parenchymal and nonparenchymal cells. The isolated endothelial Kupffer cells were quite different in their content of lysosomal structures; Kupffer cells possess many lysosomes of widely varying structure and size (see fig. 2.9), whereas lysosomes are very sparse in endothelial cells (see fig. 2.10). The few that are present are relatively small and have diameters of approximately 0.6 μ (Knook et al., 1975). Due to the properties of the lysosomes in endothelial cells and the restricted endocytic activity of these cells (Wisse, 1974), the lysosomal enzyme activities in nonparenchymal cell suspensions must be attributable mainly to the Kupffer cells.

The following enzymes were measured in suspensions of parenchymal and nonparenchymal cells: acid phosphatase (EC 3.1.3.2), β-galactosidase (EC 3.2.1.23), cathepsin D (EC 3.4.4.23), deoxyribonuclease or deoxyribonuclease II (EC 3.1.4.6), arylsulfatase B (EC 3.1.4.6), and arylsulfatase B (EC 3.1.6.1). Prior to the enzyme determinations, the cells were treated with 0.05 percent Triton X100. Therefore, the values measured represent total enzyme activities. The distribution of the enzymes over parenchymal and nonparenchymal cells is shown in table 2.3. Each enzyme is present in both cell classes, but the specific activities are much higher in nonparenchymal cells. The most striking differences are found for cathepsin D, arylsulfatase B, and the two β-galactosidase isoenzymes; these have four to thirteen times higher specific activities in nonparenchymal cells. The selective concentration of these enzymes in nonparenchymal cells strongly suggests that the enzyme content of lysosomes from parenchymal and nonparenchymal cells differs significantly.

Specific activities of a number of lysosomal enzymes were studied in liver cell suspensions from young and old rats (fig. 2.12). Lysosomal enzyme activities increase "per cell" in both parenchymal and nonparenchymal cell preparations with age. There are also increases in two of the lysosomal enzyme activities per milligram of protein in the parenchymal cells. The specific activities in nonparenchymal cells do not

Table 2.2 Cytoplasmic volume fraction of lysosomal structures in parenchymal liver tissue from young (5 months) and old (35 months) WAG/Rij rats.

	Young		
Cell type	Central	Midzonal	Peripheral
Volume (μ^3)[a]			
Hepatocytes	4230 ± 160	4100 ± 160	2710 ± 90
Nucleus	223 ± 10	183 ± 9	140 ± 8
Cytoplasm	4010 ± 160	3920 ±160	2570 ± 90
Ratio of nucleus to cell volume	0.053	0.045	0.052
Percent cytoplasmic volume of lysosomal structures[b]	0.96 ± 0.16	0.56 ± 0.12	0.45 ± 0.11

[a]Relative volumes based on diameters of hepatocytes and nuclei measured for 0.5 μ sections of liver tissue. Values represent the mean ± standard error determined on 290 micrographs. (See D. L. Knook et al. for details.)
[b]Values represent the mean ± standard deviation from the Poisson distribution.

Table 2.3 Distribution of lysosomal enzyme activities in parenchymal and nonparenchymal liver cells isolated from three-month-old female BN/Bi rats.[a]

	Act. [b]/mg protein	
Enzymes measured	Parenchymal cells	Nonparenchymal cells
Acid phosphatase	33.88 ± 3.94 (13)	53.70 ± 4.17 (7)
β-Galactosidase:		
pH 3.6	2.62 ± 0.28 (8)	10.97 ± 1.43 (5)
pH 4.5	1.97 ± 0.39 (4)	7.55 ± 1.02 (5)
Cathepsin D	8.23 ± 1.13 (6)	110.79 ± 16.48 (4)
Deoxyribonuclease II	11.30 ± 1.64 (7)	27.38 ± 7.91 (5)
Arylsulphatase B	6.63 ± 0.26 (5)	20.73 ± 3.65 (5)

[a]The values represent the mean ± standard error with the number of cell preparations in parentheses.
[b]Enzyme activities are expressed as nmoles 4-methylumbelliferone (acid phosphatase, β-galactosidase), nmoles tyrosine (cathepsin D), nmoles mononucleotide equivalents (deoxyribonuclease II) or nmoles nitrocatechol (arylsulphatase B) released per minute at 37° C.

Table 2.4 Liver lesions in old female rats.

Strain	Age in months	Number of rats	Areas of vacuolization of cytoplasm	Cystic areas	Tumorlike nodules
WAG/Rij	28–31	84	10 (12%)	8 (10%)	20 (24%)
WAG/Rij	32–35	61	17 (21%)	18 (22%)	28 (35%)
BN/Bi	28–31	18	3 (17%)	15 (83%)	0
BN/Bi	32–35	12	1 (8%)	10 (83%)	0

Old		
Central	Midzonal	Peripheral
6970 ± 320	5930 ± 210	4190 ± 200
425 ± 21	290 ± 13	245 ± 16
6545 ± 32	5640 ± 210	3955 ± 200
0.060	0.049	0.058
1.33 ± 0.19	1.65 ± 0.22	0.33 ± 0.08

change significantly. The variation in the age-related changes of the activities in parenchymal and nonparenchymal cells suggests that both cell classes have a heterogeneous lysosomal population. The results also demonstrate that the reported age changes in lysosomal activities in liver homogenates represent a mean, and often misleading, value that does not reflect the real alterations in the specific cell classes of the liver.

2.12 Spontaneous Liver Lesions

Along with the more general physiological changes with age, the aging liver has an increased tendency to develop spontaneous lesions. Therefore, the incidence of liver lesions of spontaneous origin was determined in female WAG/Rij and BN/Bi rats. In order to correlate biochemical studies with histopathology, two age groups of animals were screened, viz., a group twenty-eight to thirty-one months old, and a group aged thirty-two to thirty-five months. The liver lesions found are listed in table 2.4.

Focal areas of vacuolated and degenerated hepatocytes were found frequently in both strains of rats (fig. 2.13). Some of these areas became necrotic (fig. 2.14) and finally gave rise to a number of cysts, as is demonstrated in figures 2.15 and 2.16 (Boorman and Hollander, 1973). The cysts ranged from microscopic to those observable with the naked eye. Undoubtedly the greater percentage of the observed cysts originated in the bile duct. The focal areas of vacuolated hepatocytes and the cysts were observed only rarely in young animals. Tumorlike nodules (fig. 2.17) originating from liver parenchymal cells were observed only in the WAG/Rij rats, with a higher incidence in females than in males (Boorman and Hollander, unpublished data). There was a very clear age-related incidence of the lesions. In a large group of 290 female WAG/Rij rats, hepatic nodules were seen in only one of 21 animals that

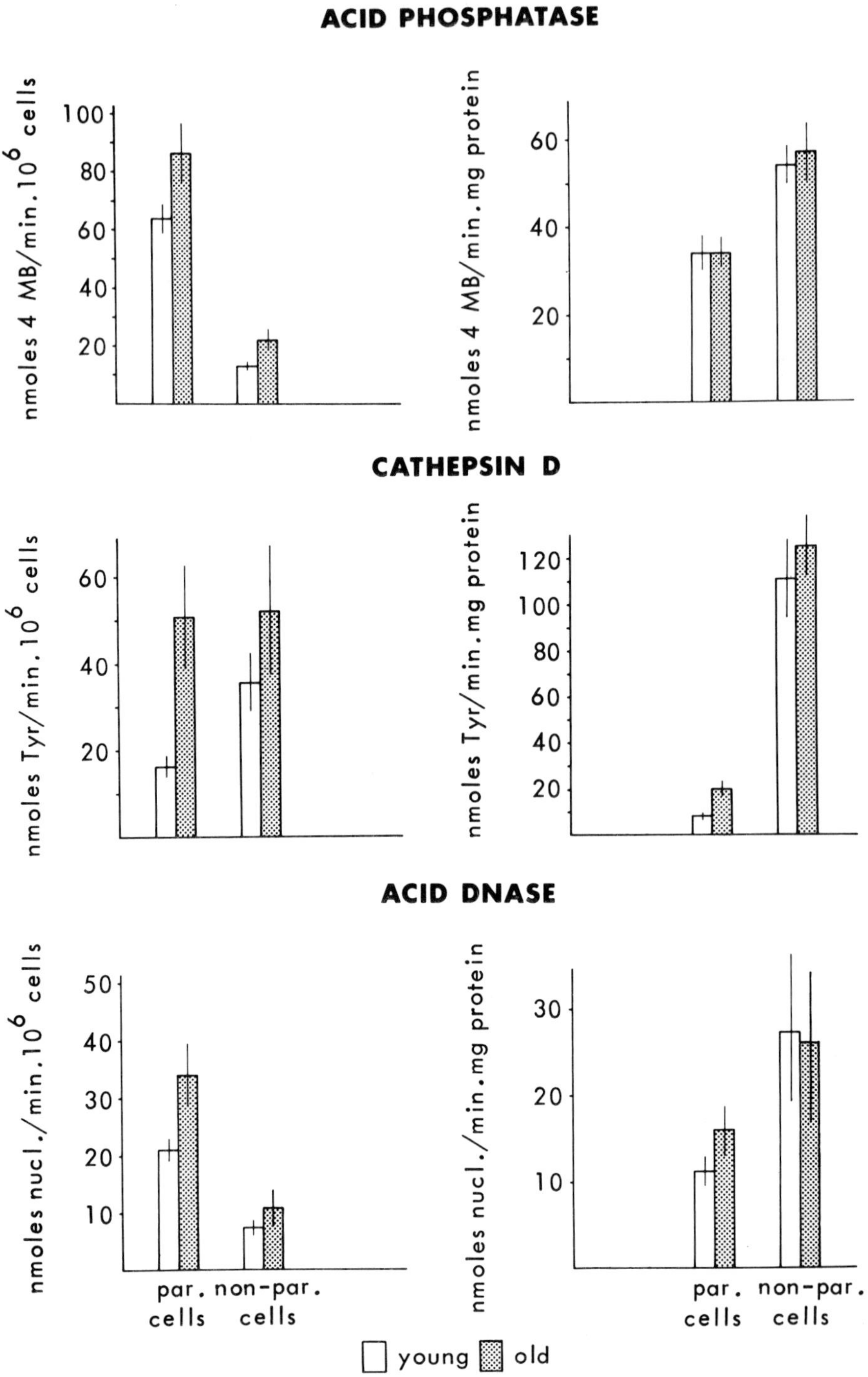

2.12 Some lysosomal enzyme activities in parenchymal and nonparenchymal liver cells from young (aged 3 months) and old (aged 30–35 months) BN/Bi rats. The activities are expressed in nmoles/min/mg of protein and in nmoles/min/10^6 cells, as described in table 2.3.

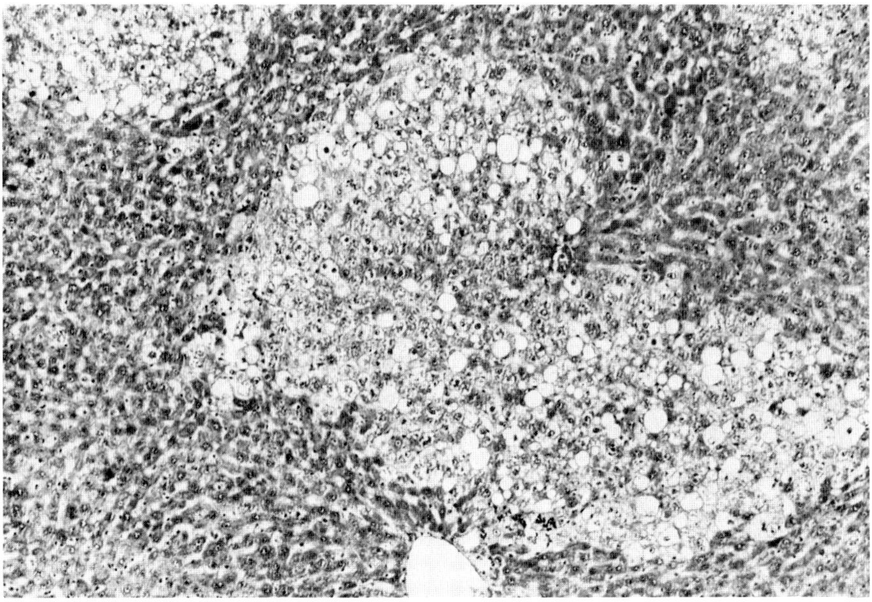

2.13 Multiple foci of highly vacuolated and degenerated hepatocytes in the liver of a 32-month-old female WAG/Rij rat (HPS, × 70). (Boorman and Hollander, 1973, *J. Geront.* 28:152–159.)

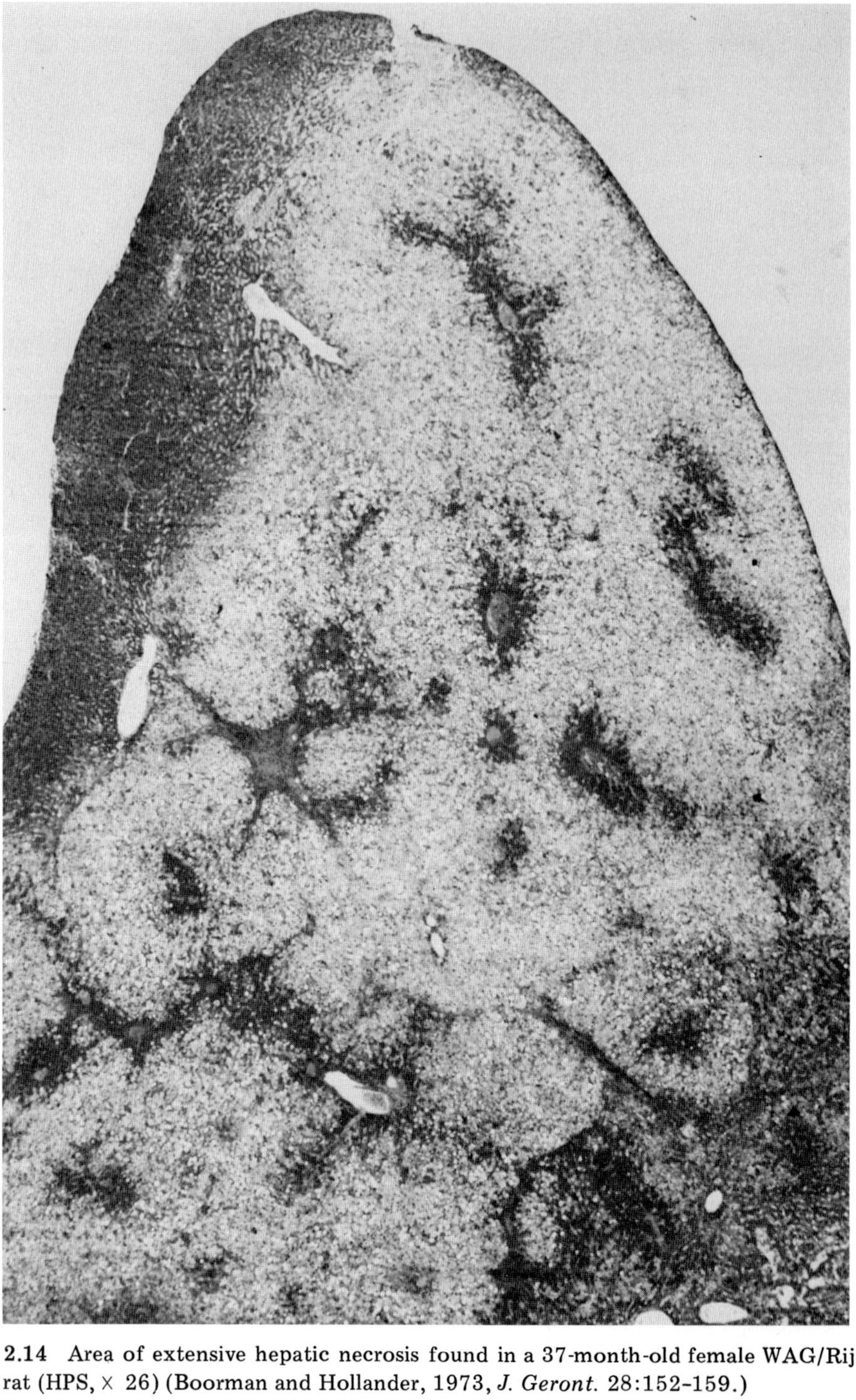

2.14 Area of extensive hepatic necrosis found in a 37-month-old female WAG/Rij rat (HPS, × 26) (Boorman and Hollander, 1973, *J. Geront.* 28:152–159.)

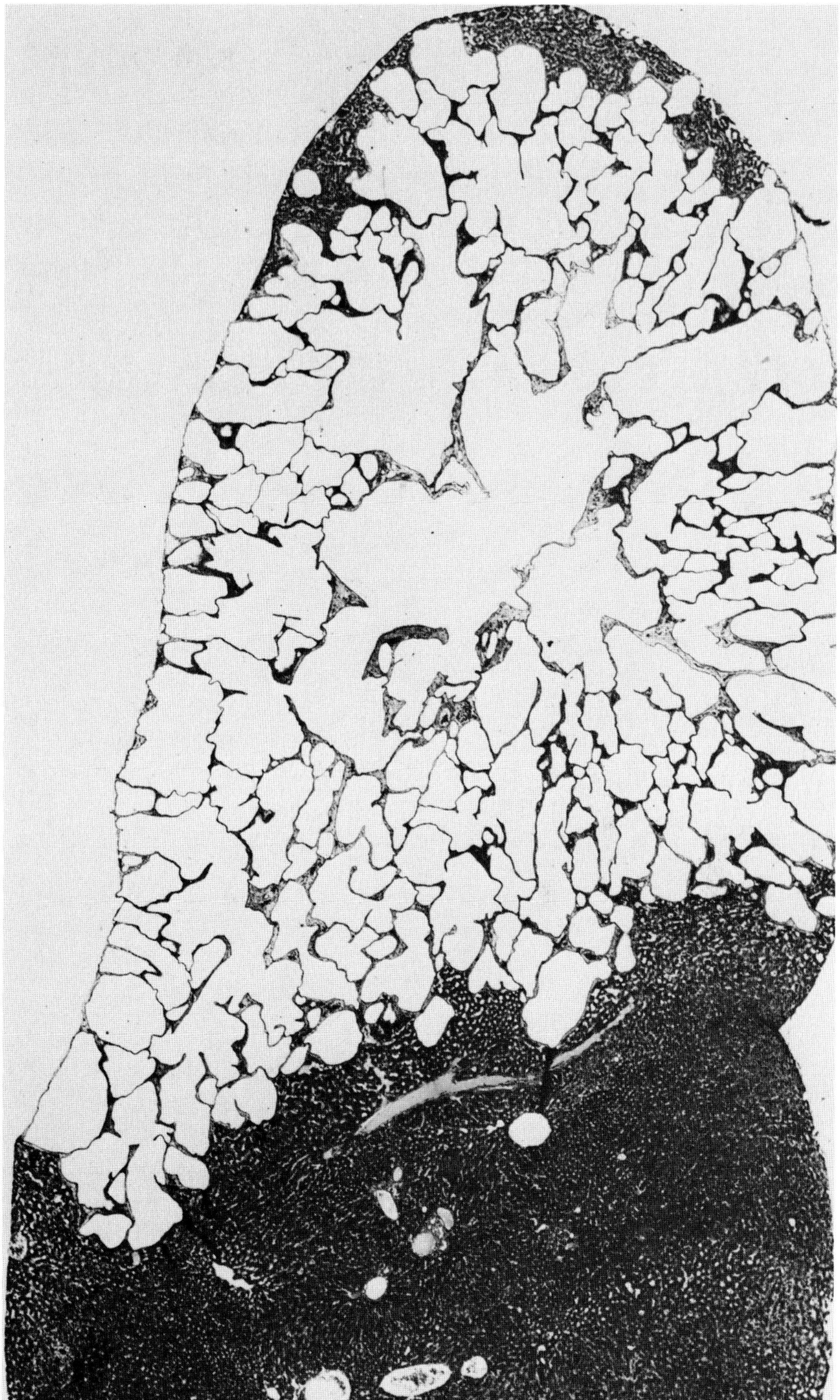

2.15 Cystic area, believed to be a sequel to the hepatic necrosis, in a 33-month-old female WAG/Rij rat (HPS, × 31).

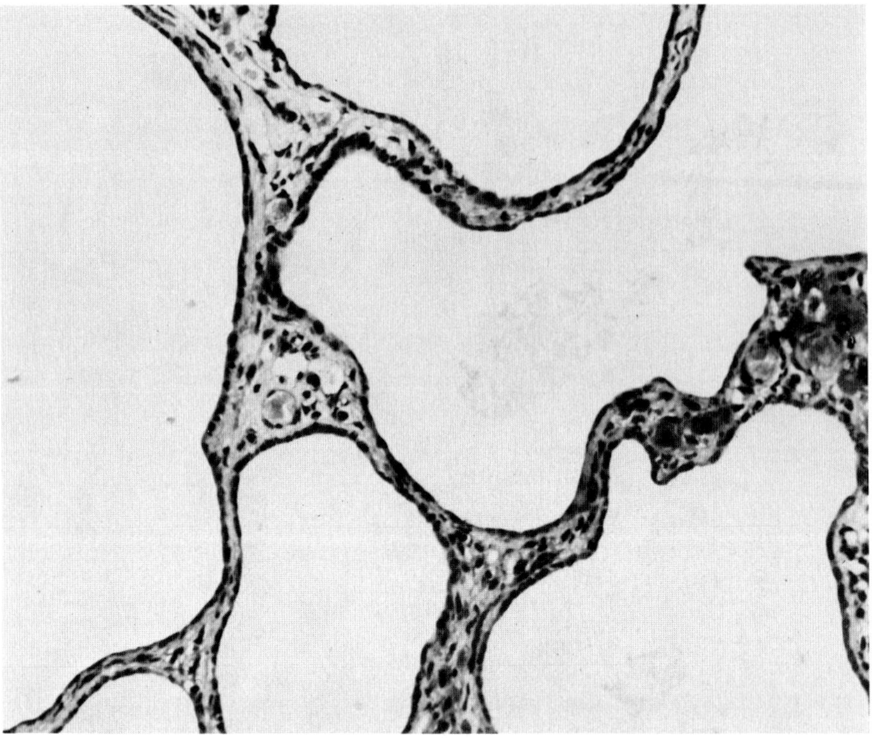

2.16 Detail of Figure 2.15 showing cyst lined by flattened cells (HPS, × 260).

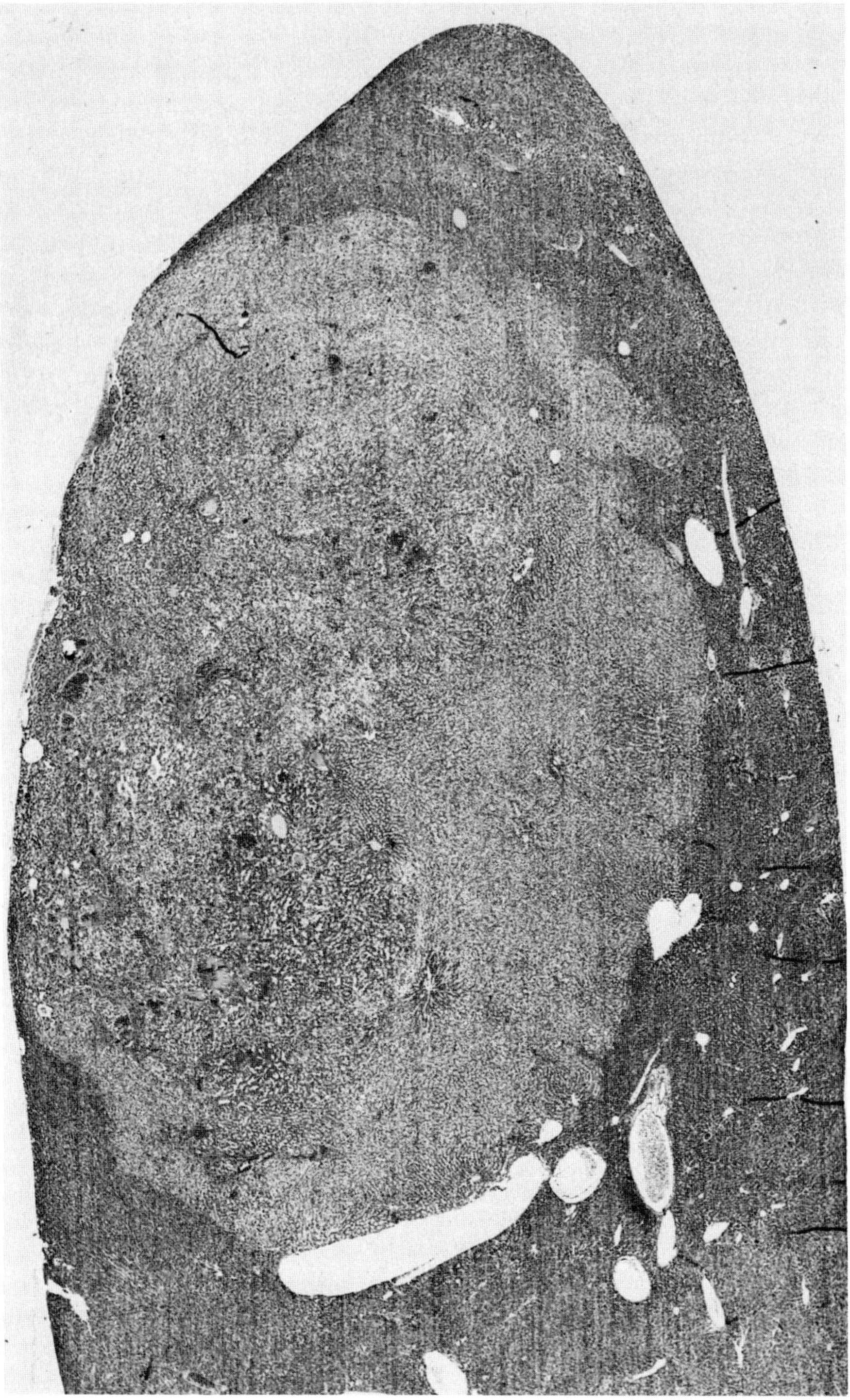

2.17 Tumorlike nodule in the liver of a 26-month-old female WAG/Rij rat (HPS, × 20).

died under two years of age; however, they were very common in rats over three years of age (Boorman and Hollander, 1973). The lesions were difficult to evaluate histologically. Mitotic figures and vacuolated and degenerating cells were often seen in the same area. Liver parenchyma adjacent to these nodules were usually compressed. Some of the nodules reached a diameter of 2–3 cm. However, because no local invasion was observed and no distant metastases were found, we hesitate to call these tumorlike nodules carcinomas. The histological evaluation and the classification of these nodules requires further work.

When using an intact organ system, a cell-free homogenate, or isolated cell suspensions for aging studies, one should never forget to take into account the influence of possible pathological lesions of the organ. Furthermore, not only parenchymal cells but various distinct cell populations also may contribute to the age-related biochemical changes in the liver.

2.13 More Recent Developments

Since this paper was written, the model system offered by isolated parenchymal and nonparenchymal liver cells for the study of developmental and aging changes has been further explored. As a result the following papers have been published:

Brouwer, A., Van Bezooijen, C. F. A., and Knook, D. L., 1977. Respiratory activities of hepatocytes isolated from rats of various ages.

Knook, D. L. 1977. Model systems for studies on the cellular basis of organ aging.

Knook, D. L., and Sleyster, E. Ch. 1976a. Separation of Kupffer and endothelial cells of the rat liver by centrifugal elutriation.

Knook, D. L., and Sleyster, E. Ch. 1976b. Lysosomal enzyme activities in parenchymal and nonparenchymal liver cells isolated from young, adult and old rats.

Van Bezooijen, C. F. A., et al. 1976. Bromsulfophthalein uptake by isolated liver parenchymal cells.

Van Bezooijen, C. F. A., et al. 1976. Albumin synthesis by liver parenchymal cells isolated from young, adult and old rats.

Van Bezooijen, C. F. A., Grell, T., and Knook, D. L. 1977. The effect of age on protein synthesis by isolated liver parenchymal cells.

Recent information on morbidity and mortality data of rats at the Institute for Experimental Gerontology is provided in the following review:

J. D. Burek (in press). Pathology of aging rats. A morphological and experimental study of the age-associated lesions in aging BN/Bi, WAG/Rij and (WAG $\times$ Rij)F_1 rats. Thesis, University of Utrecht.

2.14 Conclusions

We have briefly reviewed the different roles of the various cell types in the rat liver during development and aging.

The volume fractions and relative numbers of parenchymal and sinusoidal lining cells (endothelial, Kupffer, fat-storing cells) change greatly during prenatal and postnatal development.

Parenchymal cells undergo a morphological differentiation during the early postnatal state and a shift to larger polyploid cells later in life. This shift to polyploidy cannot be considered an aging phenomenon.

The results obtained upon the determination of endogenous respiration and of the lysosomal enzyme content in suspensions of isolated parenchymal cells and nonparenchymal cells (mainly sinusoidal lining cells) demonstrate the benefit of this system in studies in vitro of development and aging.

The major liver lesions occurring with age are focal areas of vacuolated hepatocytes, cysts, and tumorlike nodules.

2.15 Acknowledgment

The authors gratefully acknowledge the assistance of C. F. A. Van Bezooijen, M. J. Van Noord, A. L. Nooteboom, E. C. L. Sleyster and F. G. Westerhuis in the preparation of the manuscript. Figures 2.2 and 2.3 were kindly supplied by E. Wisse, Laboratory for Electron Microscopy, Leiden, The Netherlands.

2.16 References

Abercrombie, M., and Harkness, R. D. 1951. The growth of cell populations and the properties in tissue culture of regenerating liver of the rat. *Proc. Roy. Soc. S.B.* 138:544-561.

Alfert, M., and Geschwind, I. I. 1958. The development of polysomaty in rat liver. *Exp. Cell Res.* 15:230-270.

Boorman, G. A., and Hollander, C. F. 1972. Neoplasia and degenerative lesions in the ageing BN/Bi rat. In *Annual Report of REP-TNO*, pp. 176-177.

Boorman, G. A., and Hollander, C. F. 1973. Spontaneous lesions in the female WAG/Rij (Wistar) rat. *J. Geront.* 28:152-159.

Brouwer, A., Van Bezooijen, C. F. A., and Knook, D. L. 1977. Respiratory activities of hepatocytes isolated from rats of various ages. *Mech. Age. Dev.* 6:265-269.

Bulos, B., Shukla, S., and Sacktor, B. 1972. Effect of thyroid hormone on respiratory control of liver mitochondria from adult and senescent rats. *Arch. Biochem. Biophys.* 151:387-290.

Burek, J. D. (in press). Pathology of aging rats. A morphological and experimental study of the age-associated lesions of aging BN/Bi, WAG/Rij and (WAG × Rij)F_1 rats. Thesis, University of Utrecht.

Chen, J. C., Warshaw, J. B., and Sanadi, D. R. 1972. Regulation of mitochondrial respiration in senescence. *J. Cell. Physiol.* 80:141-148.

Croisille, Y., and Le Douarin, N. M. 1965. Development and regeneration of the liver. In *Organogenesis*, eds. R. L. De Haan, and H. Ursprung. New York: Holt, Rinehart and Winston.

Dadoune, J. P. 1963. Contribution a l'étude au microscope électronique de la différenciation de la cellule hépatique chez le rat. *Arch. Anat. Micr. Morph. Exp.* 52:513-571.

Daoust, R., and Cantero, A. 1959. The numerical proportions of cell types in rat liver during carcinogenesis by 4-dimethylaminoazobenzene (DAB). *Cancer Res.* 19: 757-762.

De Leeuw-Israel, F. R. 1971. Aging changes in the rat liver. Ph.D. Thesis, Leiden.

Dvořák, M., and Mazanec, K. 1967. Differenzierung der Feinstruktur der Leberzelle in der frühen postnatalen Periode. *Z. Zellforsch.* 80:370-384.

Fabrikant, J. I. 1968. The kinetics of cellular proliferation in regenerating liver. *J. Cell Biol.* 36:551-565.

Finch, C. E. 1972. Enzyme activities, gene function and aging in mammals. *Exp. Geront.* 7:53-67.

Franke, H., and Goetze, E. 1963. Die Feinstruktur der Leberzellen von Rattenfoeten und Neugeborenen in verschiedenen Entwicklungsstadien. *Acta Biol. Med. Germ.* 11:424-432.

Gold, P. H., Gee, M. V., and Strehler, B. L. 1968. Effect of age on oxidative phosphorylation in rat. *J. Geront.* 23:509-512.

Grant, H. C. and Rees, K. R. 1957. The precancerous liver; correlations of histological and biochemical changes in rats during prolonged administration of thioacetamide and "butter yellow." *Proc. Roy. Soc. S. B.* 148:117-136.

Greengard, O., Federman, M., and Knox, W. E. 1972. Cytomorphometry of developing rat liver and its application to enzymic differentiation. *J. Cell Biol.* 52: 261-272.

Henneberg, B. 1937. *Normentafeln zur Entwicklungsgeschichte der Wanderratte (Rattus norvegicus Ersleben).* Jena: Gustav Fischer.

Herzfeld, A., Federman, M., and Greengard, O. 1973. Subcellular morphometric and biochemical analysis of developing rat hepatocytes. *J. Cell Biol.* 51:475-483.

Hollander, C. F. 1976. Current experience using the laboratory rat in aging studies. *Lab. Animal Sci.* 26:320-328.

Knook, D. L. 1973. Age-related changes in isolated cells of rat liver. In *Gerontology, Proceedings of the 9th International Congress*, Kiev, 1972, eds. D. F. Chebotarev, V. V. Frolkis, and A. Ya. Mints, vol. 2, *Symposia*, pp. 94-96. Kiev: Institute of Gerontology, USSR Academy of Sciences.

Knook, D. L., 1977. Model systems for studies on the cellular basis of organ ageing. *Akt. Gerontol.* 7:1-9.

Knook, D. L., and Sleyster, E. Ch. 1976a. Separation of Kupffer and endothelial cells of the rat liver by centrifugal elutriation. *Exp. Cell Res.* 99:444-449.

Knook, D. L., and Sleyster, E. Ch. 1976b. Lysosomal enzyme activities in parenchymal and nonparenchymal liver cells isolated from young, adult and old rats. *Mech. Age. Dev.* 5:389-397.

Knook, D. L., Sleyster, E. Ch., and Van Noord, M. J. 1975. Changes in lysosomes during aging of parenchymal and nonparenchymal liver cells. In *Impairment of Cellular Functions during Aging in Vivo and in Vitro*, eds. E. Holeckova and V. J. Christofalo, pp. 155-169. New York: Plenum Press.

Metcalf, D., and Moore, H. A. S. 1971. *Haemopoietic cells.* Amsterdam: North-Holland Publishing Company.

Nadal, C., and Zajdela, F. 1966. Polyploidie somatique dans le foie de rat. 1. Le rôle des cellules binucléés dans la genèse des cellules polyploides. *Exp. Cell Res.* 42:99-116.

Nicholas, J. S. 1949. Experimental Methods and Rat Embryos. In *The Rat in Laboratory Investigation*, eds. E. J. Farris and J. Q. Griffith, pp. 51-67. Philadelphia: J. B. Lippincott Company.

Phillips, M. J., Unakar, N. J., Doornewaard, G., and Steiner, J. W. 1967. Glycogen depletion in the newborn rat liver. An electron microscopic and electron histochemical study. *J. Ultrastruct. Res.* 18:142-165.

Sibatani, A., and Fukuda, M. 1953. Increase in number of nuclei and mitotic activity of parenchymal cells of the rat liver in postnatal growth. *Cytologia* 18: 80-85.

Spycher, M. A. 1967. Zur Frühembryogenese der Leber. Elektronenmikroskopische Untersuchungen an der embryonalen Rattenleber. *Path. Microbiol.* 30:303-352.

Stowell, R. E. 1952. Use of histochemical and cytochemical techniques in problems of pathology. *Lab. Investigation* 1:210-230.

Striebich, M. J., Shelton, E., and Schneider, W. C. 1953. Quantitative morphological studies on the livers and liver homogenates of rats fed 2-methyl- or 3′-methyl-4-dimethylaminoazobenzene. *Cancer Res.* 13:279-284.

Swartz, F. J. 1956. The development in the human liver of multiple desoxyribose nucleic acid (DNA) classes and their relationship to the age of the individual. *Chromosoma* 8:53-72.

Van Bezooijen, C. F. A., Van Noord, M. J. and Knook, D. L. 1974. The viability of parenchymal cells isolated from young and old rats. *Mech. Age. Dev.* 3:105-117.

Van Bezooijen, C. F. A., Grell, T., and Knook, D. L. 1976a. Bromsulfophthalein uptake by isolated liver parenchymal cells. *Biochem. Biophys. Res. Commun.* 69: 354-361.

Van Bezooijen, C. F. A., Grell, T., and Knook, D. L. 1976b. Albumin synthesis by liver parenchymal cells isolated from young, adult and old rats. *Biophys. Biochem. Res. Commun.* 71:513–519.

Van Bezooijen, C. F. A., Grell, T., and Knook, D. L. 1977. The effect of age on protein synthesis by isolated liver parenchymal cells. *Mech. Age. Dev.* 6:293–304.

Weibel, E. R., Stäubli, N., Gnägi, H. R., and Hess, F. A. 1969. Correlated morphometric and biochemical studies on the liver cell. I. Morphometric model, stereological methods, and normal morphometric data for rat liver. *J. Cell Biol.* 42:68–91.

Wilson, P. 1973. Enzyme changes in aging mammals. *Gerontologia* 19:79–125.

Wisse, E. 1972. An ultrastructural characterization of the endothelial cell in the rat liver sinusoid under normal and various experimental conditions, as a contribution to the distinction between endothelial and Kupffer cells. *J. Ultrastruct. Res.* 38: 528–562.

Wisse, E. 1974. Observations on the fine structure and peroxidase cytochemistry of normal rat liver Kupffer cells. *J. Ultrastruct. Res.* 46:393–426.

2.17 Discussion

P. Newberne I'd like to ask one question right away while I'm thinking of it. Have you looked at the biochemistry of the cells separated from a nodule as in your figure and compared it to nonnodular areas?

Knook That is, of course, possible with large nodules, but we have not done so. The problem is to determine viability of the isolated cells.

Grice Have you used the rats that develop liver nodules in old age for cancer studies?

Knook No. All those pathological studies we have done have been of spontaneous tumors only. It turns out that most institutions performing carcinogenesis studies don't take into account all these spontaneous lesions, but all of these strains have been used for cancer studies in the Netherlands.

Yoon Do you consider that the nodules developed from the focal degenerative changes? Or are they completely separated from those changes? I am asking because I have studied the nodules, and in young rats I found focal degeneration and later on I saw nodules which increase with age. So I thought this showed some progression with age.

Knook From this study it is difficult to determine the stages of development.

Butler Do the cystic areas occur in the same liver as the tumorlike nodules, and do you have any other evidence of infarction, necrosis, within these same livers?

Knook The lesions may occur in the same liver, and only in one is there evidence of other lesions.

Bannasch Could you go a bit more into detail about your criteria to differentiate endothelial cells and Kupffer cells? Is your criterion based only on lysosomal structures, and is it only morphological?

Knook The morphological differences between the Kupffer cells and the endothelial cells have been studied by Wisse (*J. Ultrastruct. Res.* 46:393, 1974). Lysosomes in Kupffer cells vary greatly in size, and peroxidase activity is present only in Kupffer cells. We are using isolated cells and are determining which are Kupffer and endothelial cells. We are interested in the association with lipofuscin which is slightly increased.

Becker We know that in aging rats there is a visible increase in what have been called the residual bodies. These bodies are lysosome derived and still contain lysosomal enzymes. Is it possible that the increase you are seeing is this accumulation of bodies containing pigments which cannot be eliminated?

Bannasch Is glycogen present in the nodules, and did you look for glycogen in vacuolated cells?

Knook We did not find large amounts of glycogen in the liver or the vacuolated cells.

P. Newberne In our experience the smaller lesions with vacuolated cells contain glycogen, but this may be lost.

Harris Are the fat-storing cells sinusoidal cells? I've seen cells of this type in fat stains of the rat liver with no fat anywhere else. There is not a great deal of fat in them, but they stain and are very striking when they are present. I always thought they were Kupffer cells.

Knook We found them mainly in suspensions from old rats but only in small numbers. They are completely filled with fat and represent "cleft cells," which are not the sinusoid lining cells but lie in the space of Disse.

Sigler You consider that the large cystic areas are bile ducts. Can you elaborate on this? Possibly they are lymphatics. By chemical analysis, the fluid contents are rich in protein and have a high lymphocytic count.

P. Newberne I believe there is more than one kind of cyst. Certainly some are lymphatic. There are bile duct structures that proliferate and dilate. And there is another kind that we don't understand as well.

Knook At least two-thirds are bile ducts.

Laqueur I have seen these same kinds of lesions in old Fischer rats at fifteen to eighteen months which contain protein, have no increase in connective tissue, and may be confluent.

Baldwin Are the lesions found in these old animals transplantable?

Knook We have not transplanted them.

Weisburger Have you observed other types of spontaneous tumors?

Knook There is a high incidence of tumors of the kidney and urogenital system in the brown Norway rat which could be used as model systems for these types of tumors.

P. Newberne Dr. Hollander has recently shown me some material that indicated the liver was the one area that was not affected at even thirty-six months. There was, however, a whole spectrum of other kinds of spontaneous tumors.

Weisburger The brown Norway rats that we have at the NIH have a high incidence of spontaneous bladder tumors.

Becker You said that there were pituitary adenomas in the aging male of this special strain. Are they functional in any way?

Knook Yes, Boorman has studied this, and they are functional.

Becker The reason I ask is that although the females may not show the adenomas, there may be a hypersecretion of growth hormone. We know that growth hormone induces and encourages carcinoma and adenoma formation, so there may be an association here between pituitary secretion and the lesions you see in the liver.

Leffert You said something about the possible involvement of pituitary hormones in elderly rats; a great deal of my talk will be concerned with hormones, and I would ask you whether or not additional anterior pituitary hormones like TSH and growth hormone could have a pancreatic effect and could be mediating insulin release? There is good evidence that insulin and glucagon in in vitro studies are involved in hepatocellular regeneration.

Becker Yes. I agree with you. It has been reported that continual hypersecretion of some of the pituitary hormones may be stimulatory. I was envisioning these old rats pouring out a series of anterior pituitary hormones, so that specific identification of the type of pituitary adenoma could be of tremendous interest. It might give us some information as to which specific cell and which specific hormone or hormonal combination is involved. And this would, by the way, make this animal a very complicated animal to use for chemical carcinogenesis studies.

3 Characterization of Hepatic Nodules

Frederick F. Becker
Department of Anatomic and Experimental Pathology
M. D. Anderson Hospital
Tumor Institute
6723 Bertner
Houston, Texas 77030

3.1 Introduction

A premalignant lesion can be defined as one from which malignant tumors arise. This concept need not include a sense of irrevocability, however, and perhaps should not exclude a sense of reversibility. The exact mechanisms by which such lesions arise and those factors that further their evolution to malignancy are unknown, despite the devotion of a considerable effort to the identification of such lesions in tissues exposed chronically to chemical carcinogens. Several benefits would result from the identification of such lesions:

1. Recognition of similar lesions in human tissues might alert us to the possibility of a chemical etiology for a given lesion.
2. In both experimental and human tissues, attempts to reverse carcinogenic evolution could be applied prior to and subsequent to the appearance of such premalignant lesions.
3. The availability of such lesions would allow us to focus our analyses and thus avoid the danger of "dilution" inherent in utilizing an entire organ, which includes a quantity of tissue(s) not evolving toward malignancy. It would also help us to understand better the multiphasic sequence of metabolic and biologic events leading to the appearance of malignancy.
4. The characterization of such lesions might enable us to understand better the characteristic(s) of the malignant cells. It must be pointed out that, to date, we have been unable to define the feature(s) acquired by cells during their exposure to carcinogens that enable(s) them to escape the normal control mechanisms, to grow, to invade, and to destroy. It is possible that our identification of a premalignant lesion will enable us to identify such crucial characteristics.

This chapter reflects data available at the time of the workshop.

3.2 Hepatic Nodules as Premalignant Lesions

Perhaps the most intensively studied carcinogenic sequence is that which results in the liver of rats during and after the ingestion of chemical carcinogens. Almost invariably, a successful dietary regimen results in the appearance of aggregates of morphologically and metabolically altered hepatocytes, termed hepatic nodules (Farber, 1973; Firminger, 1955; Reuber and Firminger, 1963). The nodules apparently result from a combination of cell division and cell hypertrophy and vary in size from 1 mm to 3 cm. They are considered to play a significant role in the eventual appearance of malignant hepatocellular carcinomas. Supporting evidence can be summarized as follows:

1. Almost every hepatocarcinogenic agent, when administered chronically, results in the appearance of such nodules (Farber, 1973).
2. In numerous instances, focal aytpical cells have been identified histologically within such nodules prior to the appearance of gross hepatomas.
3. A number of characteristics noted in malignant cells but not in normal interphase hepatocytes have been detected in the cells of these nodules.

Perhaps the strongest evidence supporting this thesis, however, was originally noted by Reuber (1965) when he reported that some hepatic nodules in rats disappeared when subcarcinogenic dosing with a chemical agent was terminated. Teebor and I obtained confirming data from experiments for which we devised an intermittent feeding regimen that was highly productive of hepatomas (Teebor and Becker, 1971). The dietary schema consisted of feeding a high dose of 2-N-fluorenylacetamide (2-FAA) for three weeks followed by a single week of normal diet. After three such "cycles" had been administered, the rat livers demonstrated a significant degree of gross nodularity. However, few if any of these rats eventually demonstrated hepatomas, despite the appearance of a significant number of hepatomas in rats fed a comparable total dose on a continuous regimen. When one additional feeding cycle was added, the nodularity of the liver increased and there was a high incidence of hepatocellular carcinoma.

Serial laparotomy of animals in this study demonstrated that the vast majority of the nodules induced by the three-cycle feeding regimen disappeared within three months after carcinogen feeding was discontinued. In contrast, a high percentage of the nodules present at the termination of the fourth feeding cycle persisted until cancer appeared within those livers. No liver in which total regression of the nodules occurred developed a tumor. Thus, one could state more critically that it was the persistent hepatic nodule that was at high risk for malignancy. Our findings further supported the concept that has been strongly espoused by Farber (1973) and others (Becker, 1971) that the process of nodule formation is a dynamic, multiphasic sequence in which progressive alterations are induced in hepatocytes by the continuing presence of the carcinogen. Thus the eventual premalignant cell, which could be defined simply as having an irrevocable impulse toward malignant evolution, would result from a continuing action of carcinogen upon a continually altering cell population.

It is of crucial importance in any discussion of hepatic nodules to emphasize their diversity. Reuben (1965) and others have suggested that the initial nodularity seen in livers exposed to carcinogen results from regenerative activity in response to the death of normal hepatocytes. The reversibility of many hepatic nodules following cessation of a subcarcinogenic dose supports this concept. However, there have been several pieces of evidence that indicate that the cells of reversible nodules are different from normal hepatocytes and certainly different from normal regenerative hepatocytes (Bannasch, 1968; Becker, 1969; Becker et al., 1972; Kitagawa and Sugano, 1973). The concept may be more applicable to livers exposed to necrogenic carcinogens than to those which are less overtly damaging. In any case, according to this schema, the regenerative nodule is eventually replaced, possibly by continuing cell division, and results in the persistent nodularity characteristic of the premalignant liver. In a manner as yet undefined, further alteration, with or without the intervention of carcinogen, may select the few nodules that will eventually become liver cell cancer.

Regardless of the correctness of these concepts, the hepatic nodule represents an important tissue in our analysis of chemical carcinogenesis. If offers us the opportunity of examining a carcinogen-altered tissue at various stages in its progression toward malignancy. Equally significant is our ability to obtain sufficient material from a single nodule for analysis and for comparison with nodules of other stages and with nonnodular parenchymal areas (Merkow et al., 1967; Teebor and Becker, 1971).

It must be pointed out that the nodules, even the persistent nodules that are at high risk for malignancy, cannot be construed as the ultimate tissue. At least two major problems exist in our interpretation of results regarding these nodules. The first is the existence within the nodule of significant subpopulations of cells that differ strikingly from the majority. Second, and a corollary of the first problem, is the dynamic nature of alterations within these nodules—a feature that may or may not be understood by examining the persistent nodule. Numerous studies have suggested the presence of such subpopulations within the hepatic nodules.

3.3 Characterization of Hepatic Nodules

Merkow et al. have demonstrated, by electron microscopic examination, cells within the nodules that possess unique cytostructure (Epstein et al., 1967; Merkow et al., 1971). These cells demonstrate histologic atypia with varied nuclear size and structure (Becker et al., 1971). Equally interesting have been the studies of Daoust and his colleagues demonstrating subpopulations of cells with altered distribution of RNA within the nodules (Daoust and Calamai, 1971; LePage et al., 1973). The experiments of Kitagawa demonstrate strikingly the presence of varied cell populations and, in addition, delineate the dynamic quality of the progression of the growth of the nodules (Kitagawa and Sugano, 1973). In recent reports Kitagawa has described the variability of cells of the nodule throughout exposure to carcinogen. In the initial stages of nodule formation, cells are present that are actively engaged in mitotic activity and are enzymatically "immature"; subsequent examination demonstrates that these cells divide and mature, in terms of normal or excess enzyme content. Such cells then "blend" into the normal architecture of the liver. At this stage the residual nodules are comprised of enzymatically immature cells that continue to divide progressively. Further evidence for the dynamic quality of the hepatic nodules has been offered by Daoust, who has reported that focal cell aggregates demonstrate a deficiency in RNAse prior to the acquisition of altered RNA distribution (Fontainiere and Daoust, 1973). The eventual, altered foci represent the hyperbasophilic areas that appear in nodules, and they may represent the final sequential stage prior to overt malignancy. Work from my own laboratory has also contributed to the accumulated evidence for the presence of subpopulations of cells within the nodules and a continuing evolution of the nodules' properties. It has been demonstrated that the cells of these nodules exhibit a continuing, excessive mitotic activity, compared to the nonnodular liver (Becker and Klein, 1971; Karasaki, 1969). This phenomenon continues long after the cessation of carcinogen (Becker and Klein, 1971). In addition, the majority of the cells of these nodules remain responsive to the mitotic stimulus of 70 percent hepatectomy (Becker and Klein, 1971; Kitagawa, 1971). The normal dividing hepatocyte is exquisitely sensitive to the inhibitory property of the enzyme L-asparaginase (Becker and Broome, 1967). A combination of 70 percent hepatectomy and L-asparaginase was used to demonstrate that the enzyme inhibited the mitotic response of the majority of the cells of the nodule; however, small foci of cells were unaffected and divided (Becker and Klein, 1971). Although this evidence cannot be considered absolute, it suggests the presence of clonal subpopulations.

In a parallel series of experiments, the capacity of such nodules to produce plasma protein was tested and compared to that of normal and regenerating liver (Becker et al., 1972). In this approach, small fragments of a single nodule or a comparable aliquot of nonnodular tissue are incubated in vitro in the presence of radioactively labeled amino acids. Subsequent analysis by immune-electrophoresis radioautography enabled us to estimate the rates of synthesis of a broad spectrum of normal plasma proteins. Protein synthesis was found to be generally suppressed in the nodules as a result of a full carcinogenic regimen; a few proteins were produced in near normal quantities, but the production of most was diminished or absent. This selective effect indicated that the process was not a simple alteration of amino acid transport, pool size, or a nonspecific "translation attack." Later on, however, when persistent nodules were examined, a much more varied pattern of plasma protein synthesis emerged. First, the pattern was different for each nodule examined. Furthermore the level of synthesis varied within each nodule. The production of certain plasma proteins was still suppressed; others were again being produced at normal levels, and in several instances plasma protein synthesis was considerably enhanced. Thus, the functional capacity of the nodules exhibits sequential alteration, individual variation, and significant difference from normal tissues, as late as six months after the cessation of carcinogen feeding. The finding of variation among persistent nodules must make us cautious in interpreting our findings. In addition to differences in plasma protein synthesis between nodules at this stage, several laboratories (Farber, 1973; Teebor and Becker, 1971) have demonstrated that many persistent nodules fail to evolve as tumors even when followed through the life span of the rat. After twelve to fourteen months of an experiment, almost no malignancies are found despite the presence of nodules.

Another major area of investigation of chemical hepatocarcinogenesis has focused upon the metabolic apparatus of the target cells; a variety of results, depending somewhat upon the conditions utilized, has emerged. Thus it is of absolute importance in comparing such results to keep in mind the carcinogen used, the dietary regimen, the stage of exposure, the strain and age of rat, and many other modifying conditions.

The most striking features of most metabolic studies have been the suggestions that the hepatic nodules may demonstrate aberrant enzyme response to various stimuli (Horton et al., 1973; Poirier and Pitot, 1969), and that the nodules may possess "fetal enzymes" (Potter, 1973; Yanagi et al., 1974). The former relationship has been examined in depth by many investigators attempting to demonstrate a correlation between the form of aberration in the control of enzyme activity and the development of malignant tumors. The search for fetal enzymes had the same purpose, that is, to demonstrate that tumors and nodules possess a reemerging fetal metabolic pattern. Although both approaches have identified similarities between the nodules and the tumors that develop from them, in some instances these relationships did not occur (Bannasch, 1968; Teebor and Seidman, 1970). In no instance has the metabolic process so delineated been adequate to explain malignant behavior. Thus, at present, these changes must be considered only to demonstrate alteration in metabolic regulation induced by the carcinogen; they may or may not shed light on the basic nature of the oncogenic process.

Several recent papers have reported the induction of the fetal plasma protein, alpha-fetoprotein (α_1 F), during the exposure of the liver to chemical carcinogens (Akazaki, 1973; Kroes et al., 1972 and 1973). These studies were stimulated by the reported synthesis of this protein by a large percentage of chemically induced liver tumors. Alpha-fetoprotein was detected in the serum at the time of appearance of the hepatic nodules, and it was hypothesized that its presence was evidence of the premalignent nature of these nodules. Two problems arose regarding this hypothesis. First, a significant percentage of such tumors appear not to produce α_1 F (despite the application of extremely sensitive radioimmunoassay) (Becker et al., 1973). Second, the methods used to determine circulating α_1 F were relatively insensitive. In collaboration with S. Sell, we have applied an extremely sensitive radioimmunoassay to the determination of circulating α_1 F during exposure to 2-FAA (Becker and Sell, 1974). A significant elevation of α_1 F began to appear almost immediately, even at exposures as low as 0.001 of a carcinogenic dose. No tissue alteration was demonstrable at that time, and no significant, further increase in the circulating level was noted when hepatic nodules appeared. Thus, although it remains possible that some subpopulation of hepatic nodule cells produces α_1 F, it is difficult at this time to impart any significnace to this relationship. It is our contention that the early onset of synthesis is indicative of a specific derepression induced by carcinogens (chemical analogues have failed to produce such elevation) and that this phenomenon represents another piece of evidence for the multisequential effect of chemical carcinogens.

3.4 Recent Studies

In parallel studies, my laboratory has developed techniques that permit cytogenetic analysis of the chromosome composition of solid hepatic tissues (Becker et al., 1971). This approach made it possible to detect chromosomal abnormalities in the cells of hepatic nodules induced by 2-FAA. However, although the mitotic rate of the cells of the nodule is elevated compared with that of normal tissue, it is lower than required for examination of spontaneous cell division. We therefore induced high mitotic rates by performing 70 percent hepatectomies on the experimental animals (Becker and Klein, 1971). The cells were then examined cytogenetically and were compared with normal regenerating hepatocytes and circulating lymphocytes.

Some chromosomal gaps and breaks were evident in cells derived from nodules during early exposure to carcinogen (three cycles), but these features were not detected following a complete dietary regimen (four cycles) or in the persistent nodule. There were two possible explanations: either the chromosomal alterations had led to a failure of these cells to propagate or the damage had healed and did not appear as detectable lesions in later instances. Of equal interest was a dramatic arrest of maturation of cell ploidy. Thus, although age-paired rats demonstrated a progressive increase in the percentage of tetraploid cells, the cells of the hepatic nodules did not increase similarly. This persistent diploidy might represent further evidence for clonal selection of nodules' cells (Scherer and Hoffmann, 1971; Webber and Stich, 1965), or it may have broader implications related to the reemergence of fetal characteristics. Preliminary studies of the cells of animals exposed to azo carcinogens have revealed much more dramatic chromosome alterations (A. Horland, S. R. Wolman, and F. F. Becker, unpublished data). Alterations in Giesma banding patterns have been detected as well as a tendency toward alterations in chromosome number. Whether these changes represent irreversible and significant trends is yet to be determined, but the differences between 2-FAA and azo dye carcinogens may be related to their different capacities for cytotoxicity. Azo carcinogens are far more toxic to hepatocytes than are aromatic amides at comparable tumor-inducing doses.

A persistent and widely reported problem has been an inability to transplant hepatic nodules successfully (Farber, 1973; Reuber, 1965). The semantic problem—that tissues that grew after transplantation were termed liver cancers (hepatomas)—does not appear to be the only explanation. Thus fragments of nodules planted in omentum, peritoneum, subcutaneously, or in thoracic air sac resorb completely and demonstrate no persistent tissue (F. F. Becker and M. J. Folkman, unpublished data). It is well known that many tissues will persist at a diameter of 1–2 cm, nourished by diffusion of nutrients (Folkman, 1974). The failure of nodular tissue to persist, despite the use of highly inbred rats, may possibly result from some antigenic alteration induced by chemical carcinogens. A recent study by Farber suggests that an unusual antigen exists in the cells of such nodules, but this finding needs further clarification (Farber, 1974). It was apparent from our own studies of transplantation of nodules (Becker and Folkman, unpublished data) that the cells of transplanted nodules did not possess the important ability to produce tumor angiogenic factor, as described by Folkman (1974). That the cells of hepatic nodules could be clearly differentiated from those of liver cell tumor was demonstrated by the rapid growth of tumors and the ability of these neoplasms to induce vascular proliferation (Becker and Folkman, unpublished data). Interestingly, in a single instance a nodule regressed completely, by gross examination, and subsequent rapid growth was detected after a lag period of one month. This tissue proved to be a typical liver cell tumor. It demonstrated an enormous capacity to induce vascular proliferation and possibly arose from a single malignant clone.

3.5 Conclusions

In biological research or in the investigation of pathological lesions, we must often utilize the *best available* tissue rather than the *best tissue.* Although cultured mammalian cells or lower organisms may add immensely to our understanding of mechanisms of cell damage and carcinogenesis, more frequently they offer clues that must be pursued in vivo. The hepatic nodule—in particular, the persistent hepatic nodule—would be a *best tissue* if it were composed of a clone of cells in which each cell had equal potential for malignancy, via the same mechanisms. Although this is a possibility (Webber and Stich, 1965; Scherer and Hoffman, 1971), studies of subpopulations within nodules indicate that it is unlikely. For the moment then, the hepatic nodule is the *best available* tissue to which we can apply new analytical techniques and upon which we can test new oncological concepts. By this approach we may be able to define the population of premalignant cells within these structures and go on to identify the mechanisms by which such cells appear and progress to malignancy.

3.6 Acknowledgment

Portions of this work were supported by Grant No. 12141 from the National Cancer Institute and Contract E72-3258, National Cancer Institute.

3.7 References

Akazaki, K., et al., eds. 1973. *Malignant Diseases of the Hemapoetic System. Gann Monograph on Cancer Research*, no. 15. Baltimore, Md.: University Park Press.

Bannasch, P. 1968. *The Cytoplasm of Hepatocytes during Carcinogenesis. Recent Results in Cancer Research*, vol. 19. New York: Springer-Verlag.

Becker, F. F. 1969. Structural and functional correlation in the regenerating liver. In *Biochemistry of Cell Division*, ed. R. Basgera, p. 113. Springfield, Ill.: C. C. Thomas Co.

Becker, F. F. 1971. Cell function: its importance in chemical carcinogens. *Fed. Proc.* 30:1736-1741.

Becker, F. F., and Broome, J. D. 1967. L-Asparaginase: inhibition of early mitosis in regenerating rat liver. *Science* 156:1602-1603.

Becker, F. F., Fox, R. A., Klein, K. M., and Wolman, S. R. 1971. Chromosome patterns in rat hepatocytes during N-2-fluorenylacetamide carcinogenesis. *J. Nat. Cancer Inst.* 46:1261-1269.

Becker, F. F., and Klein, K. M. 1971. The effect of L-asparaginase on mitotic activity during N-2-fluorenylacetamide hepatocarcinogenesis: subpopulations of nodular cells. *Cancer Res.* 31:160-173.

Becker, F. F., Klein, K. M., and Asofsky, R. 1972. Plasma protein synthesis by N-2-fluoroacetamide—induced primary hepatocellular carcinomas and hepatic nodules. *Cancer Res.* 32:914-920.

Becker, F. F., Klein, K. M., Wolman, S. R., Asofsky, R., and Sell, S. 1973. Characterization of primary hepatocellular carcinomas and initial transplant generations. *Cancer Res.* 33:3330-3338.

Becker, F. F., and Sell, S. 1974; Alpha-1-fetoprotein in N-2-fluorenylacetomine hepatocarcinogenesis. *Cancer Res.* 34:2489-2494.

Daoust, R., and Calamai, R. 1971. Hyperbarophilic foci as sites of neoplastic transformation in hepatic parenchyma. *Cancer Res.* 31:1290-1296.

Epstein, S. M., Ito, N., Merkow, L., and Farber, E. 1967. Cellular analysis of liver carcinogenesis: the induction of large hyperplastic nodules in the liver by 2-fluorenylacetamide or ethionine and some aspects of their morphology and glycogen metabolism. *Cancer Res.* 27:1702-1711.

Farber, E. 1973. Hyperplastic liver nodules. In *Methods in Cancer Research*, vol. 7, ed. H. Busch, pp. 345-375.

Farber, E. 1974. Pathogenesis of liver cancer. *Arch. Path.* 98:145-148.

Firminger, H. I. 1955. Histopathology of carcinogenesis and tumours of liver in rats. *J. Nat. Cancer Inst.* 15:1427-1442.

Folkman, M. J. 1974. Proceedings: tumor angiogenesis factor. *Cancer Res.* 34: 2109-2113.

Fontanière, B., and Daoust, R. 1973. Histochemical studies on nuclease activity and neoplastic transformation in rat liver during diethylnitrosamine carcinogenesis. *Cancer Res.* 33:3108-3111.

Horton, B. J., Horton, J. D., and Sabine, J. R. 1973. Metabolic controls in precancerous liver. V. Loss of control of cholesterol synthesis during feeding of the hepatocarcinogen 3′-methyl-4-dimethylaminoazobenzene. *Eur. J. Cancer* 9:573-576.

Karasaki, S. 1969. The fine structure of proliferating cells in preneoplastic rat livers during azo-dye carcinogenesis. *J. Cell. Biol.* 40:322-335.

Kitagawa, T. 1971. Responsiveness of hyperplastic lesions and hepatomas to partial hepatectomy. *Gann* 62:217–224.

Kitagawa, T., and Sugano, H. 1973. Combined enzyme histochemical and radioautographic studies on areas of hyperplasia in the liver of rats fed N-2-fluorenylacetamide. *Cancer Res.* 33:2993–3001.

Kroes, R., Williams, G. M., and Weisberger, J. M. 1972. Early appearance of serum fetoprotein during hepatocarcinogenesis as a function of age of rats and extent of treatment with 3-methyl-4-dimethylaminoazobenzene. *Cancer Res.* 32:1526–1532.

Kroes, R., Williams, G. M., and Weisburger, J. H. 1973. Early appearance of serum-fetoprotein as a function of dosage of various hepatocarcinogens. *Cancer Res.* 33: 613–617.

LePage, R., Moulin-Camus, M. C., De Lamirande, G., and Daoust, R. 1973. Quantitative estimations of RNA sensitive to mild RNAse treatment in sections of normal, regenerating and neoplastic rat livers. *Cancer Res.* 33:2609–2614.

Merkow, L. P., Epstein, S. M., Caito, B. J., and Bartus, B. 1967. The cellular analysis of liver carcinogenesis: ultrastructural alterations within hyperplastic liver nodules induced by 2-fluorenylacetamide. *Cancer Res.* 27:1712–1721.

Merkow, L. P., Epstein, S. M., Slifkin, M., and Pardo, M. 1971. Ultrastructural alterations within hyperplastic liver nodules induced by ethionine. *Cancer Res.* 31:174–178.

Poirier, L. A., and Pitot, H. C. 1969. Dietary induction of some enzymes of amino acid metabolism during azo dye feeding. *Cancer Res.* 29:475–480.

Potter, V. R. 1973. Biochemistry of Cancer. In *Cancer Medicine,* eds., J. F. Holland and E. Frei, p. 178. Philadelphia: Lea & Febiger.

Reuber, M. D. 1965. Development of preneoplastic and neoplastic lesions of the liver in male rats given 0.025 percent N-2-fluorenyldiacetamide. *J. Nat. Cancer Inst.* 34:697–723.

Reuber, M. D., and Firminger, H. I. 1963. Morphologic and biologic correlation of lesions obtained in hepatic carcinogenesis in A × C rats given 0.025 percent N-2-fluorenyldiacetamide. *J. Nat. Cancer Inst.* 31:1407–1429.

Scherer, E., and Hoffmann, M. 1971. Probable clonal genesis of cellular islands induced in rat liver by diethylnitrosamine. *Eur. J. Cancer.* 7:369–371.

Teebor, G. W., and Becker, F. F. 1971. Regression and persistance of hyperplastic hepatic nodules induced by N-2-fluorenylacetamide and their relationship to hepatocarcinogenesis. *Cancer Res.* 31:1–3.

Teebor, G. W., and Seidman, I. 1970. Retention of metabolic regulation in the hyperplastic hepatic nodule induced by N-2-fluorenylacetamide. *Cancer Res.* 30:1095–1101.

Webber, M. M., and Stich, H. F. 1965. Combined effects of X-irradiation and 3-methyl-4-dimethylaminoazobenzene on liver cell population. *Canad. J. Biochem.* 43:811–815.

Yanagi, S., Makiura, S., Akai, M., Matsumura, K., Hirao, K., Ito, N., and Tanaka, T. 1974. Isozyme patterns of pyruvate kinase in various primary liver tumors induced during the process of hepatocarcinogenesis. *Cancer Res.* 34:2283–2289.

3.8 Discussion

Leffert How long must you give the L-asparaginase, and at what times is there inhibition of mitosis? Did you look at asparagine synthetase in the nodules?

Becker When given from the time of partial hepatectomy to about eight hours it is effective but after twelve hours ineffective. However, it only delays the response and does not abate the response. On multiple doses, the mitotic response breaks through so that we would expect the DNA synthetic response to be delayed in the normal liver and in the nodule by about eight hours. Following partial hepatectomy, asparagine synthetase is induced from zero, and the liver has a very substantial asparagine level. Therefore, asparaginase sensitivity of the regenerating liver is a relative need for more asparagine at that time, and it can overcome the need itself even in the presence of the L-asparaginase. We have not looked at asparagine synthetase in the nodules. The nodules responded to 70 percent hepatectomy with a mitotic rate of a 60-gram rat—a young rat—as against the rat that is 300 grams.

We studied a number of our own and some of Morris's tumors and found that sensitivity to L-asparaginase is inversely proportional to growth rate; that is, if you take a tumor which takes nine months to grow, it has a sensitivity to L-asparaginase similar to that in normal liver. Its overall DNA synthesis is suppressed simply by asparaginase. Fast-growing tumors are resistant, but the tumors are diverse and one just cannot make bold statements about them.

Leffert In your plasma protein studies, have you investigated the possibility that the cells could be synthesizing but not secreting plasma protein?

Becker We are analyzing for the alpha-1-antitrypsin component. If you put rat sera that contain α_1F on electrophoretic gel, then you find α_1F and alpha-1-antitrypsin. They are both lipoproteins. Alpha-fetoprotein, which we have tested, has no antitrypsin activity. Our impression is that when we put tumors into our incubates, they appeared to secrete protein. They had no secreting difficulty. There is other evidence that α_1F is really secreted by the tumor. Sell has shown that when he innoculated 10^6 cells of a tumor which produces enormous levels of α_1F, he could detect an elevation of α_1F in the serum before he could detect the tumor. Others have suggested that these proteins are not secreted by a tumor but that the tumor necroses and releases them. That is not really likely, as a tumor undergoes necrosis in the center of a spherical area. Although this area is probably vascularized, it is not well vascularized. The chances of getting a formed plasma protein out of this necrotic mass I consider very unlikely.

Leffert Less than 20,000 fetal cells can produce detectable α_1F.

Weisburger Dr. Becker mentioned the differing response of the various strains to acetylaminofluorene. We have compared the induction of different neoplasms and find that breast cancer varies all the way from zero to about 65 percent, liver cancer from zero to some 80 percent. For ear duct and the intestinal tumors there is great variation. Some strains have high incidence and the rest very low. Bladder cancer also shows a tremendous variation in response. All this is to just one carcinogen. It is very difficult to believe that there is such a tremendous difference in the metabolism to explain this variable effect. We have also studied the effect of sex on the response to diacetylaminofluorene in the Irish rat. We find that the female rat had a very low incidence of liver cancer and the males a high incidence. When the female rats were castrated and given testosterone, they reacted like the intact males, and when the males were castrated and given stilbesterol, they reacted almost like the intact female.

Goyings Dr. Weisburger's data shows a rather significant difference in the distribution of tumors with a common carcinogen. There ought to be some implications about this with respect to the type of model system that we should select for carcinogenic testing. Do you have any comments in regard to that?

Weisburger With respect to selecting a rat for carcinogenic tests, I would not suggest a Long-Evans rat, for example, because they are very resistant. Some kinds of rats show very high spontaneous tumor rates, which complicates the evaluation of the results. We have no ideal rat, and we are still searching for this animal.

Becker Now, if you used the Fischer Rat, they are very sensitive to 2-FAA. We started to use Fischers, just to see what would happen in our α_1F experiment. We gave a day or two of our dose, and within two days, where others had begun to show a statistically significant elevation of α_1F, the male Fischer rats had higher levels than those with tumors.

Merkow In that 3-by-3 and 4-by-3 cycle, how long did you wait, or how old is the oldest rat in the 3-by-3 cycle? Did you let them live as long as they could and then autopsy them to find out what the incidence of hepatomas or carcinomas was? The second part of my question is, do you have any explanation or suggestion as to why the 3-by-3 and 4-by-3 produce such great differences in the incidence of hepatocarcinomas at the end?

Becker We have followed them until death. Many of them live over two years. In fact, sometimes if you give a low dose of carcinogen, they live longer than the normal animal.

We have done the experiment twice; in one series (fed 3-by-3 cycle) when they finally all died off, we had an aggregate percentage of around 4 or 5 percent of the remaining livers with persistent nodules, 4 percent hepatocarcinomas. In the second experiment we had about 9 percent persistent nodules after this regimen and 4 percent tumors.

We do not know any reason why one more feeding cycle reinforces or impresses or makes irrevocable the condition of premalignancy. The only thing that we have been able to detect is this plasma protein difference. There is one possibility, based on the work on partial hepatectomy plus carcinogens. We did one thousand weanling rats with five carcinogens, giving the carcinogens at five points in the mitotic cycle, and we did not get one hepatoma. Now, throughout exposure to carcinogen, mitotic activity is raised in these nodules.

It may be that during this one last burst of DNA synthesis transcription or alteration has occurred, and the chromatin has changed and nonhistone proteins have shifted. We know there is more euchromatin which may be more vulnerable. It could be something very simplistic like one more wave of mitosis. It is a commonly held view that cell division is imperative in the process of chemical carcinogenesis no matter what tissue you are talking about. This may not be so of some highly potent carcinogens, but I cannot answer that. In most systems I have described, cell division is absolutely characteristic.

Butler You said these early nodules are diploid. Is there ever enough nonnodular liver to study the ploidy?

Becker Yes, but there is no way of doing it because you cannot get mitotic figures. It does not respond to 70 percent hepatectomy. You need a certain number of mitotic figures to karyotype. So the way we do this is to partially hepatectomize these animals, and then we study the mitotic figures that result in the nodules, but the nonnodular tissue never responds because it is under the suppressive effect. Now, two months after the carcinogen ceases, you can begin to study it.

Butler Dr. Neal at Carshalton has been looking at this by fractionating nuclear populations following aflatoxin treatment. One does not see the normal maturation from diploid to tetraploid nuclei, so that a nonnodular liver with diploid population of parenchymal cells results. When the diet is stopped the changed distribution of ploidy persists for two years. When the carcinoma is present, it appears to be diploid as well.

Becker I think that is terribly important, because, again, it shows that the effect is diffuse.

P. Newberne With 2-FAA the liver goes through the stages you have described and on to hepatocarcinoma. Now, what is your comment on giving Compound B going through this stage and getting to the nodule stage? Do you say it might be a carcinogen? We have, in fact, people who are saying that.

Becker I have already done it actually, in this respect: we gave acetylaminofluorene, 3 × 3 (I might add that control animals given only that developed no carcinomas in this particular run, although we only followed them for fourteen months. We then took the same regimen, 2-FAA 3 × 3, and we gave a single dose, 5 mg/kg, IP, of dimethylnitrosamine. I understood that this single dose produces no hepatocarcinomas, and we had controls which did not get carcinoma. When we gave this dose of DMN to these animals, we got 100 percent carcinomatous transformation at about the same time. The nodules almost all remained persistent through the life span of the animal and many livers had four and five foci of malignant tumor.

The answer is, it is not that any dose of any carcinogen will make them persist and keep them persistent and product tumors. If the nodules do stay persistent, you seem to get tumors.

Bannasch It is difficult for me to comment about your investigations as I have not seen the slides of your nodules, but I feel that reversible or persistent nodules are basically two different situations. I do not know whether it is valid to do comparative studies on reversible and persistent nodules if you want to investigate carcinogenesis. We should avoid it because we have so many cellular reactions which are either toxic or regenerative in nature. I think these changes cannot be a very early stage of carcinogenesis because they are reversible. Toxic and degenerative changes can be avoided if we lower the dose.

Becker If the dose of 2-FAA is reduced to a level where nodules are not produced, hepatocarcinomas also are not induced. On continuous feeding of 0.05 percent 2-FAA, you get a high mortatity and a very high percentage of tumors at about six months and also enormous nodularity. At 0.015 percent you get very little nodularity even after 150 days and less than 5 percent hepatocarcinomas. At 0.01 percent we see no hepatocarcinomas or nodules.

I disagree with you about using those nodules. The reversible nodule is my control tissue. Our choices for control tissues are normal liver, 70 percent hepetectomized liver to control for the known mitotic activity, and normal liver exposed to the carcinogen but prior to nodularity. After the carcinogen you have a population in which the toxic effects of the agent probably predominate because the proliferative effects are not yet striking. But in the nodules you have a tissue altered from the normal, and they show certain characteristics of having been influenced by the carcinogen. They resulted from cell division which took place in the continuous presence of carcinogen. Initially if you feed the carcinogen for two days and then hepatectomize the rat, no cell division is seen. The normal liver is severly inhibited by 2-FAA. It is only when the nodules appear that the tissue becomes responsive to the stimulus of 70 percent hepatectomy. The first tissue that shows any response to hepatectomy is the basophilic focus which escapes the mitotic inhibitory effect of 2-FAA. The surrounding tissue shows no mitosis.

Bannasch Can you say anything about the relation of the reversible and the persistent nodules under the influence of different doses?

Becker Reuber and Firminger (*J. Nat. Cancer Inst.* 31:1407, 1963) described it with a dose of 0.025 percent. They described reversibility of nodules with the subcarcinogenic dose at 0.025 percent, and persistence with 0.025 percent given longer. When 3′-methyl-dimethylaminoazobenzene (DAB) is fed for four weeks and stopped, the nodules disappear and no carcinomas develop. If 3-methyl-DAB is fed for more than four weeks, the nodules persist and carcinoma results. When we have used a lower dose, we get nodules which reverse and nodules which persist. It just depends on where you stop treatment. Our nodules are histologically and ultrastructurally identical to those that others have published. The nodules are composed of large eosinophilic cells with a foamy cytoplasm, a high cytoplasmic-nuclear ratio, and most of the nuclei are of diploid size and diploid in terms of spectrophotometric analysis. They have alterations in glycogen staining.

Butler You have suggested that if you induce a carcinoma the liver goes through a phase of nodule formation.

Becker No. What I said is that the nodule is a phase of evolution toward carcinoma. You do not get nodules when you give some carcinogenic agents following that 70 percent hepatectomy. Does aflatoxin produce nodules?

Rogers Yes, certainly. Nodules, but not cirrhosis.

Butler In the system we use with aflatoxin, we feed the compound for six weeks, then stop, which will induce 100 percent carcinoma, but we do not got through this nodular phase during the six weeks feeding.

Becker I am saying that when you have nodules, in this situation that I have described, they are related to cancer. There are many pathways to cancer.

Butler I am interested to know where you consider you have induced carcinoma. I would think that in our material at six weeks there is an irreversible change and the carcinoma is present. I have no idea how to identify it, but I do not presume that there are multiple phases after that.

Becker Well, let me ask you the question before I even respond to that. How long is it before you can identify the smallest malignancy, one that would transplant?

Butler This is what we are starting to do, but do not necessarily accept transplantation as evidence of malignancy.

Becker Let me just point out one thing, because I do disagree with you. I do not think you have cancer. I think that there are many things that happen. If you give most of these carcinogenic regimens, there is always a substantial lag period before any detectable change that you can identify as malignancy appears.

Butler This depends on how you interpret a lag period. I would suggest that the techniques are inadequate or insensitive. When one considers the kidney system we described (Hard and Butler, *Cancer Res.* 30:2806, 1970) nothing appeared to happen for six to eight weeks, but if you look at the kidneys closely, you can find morphologically transformed cells in four days, which will culture by twenty-four hours. So I suspect in other systems that there is no lag period.

Becker You may be absolutely right, but there is often a three- to four-month lag period before anything is detectable. Now what is more important is your definition of malignancy. If you define a malignant cell as one that is growing and accumulating mass as against a dormant cell that will grow or accumulate mass, then that is not true. But if you are willing to accept a cell which has been changed and will eventually manifest itself with growth as a malignant cell, then you are absolutely right. It is just a matter of definition. I do not define it as "malignant" until it starts to accumulate as a cell, that is, until one cell becomes two and four and you get a mass of cells. If is is dormant, it could be malignant or it might not be. It's not manifesting anything.

Butler I am not suggesting they are sitting there dormant. There are just so few of them and you have difficulty in finding them.

Becker That I agree with. But it then takes them three to five months to grow to where they are identifiable.

J. Newberne The dose response you illustrated so nicely is what some of us in the pharmacology world are always interested in. I think you may have demonstrated what I would like to call "no effect level." Would you allow yourself to call this a "no effect level" of carcinogen?

Becker Not without being abused by my colleagues. I would state fairly strongly that, under the conditions I have described, for this agent and these rats, we can give them many doses that produce effects in the liver and yet, within the life of the rat, never manifest themselves as hepatocarcinoma? and I am not particularly concerned by the argument that it would if the rat lived longer, because that is not the way the world goes. If I could be guaranteed that I would not have a carcinoma within my lifetime, I would be perfectly satisfied to admit I might have one afterward. So, what I can state—and I think it is true for everyone involved in carcinogenesis, I am sure even with the nitrosamines and aflatoxin—we could always select a dose that, given the lifetime of the animal, will not produce a carcinoma in a statistically significant number. Carcinogens have been administered at doses that never produce changes, in terms of carcinoma. So I would say "yes, that is absolutely true." In fact, I think the best evidence is that we are not all walking around with masses of carcinoma, since we are exposed to things all the time.

P. Newberne That is a very important point. I believe in a dose response in carcinogenesis, just as in toxicology. We have done some experiments recently with aflatoxin at very low levels of 1 ppb and up to 100 ppb, and we have tumors at 1 ppb in two or three animals; I think there are three animals out of the hundred. So I do not know that we can say that that is the "no effect level," but it is very very low. Now what about the other ninety-seven? Some developed nodules; others did not develop anything that I can see histologically. You have pointed out here that at very low levels of 2-FAA an effect is produced.

Becker We can give levels of acetylaminofluorene that produced thousand-fold or hundred-fold increases in α_1F and yet did not alter the cells histologically, by electron microscopy. There was no elevation of circulating serum glutamic oxalacetic or pyruvic transaminases, no alteration in glycogen pattern, no alteration in the basophilia or the ribonucleic acid pattern. The only change we find is an elevation of α_1F. 3′-Methyl-DAB at very low doses also triggers off a very high level of α_1F. You have to be a little more careful with 3-methyl-DAB because one gets a tremendous stimulation of mitosis which, as Dr. Leffert points out, releases α_1F.

Sigler Assuming you have various dose levels in carcinogenic tests and you have livers with bumps at all levels, even some of the controls, but compared to the life span of the controls, these are not life threatening. Is the compound really a carcinogen?

Becker I think a carcinogen is a carcinogen, but a carcinogenic dose is different. We have animals given a carcinogenic dose which we followed until their death; they have persistent nodules, but they never develop hepatocarcinoma.

4 Sequential Cellular Alterations during Hepatocarcinogenesis

P. Bannasch

Abteilung für Cytopathologie
Institut für Experimentelle Pathologie am
Deutschen Krebsforschungszentrum
Heidelberg, Germany

4.1 Introduction

The transformation of a normal liver cell into a cancer cell does not occur suddenly; rather it is a slow process that probably develops stepwise. A direct analysis of all the steps of this transformation should be possible, since there are many suitable methods that yield a high incidence of liver cell carcinomas in experimental animals. The fact that most tumors develop focally presents a problem, however. Another difficulty is that all hepatocarcinogens may kill many of their target cells while producing neoplastic transformations in others. This fact makes it nearly impossible, for instance, to distinguish biochemically, in homogenates of livers that had been exposed to a carcinogen up to the time of their examination, between carcinogenic and necrogenic effects.

The clarification of the morphological and biochemical changes during cellular transformation presupposes that the precancerous cell itself can be identified. During the past two decades, cytomorphological and cytochemical methods have been used for this purpose with some success. Our own investigations in this field were made on about 500 rats that received in their drinking water nitrosomorpholine (NNM)—a cyclic nitrosamine—or thioacetamide (TAA) in concentrations of between 5 and 50 mg/100 ml for different periods. (For detailed experimental data see Bannasch, 1968; Bannasch, 1969; Bannasch, 1974a; Bannasch, 1974b; Bannasch and Angerer, 1974; Bannasch and Miller, 1964; Bannasch and Reiss, 1971; Bannasch et al., 1972 and 1974.)

Under the influence of hepatocarcinogens, the liver cells show many alterations that can be classified as reversible, persistent, or progressive, depending on their further evolution after discontinuance of the carcinogen in so-called "stop experiments" (Bannasch et al., 1972). These changes result in characteristic cytotoxic patterns differing mainly in the quantity or structure of glycogen, the endoplasmic reticulum, and the ribosomes. No doubt there are also structural alterations of other cytoplasmic and nuclear components (see Bannasch, 1975), but with the exception of the enlargement of the nucleus and the nucleolus in late stages of hepatocarcinogenesis, other alterations seem to be rather inconstant and will therefore be disregarded here.

4.2 Reversible Cytotoxic Alterations

Reversible alterations have been well documented and are characterized by a predominantly centrilobular loss of glycogen (fig. 4.1) and by a concomitant dispersal of the basophilic bodies (Bannasch, 1968 and 1975). The electron microscope shows that this alteration of the ergastoplasm starts with a scattering of the granular cisternae throughout the cyloplasm (fig. 4.2). As judged by the cytochemically demonstrable activity of the membrane-bound enzyme glucose-6-phosphatase, initially the function of the scattered membranes appears to be normal (Bannasch and Angerer, 1974). The product of the cytochemical reaction is clearly visible in the form of dense lead deposits throughout the granular cisternae and the nuclear envelope (fig. 4.2). However, as a consequence of the disorganization of the ergastoplasm, the reaction sites come to be distributed all over the cytoplasm. Later on, the granular cisternae show dilatation, fragmentation, and partial loss of their ribosomes. At this stage the activity of the glucose-6-phosphatase may be reduced. Finally a vesicular transformation of the whole endoplasmic reticulum occurs together with a considerable reduction of the membrane-bound and the free ribosomes (Theodossiou et al., 1971).

Parallel radioautographic studies by several authors indicate that these changes represent the morphological equivalent of an impaired protein synthesis (Smuckler and Arcasoy, 1969).

Most cells thus affected undergo coagulation necrosis (fig. 4.3a) if the administration of the carcinogen is continued (Bannasch, 1968). At medium and high concentrations of the carcinogen, the necrosis elicits a compensatory proliferation of mesenchymal cells and bile ducts; fibrosis, cirrhosis, and, in severe cases, also cholangiofibrosis follow (Bannasch, 1968; Bannasch and Reiss, 1971). When the administration of the carcinogen is stopped, the loss of glycogen as well as the disorganization of the ergastoplasm and the reduction, in the center of the lobule, of the enzymatic activity are reversed within two to four weeks (Bannasch, 1968; Theodossiou et al., 1971; Bannasch and Angerer, 1974). After administration of low doses of the carcinogen, the cytoplasmic alterations just described occur only in a few single cells and necroses are absent or very rare (Bannasch, 1968; Bannasch et al., 1974). The nearly total lack of necrotic centrilobular alterations after low doses of carcinogens explains why neither cirrhosis nor cholangiofibrosis usually develops under these circumstances.

With sublethal doses, loss of glycogen and disorganization of the ergastoplasm (fig. 4.3a) may develop almost anywhere in the liver parenchyma (fig. 4.3a) (Theodossiou et al., 1971). In many hepatocytes

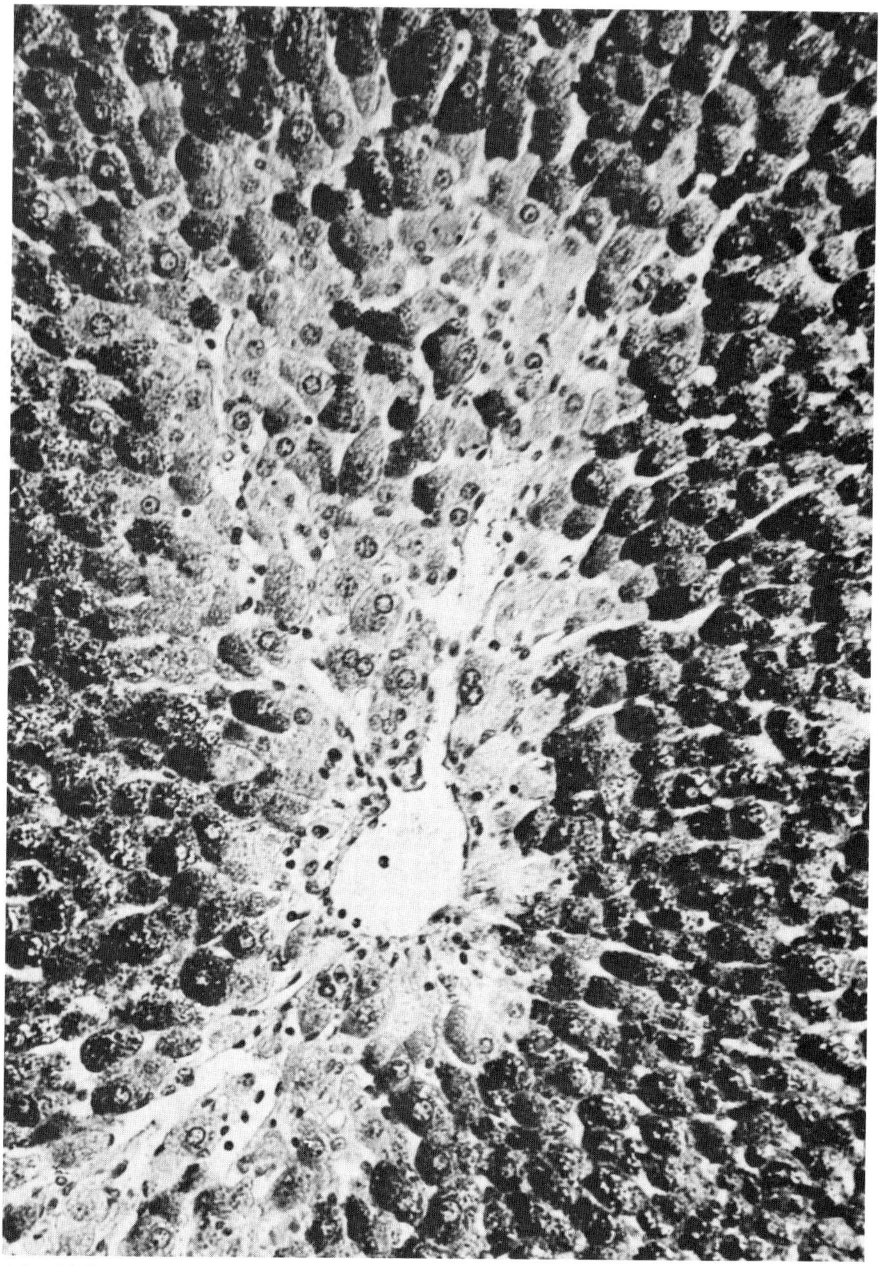

4.1 Light microscopic appearance of rat liver parenchyma after oral administration of a nitrosomorpholine (NNM) solution in drinking water (12 mg/100 ml) for 8 weeks: centrilobular loss and peripheral excessive storage of glycogen (black). Slight proliferation of mesenchymal cells in the immediate vicinity of the central vein (Tri-PAS, × 200).

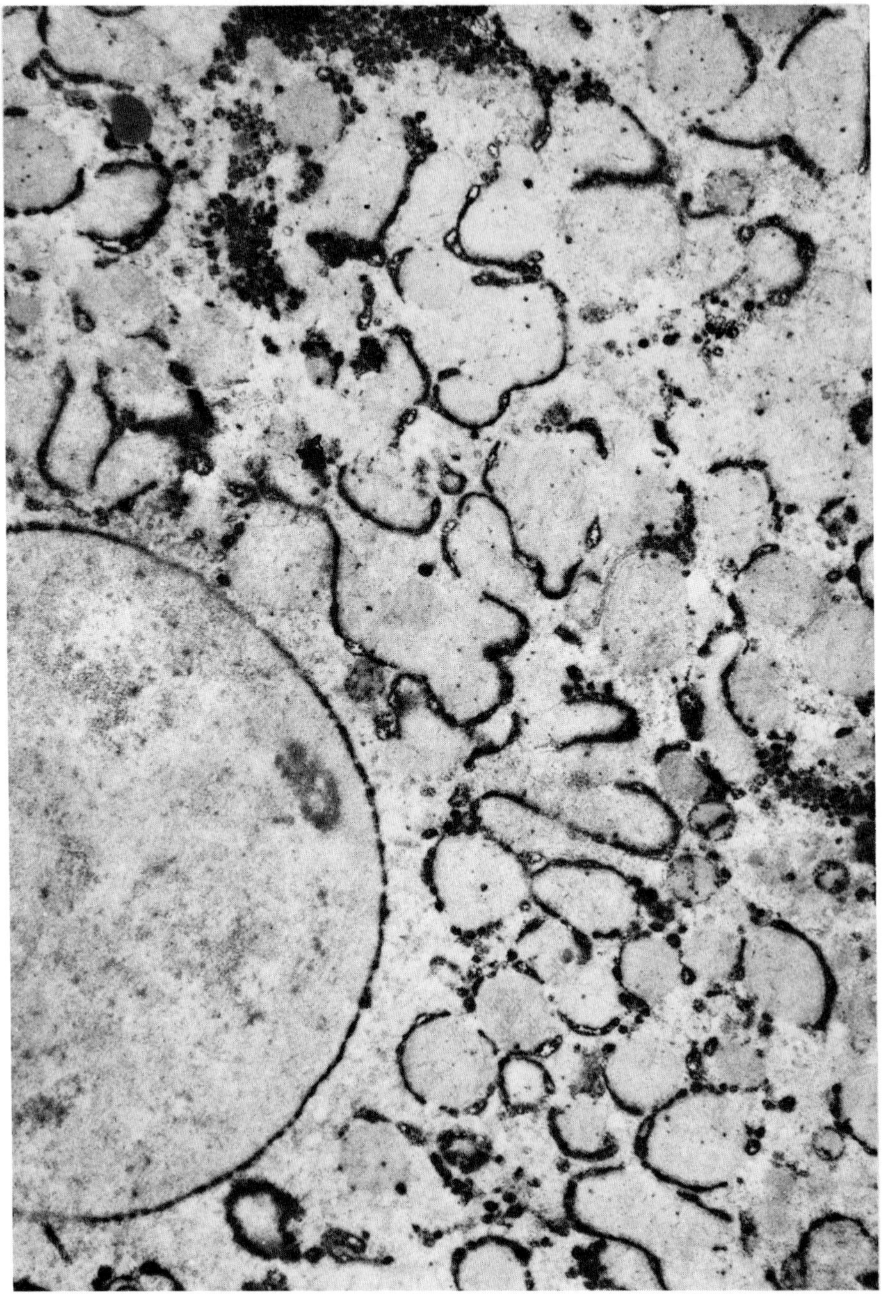

4.2 Electron micrograph of NNM-intoxicated centrilobular hepatocyte: Loss of glycogen and disorganization of the granular endoplasmic reticulum. Normal cytochemical reaction for glucose-6-phosphatase (black) throughout the scattered endoplasmic reticulum membranes and the nuclear envelope. Modified technique of Wachstein and Meisel (× 13,150). (Bannasch and Angerer, 1974, *Arch. Geschwalstforch* 43:105–114.)

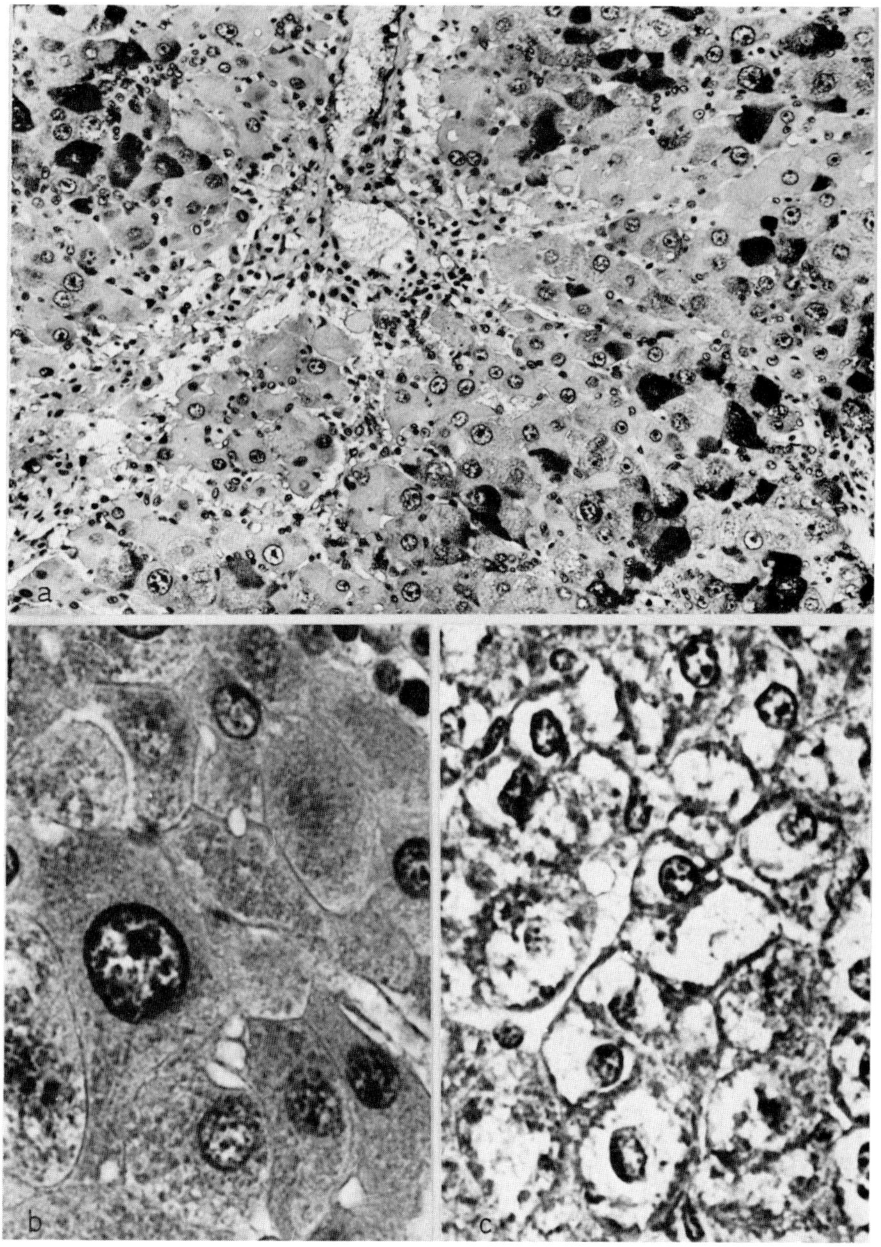

4.3 Light microscopic appearance of rat liver parenchyma under the influence (a,b) of sublethal doses of NNM (50 mg NNM/100 ml drinking water, 3 weeks and 5 weeks) after cessation and (c) of the carcinogenic treatment. (a) Loss of glycogen and enlargement of nuclei and nucleoli throughout the liver parenchyma of this section. Some hepatocytes are rich in glycogen (black) in peripheral regions of the lobule. Necroses, pronounced loss of parenchyma, and prominent mesenchymal proliferation, especially in the center of the lobule (Tri-PAS, × 180). (b) Hepatocytes with intense "diffuse" cytoplasmic basophilia ("hyperbasophilia"). Markedly enlarged cell with polyploid nucleus on the left side (H and E, × 910). (c) Reversibility of glycogen loss and cytoplasmic "hyperbasophilia" after withdrawal of the carcinogen: typical basophilic bodies and excessive storage of glycogen (clear cytoplasm) (H and E, × 760). (Theodossiou et al., 1971, *Virch. Arch. Abt. Zellpath.* 7:126–146.

these changes obviously are a sign of the severely altered metabolism of prenecrotic cells, since they are usually followed by severe parenchymal necrosis and by an intense proliferation of mesenchymal (fig. 4.3a) and ductular cells. In other hepatocytes, however, the loss of glycogen (fig. 4.3c) may be combined with a strong cytoplasmic basophilia (fig. 4.3b), which develops after a few days of poisoning and is due to a pronounced increase in free ribosomes, which sometimes are arranged in a helical pattern (fig. 4.4). The functional significance of such a marked increase in ribosomes is not entirely clear, but it may well be the morphological expression of a parenchymal regeneration that follows the extensive necrosis.

This interpretation is consistent with observations on the mitotic index (Bannasch et al., 1972; Romen and Bannasch, 1972) and the incorporation of ^{3}H-thymidine, as shown by radioautographic investigations (Heine et al., 1974). In addition to these observations on cellular proliferation, short-term experiments have shown that the early ribosomal increase is reversible (Theodossiou et al., 1971). It should, therefore, not be confused with somewhat similar alterations that appear during the later stages of the experiment and progress during the development of the hepatocellular carcinoma.

Interestingly enough, the short-term experiments with high doses of a carcinogen indicate that the pronounced early loss of glycogen is not only reversible but may even be followed by an increase in the cytoplasmic glycogen level (fig. 4.3c) after cessation of the carcinogenic treatment (Theodossiou et al., 1971).

4.3 Persistent Cytotoxic Alterations

The outstanding feature of the persistent alterations is an excessive storage of glycogen (Bannasch and Müller, 1964; Bannasch, 1968; Bannasch et al., 1972 and 1974). This accumulation of glycogen is demonstrated in the light microscope by the PAS reaction or Best's carmine stain. Following the dissolution of glycogen in the course of tissue preparation, or after treatment with diastase, the cytoplasm of the storage cells appears clear (fig. 4.3c; see also fig. 4.15b). It is considerably more abundant and may occupy an area four times the size it does in normal cells. The basophilic bodies are pushed toward the peripheral or paranuclear regions of the cell and are markedly reduced per unit volume of cytoplasm. In spite of the dislocation and relative reduction of the basophilic bodies, the fine structure of the granular reticulum almost invariably remains unchanged. The glycogen of the storage cells is found predominantly within the cytoplasmic matrix in the form of α or β particles. So far, we have not observed clear-cut differences in the structure of the accumulated glycogen in hepatocytes affected by carcinogens, but Drochmans and Scherer (1972) briefly reported that there is some evidence of a greater variability in the size

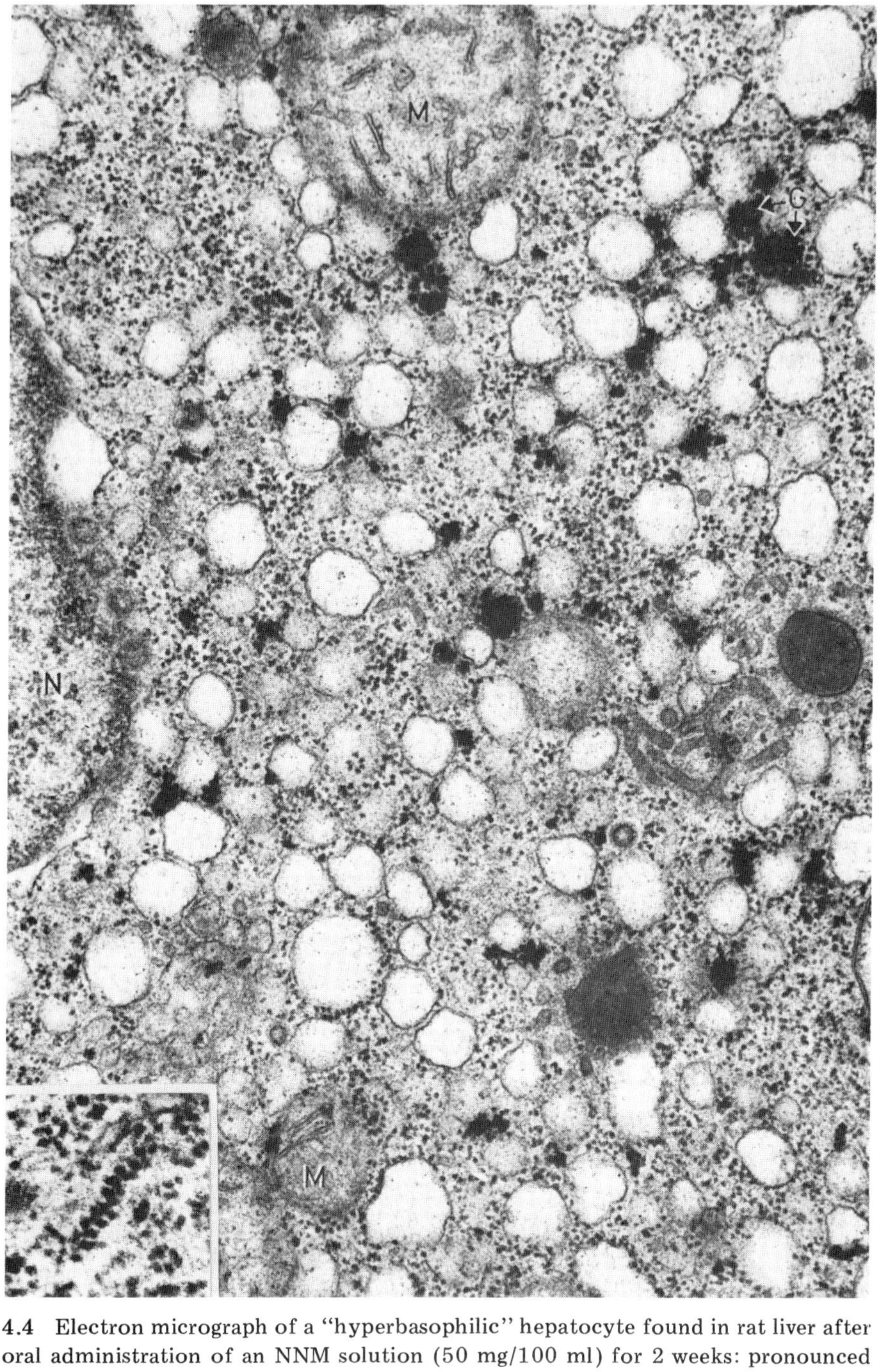

4.4 Electron micrograph of a "hyperbasophilic" hepatocyte found in rat liver after oral administration of an NNM solution (50 mg/100 ml) for 2 weeks: pronounced loss of glycogen (G), vesicular transformation of endoplasmic reticulum, and marked increase in free ribosomes. The latter are sometimes arranged in a helical pattern (inset). N = nucleus; M = mitochondria; lead hydroxide. (× 33,320; inset, × 52,920.) (Theodossiou et al., 1971, *Virch. Arch. Abt. Zellpath.* 7:126–146.)

of the glycogen particles. A substantial part of the stored glycogen sometimes is enclosed in large autophagic vacuoles and becomes finely granular, probably as the result of lysosomal degradation (Bannasch, 1968).

The development and the localization of the excessive glycogen storage depends on the dose of the carcinogen given (Bannasch, 1968). For example, if a solution of 12 mg NNM per 100 ml is given chronically, the storage of glycogen starts after one to two weeks; if poisoning is by a solution of 6/100 ml, after fourteen weeks. In both instances storage progresses as the total dose increases in the course of chronic intoxication. As a rule, the glycogen storage cells are initially localized in peripheral and intermediate parts of the lobule. Later on, the storage cells may appear also in the centrilobular region. Under certain conditions, that do not produce significant nonspecific toxic lesions in the central parts of the liver lobule—such as low dosage—excessive storage of glycogen can take place in that area even from the very beginning (Bannasch et al., 1974). When the carcinogen is withdrawn after several weeks of poisoning, the excessive storage of glycogen persists in small, or even in large, foci of the liver parenchyma for weeks and months. Thus, one may infer that the carcinogen induces a focal glycogen storage disease of the liver.

Histochemical investigations by several groups of investigators demonstrate that an unusual accumulation of glycogen appears in the liver parenchyma of different species not only after NNM but also after numerous other hepatocarcinogens, such as 2-fluorenylacetamide (Epstein et al., 1967), diethylnitrosamine (Schauer and Kunza, 1968; Friedrich-Freksa, Gössner, and Börner, 1969; Friedrich-Freksa, Papadopulu, and Gössner, 1969; Bader et al., 1971; Scherer et al., 1972; Schmitz-Moormann et al., 1972), aflatoxin (Newberne and Wogan, 1968), dimethylaminoazobenzene (Forget and Daoust, 1970), galactosamine (Lesch et al., 1973) or TAA (Bannasch et al., 1974). These observations point to the development of hepatocellular glycogenosis as an obligatory phenomenon after exposure to carcinogenic liver poisons. This conclusion, however, does not mean that every increase of glycogen in the liver cells is an indication of a precancerous cell reaction. Anaphylactic shock (Soostmeyer, 1940) or orthostatic collapse (Langer, 1965), for instance, also gives rise to increased hepatocellular glycogen levels. In contrast to the glycogenosis induced by carcinogens, however, the glycogen accumulation under these conditions is completely reversible within a short period of time. The same seems to be true for the increased hepatocellular glycogen levels observed in different species after protein deficiency (see Svoboda et al., 1966; Ericsson et al., 1966), an interesting phenomenon in view of the fact that it has been debated for some time whether protein deficiency helps to promote the development of hepatomas.

The origin of the glycogenosis induced by carcinogens is far from clear. Several authors have suggested that a glucose-6-phosphatase deficiency might play an important role (Epstein et al., 1967; Friedrich-Freksa, Gössner, and Börner, 1969; Friedrich-Freksa, Papadopulu, and Gössner, 1969; Schauer and Kunze, 1968; Schmitz-Moorman et al., 1972). This assumption seems to be supported by the cytochemical finding of foci of cells with glycogen storage, showing at the same time a reduction of enzyme activity. However, if one looks closer at the storage cells in early stages of the experiment, the peripheral regions of the cytoplasm, in which the displaced basophilic bodies are situated, exhibit a positive enzyme reaction of normal intensity (Bannasch and Angerer, 1974). In electron micrographs of such cells the reaction product can be seen throughout the granular cisternae and the nuclear envelope (fig. 4.5). These results are hardly compatible with the interpretation that there exists a causal relationship between glucose-6-phosphatase deficiency and the glycogenosis induced by carcinogens.

Two to three weeks after the emergence of the clear glycogen storage cells, striking acidophilic cells appear (Bannasch, 1968; Bannasch et al., 1974). The cytoplasmic acidophilia is due to the occurrence of a hypertrophy of the agranular endoplasmic reticulum in many storage cells (fig. 4.6) (see Bruni, 1960 and 1973; Bannasch, 1968 and 1975; Farber, 1973). There is usually a close spatial relationship between the proliferated smooth membranes and the glycogen particles. Although the smooth membranes show a typical arrangement in most cases, they may form unusual concentric lamellar complexes (fig. 4.7a) with or without glycogen. These well-known structures appear to be only a morphologic variant of the hypertrophy of the agranular reticulum (Thoenes and Bannasch, 1962; Bannasch, 1968). The reason for and the significance of the proliferation of the membranes remain obscure. Basing my view on those of Porter and Bruni (1959), I think that membrane proliferation is linked directly or indirectly to the alterations of intracellular glycogen metabolism. Such an interpretation is supported by the observation that the membrane proliferation induced by carcinogens persists for weeks and months after cessation of the carcinogenic treatment, as does the excessive glycogen storage (Bannasch, 1968; Bannasch et al., 1972 and 1974). This phenomenon contrasts with the common drug-induced hypertrophy of the agranular reticulum, which always seems to be reversible after withdrawal of the drug (see Stäubli et al., 1969).

Interestingly enough, in liver cells poisoned with NNM, the proliferated membranes initially show normal glucose-6-phosphatase activity (Bannasch and Angerer, 1974). The same holds true for the concentric lamellar membrane complexes of the agranular reticulum (fig. 4.7b). Groups of such cells might account for the foci with enhanced glucose-6-phosphatase activity, as shown with the light microscope by some investigators (Friedrich-Freksa, Gössner, and Börner, 1969; Friedrich-Freksa, Papadopulu, and Gössner, 1969; Moulin and Daoust, 1971).

Both the clear and the acidophilic cells form multiple foci during the

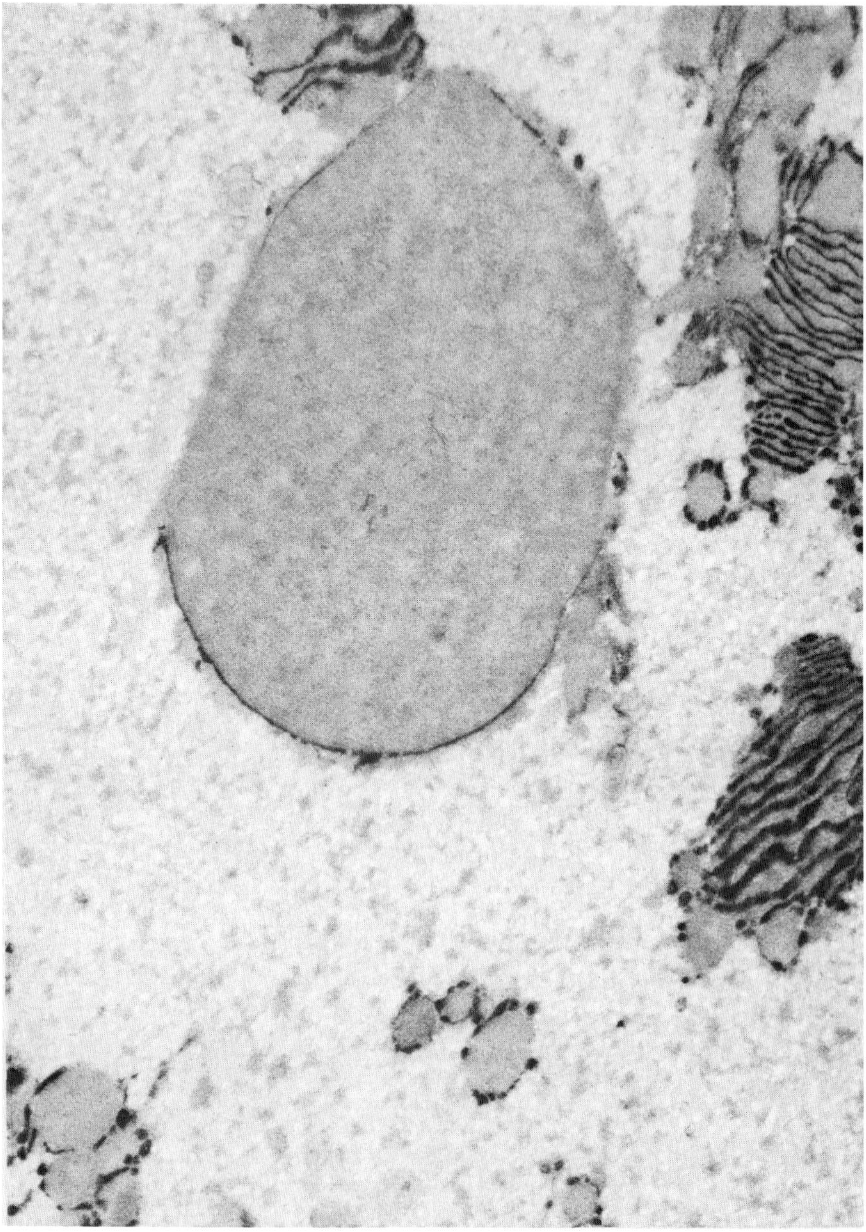

4.5 Electron micrograph of NNM-induced "clear" glycogen storage cell of rat liver parenchyma. Note large, cloudy glycogen zones and relative reduction of the basophilic bodies but normal cytochemical reaction for glucose-6-phosphatase (black) throughout the granular cisternae and the nuclear envelope. Modified technique of Wachstein and Meisel × 12,830). (Bannasch and Angerer, 1974, *Arch. Geschwulstforsch.* 43:105–114.)

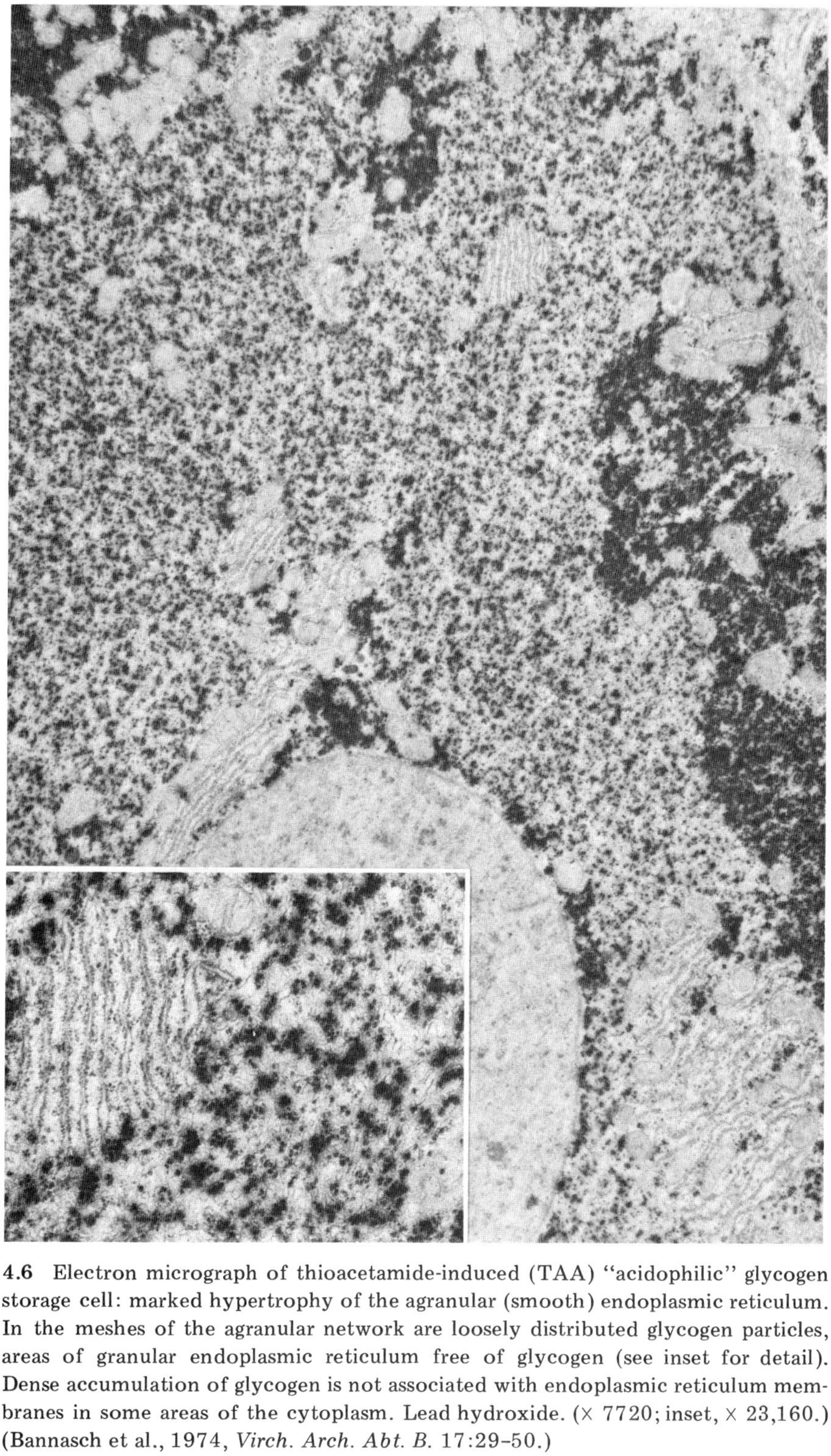

4.6 Electron micrograph of thioacetamide-induced (TAA) "acidophilic" glycogen storage cell: marked hypertrophy of the agranular (smooth) endoplasmic reticulum. In the meshes of the agranular network are loosely distributed glycogen particles, areas of granular endoplasmic reticulum free of glycogen (see inset for detail). Dense accumulation of glycogen is not associated with endoplasmic reticulum membranes in some areas of the cytoplasm. Lead hydroxide. (× 7720; inset, × 23,160.) (Bannasch et al., 1974, *Virch. Arch. Abt. B.* 17:29–50.)

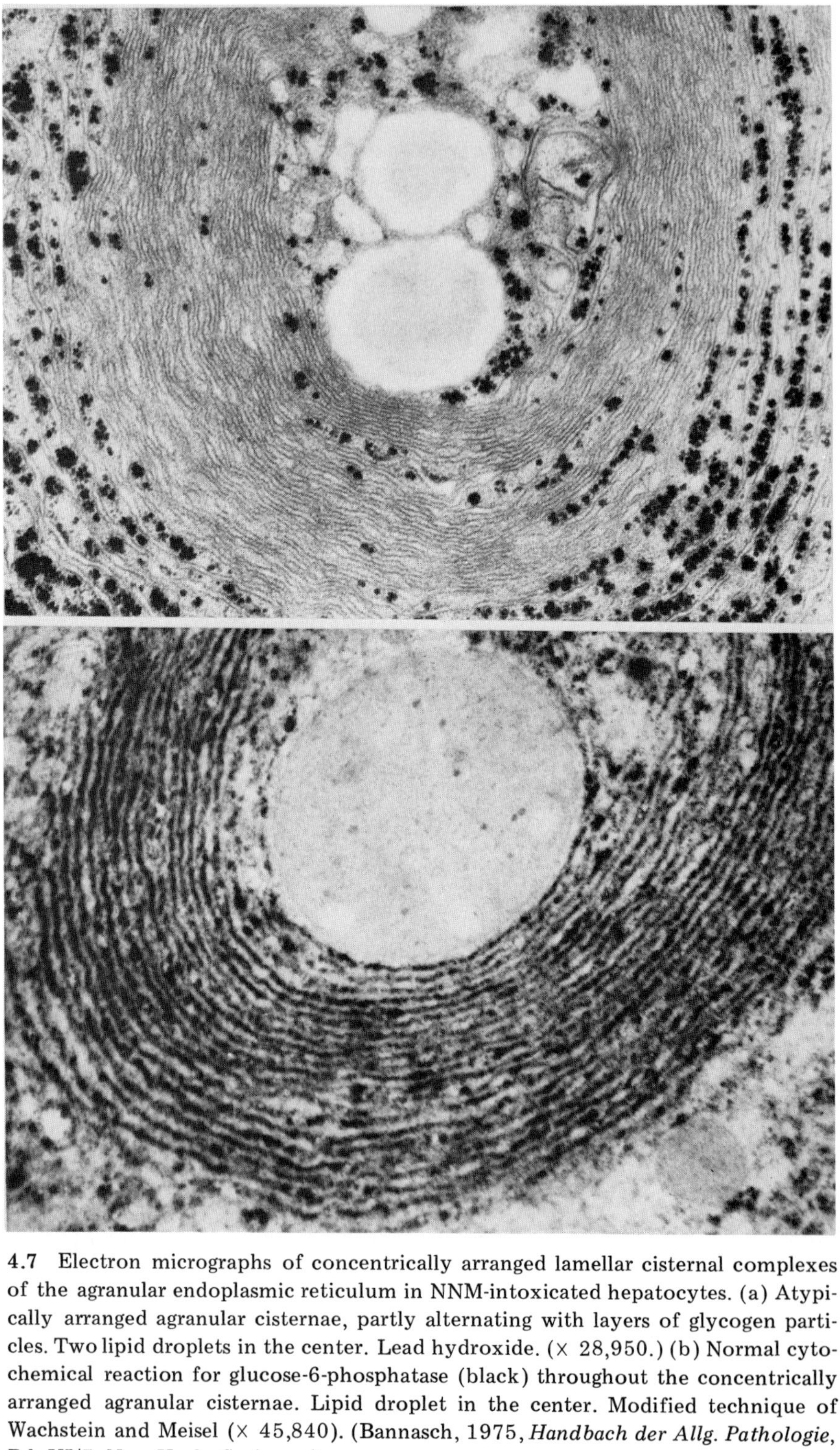

4.7 Electron micrographs of concentrically arranged lamellar cisternal complexes of the agranular endoplasmic reticulum in NNM-intoxicated hepatocytes. (a) Atypically arranged agranular cisternae, partly alternating with layers of glycogen particles. Two lipid droplets in the center. Lead hydroxide. (× 28,950.) (b) Normal cytochemical reaction for glucose-6-phosphatase (black) throughout the concentrically arranged agranular cisternae. Lipid droplet in the center. Modified technique of Wachstein and Meisel (× 45,840). (Bannasch, 1975, *Handbach der Allg. Pathologie*, Bd. VI/7. New York: Springer.)

preneoplastic phase (fig. 4.8). Such parenchymal alterations have been known, from the time of Yoshida's fundamental work, as "hyperplastische Areale," or areas of hyperplasia (Reuber, 1965). I submit now that these areas correspond to foci of glycogen storage. Whereas the rest of the liver parenchyma is completely free of glycogen after starvation or the administration of glycogen, one may still find glycogen in these foci (Farber, 1963; Steiner et al., 1964; Friedrich-Freksa, Gössner, and Börner, 1969; Friedrich-Freksa, Papadopulu, and Gössner, 1969; Scherer et al., 1972).

The persistence of hepatocellular glycogenosis after the withdrawal of carcinogens suggests that this cytopathological phenomenon is caused by an irreversible cell lesion due, in turn, to a direct action of the carcinogen on the affected cells (Bannasch, 1968 and 1975). The essential role of a regenerative cellular hyperplasia for the development of the foci of glycogen storage remains unproven. Preliminary results of radioautographic investigations with ^{3}H-thymidine in rats intoxiicated by NNM point to a considerably slower cell proliferation in the glycogen storage foci than in the surrounding liver parenchyma (Heine et al., unpublished results). This finding brings into question the validity of the term "hyperplastic area"; purely descriptive terms such as "focus of glycogen storage" and "focal glycogenosis" are preferable (Bannasch, 1968). Some authors seem to confuse the terms "hyperplasia" and "hypertrophy" when they speak of "hyperplasia" not only in order to describe an increase in the number of hepatocytes but also to characterize enlarged cells. Virchow defined the latter as hypertrophied cells. A clear distinction between these two basically different alterations of the liver parenchyma is essential for further discussions of problems of hepatocarcinogenesis.

4.4 Progressive Posttoxic Alterations

At the end of the preneoplastic phase in most animals, so-called hyperplastic nodules (fig. 4.9) develop. They are regarded as possible precursors of hepatomas (see Farber, 1973; Bannasch, 1975). The cellular composition of these nodules is often very similar to that of the foci of glycogen storage. Thus a considerable accumulation of glycogen in cells of hyperplastic nodules has been demonstrated by many investigators (see Bannasch, 1968 and 1975; Hruban et al., 1972; Farber, 1973). The same holds true for a more or less intense proliferation of the agranular endoplasmic reticulum. Moreover, a dislocation and relative reduction of the granular reticulum can often be seen. These observations strongly suggest that the nodules originate from the areas storing glycogen in excess. However, during the transformation of these areas into nodules of hyperplasia, many cells show a striking new pattern of cytoplasmic alterations, which take the place of the glycogenosis and of the concomitant structural anomalies of the endoplasmic reticulum (Bannasch,

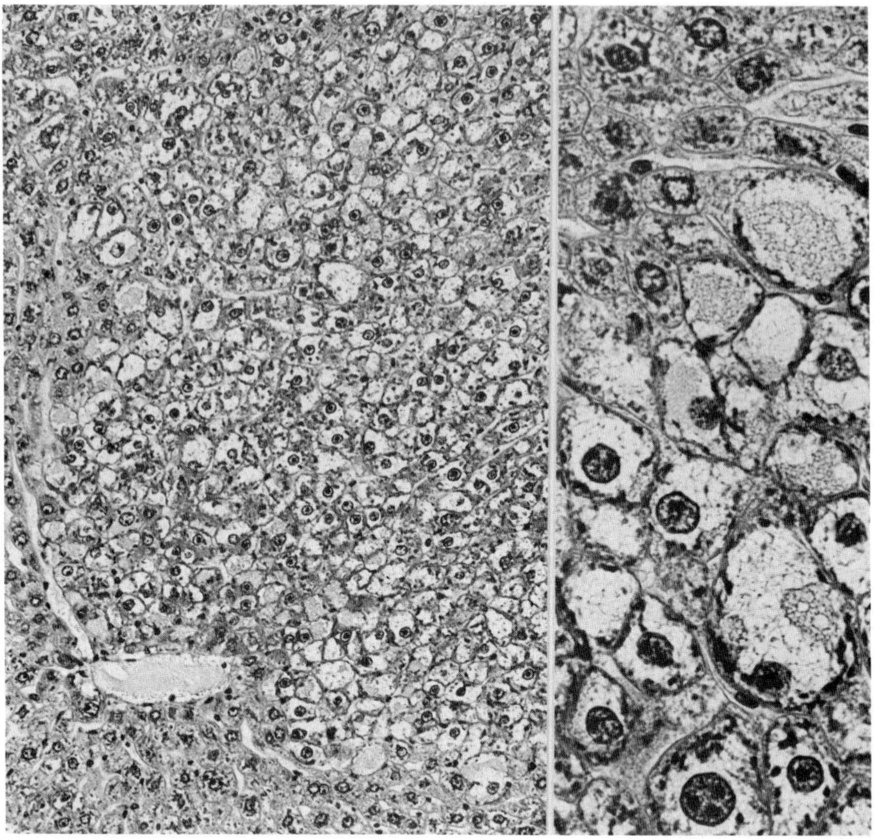

4.8 Light micrographs of TAA-induced focal glycogenosis of rat liver parenchyma (a) Large focus of glycogen storage with markedly enlarged clear and acidophilic hepatocytes (H and E, × 140). (b) Detail of the same focus showing clear and acidophilic glycogen storage cells in addition to some normal hepatocytes (H and E, × 580). (Bannasch et al., 1974, *Virch. Arch. Abt. B.* 17:29–50.)

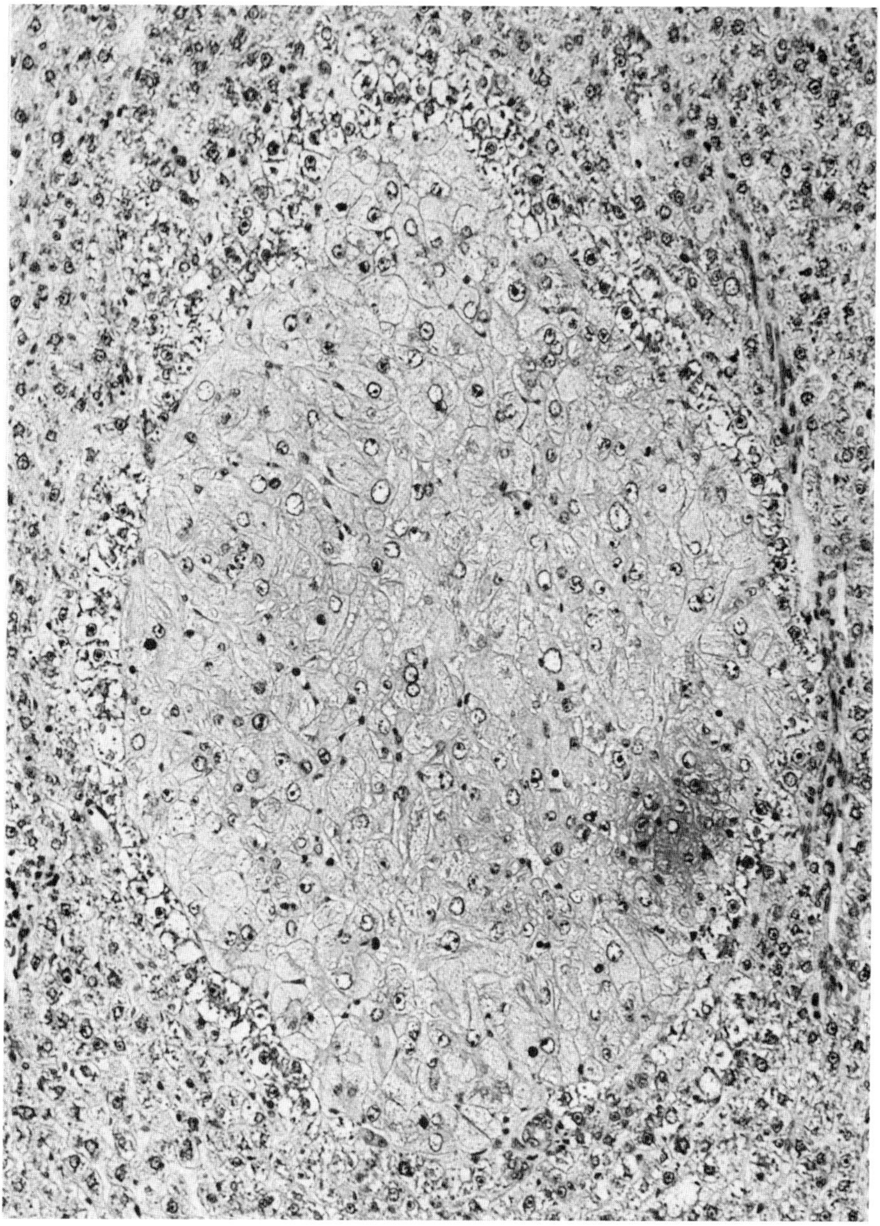

4.9 Light micrograph of a so-called hyperplastic nodule induced by NNM in the rat liver: transition from clear glycogen storage cells (narrow peripheral rim) to dark basophilic cells (small area to the right) (H and E, × 190). (Bannasch, 1975, *Handbuch der Allg. Pathologie*, Bd. VI/7. New York: Springer.)

1968). Whereas these morphological signs persist until the nodules start to develop and gradually disappear during the growth of the nodules, the new alterations are progressive (Bannasch et al., 1972). Thus, from a cytological point of view, one may distinguish two stages in the carcinogenic process: the stage of persistent changes and the stage of progressive changes. The progressive behavior of the alterations of the second stage is observed not only during exposure to a carcinogen but also many weeks or even months after its withdrawal. One might even speak of posttoxic alterations (Bannasch et al., 1972). Obviously the final cellular metamorphosis is not an immediate consequence of the intoxication but a reaction of the cell to the primary carcinogenic damage, whose nature is still unknown.

We consider the most important phenomenon of this metamorphosis to be a gradual reduction of the glycogen initially stored in excess (Bannasch, 1968; Bannasch, 1974a, b; Bannasch, 1975; Bannasch and Angerer, 1974; Bannasch and Klinge, 1971; Bannasch and Müller, 1964; Bannasch and Reiss, 1971; Bannasch et al., 1972; Bannasch et al., 1974). Sometimes cells in intermediate stages, when they still contain considerable amounts of glycogen, may take part in the formation of trabecular tumors and may even metastasize to the lungs, as observed in one case (Bannasch, 1974a). However, as a rule, glycogen is extensively depleted at the very beginning of the development of the tumor (fig. 4.10). The resulting hepatomas are usually poor in, or entirely free from, histochemically demonstrable glycogen. Similar observations have been reported by various other authors (Schauer and Kunze, 1968; Friedrich-Freksa, Gössner, and Börner, 1969; Forget and Daoust, 1970; Schmitz-Moorman et al., 1972; Lesch et al., 1973). These results are in accordance with biochemical data on transplantable tumors (see Weber, 1968).

The loss in glucose-6-phosphatase activity seems to occur only a short time before or with the reduction of glycogen (Bannasch and Angerer, 1974). As demonstrated by light and electron microscopic cytochemistry, the glucose-6-phosphatase activity of the hypertrophied smooth as well as the rough endoplasmic reticulum is definitely lost at this point. This result agrees with recent findings of Drochmans and Scherer (1972); an artifact appears to be excluded because negatively and positively reacting cells are often situated side by side. One can conclude from these observations that the cytochemically demonstrable reduction of glucose-6-phosphatase activity does not cause but merely accompanies or follows the hepatocellular glycogenosis induced by carcinogens. Obviously, a glucose-6-phosphatase deficiency is not essential for the process of neoplastic cell transformation. In rare cases development of glycogen-free microtumors that even have an enhanced activity of glucose-6-phosphatase occurs (see Bannasch and Angerer, 1974). Unfortunately, the limitations of the histochemical approach do not allow any final conclusions concerning the role of the glucose-6-

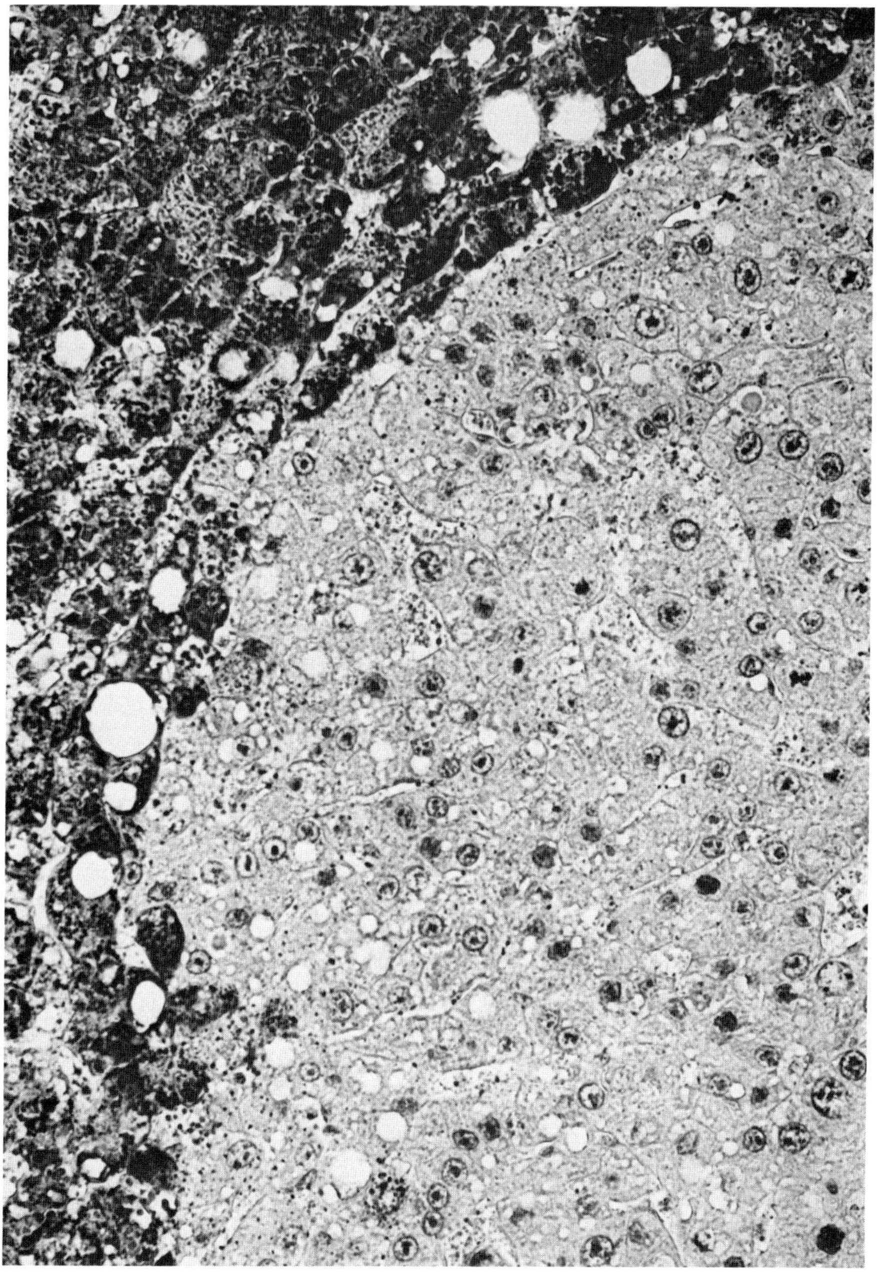

4.10 Light micrograph of the border region between an NNM-induced (rat) focal hepatic glycogenosis (to the left) and a liver cell tumor poor in glycogen (to the right). Note lipid vacuoles in both lesions and several mitoses in the glycogen-poor area (Tri-PAS, × 200).

phosphatase during hepatocarcinogenesis. On the other hand, the observation that the proliferated agranular membranes give a positive glucose-6-phosphatase reaction at first and a negative one later on permits a more general conclusion concerning the function of the endoplasmic reticulum.

Simultaneously with the reduction of the level of glycogen in the storage cells, the number of cytoplasmic ribosomes progressively increases (Bannasch, 1968). This change is shown in the light microscope by an increase in cytoplasmic basophilia (fig. 4.11a and 4.12b) long known as "chromatogenesis" (Opie, 1946). Since the glycogen-free basophilic cells are morphologically similar to hepatoma cells, many authors believe that the "foci of basophilic hyperplasia" (Opie) indicate the transition from "hyperplasia" to neoplasia (see Daoust and Molnar, 1964; Bannasch, 1968, 1975; Karasaki, 1969; Daoust and Calmai, 1971). In appropriate sections one can distinguish all intermediate stages between clear and acidophilic glycogen storage cells on the one hand and glycogen-free basophilic hepatoma cells on the other (figs. 4.11a, and 4.12). Frequently a transitory accumulation of fine or coarse fat droplets occurs and gives rise to vacuoles (fig. 4.11b) in the customary tissue preparations (Bannasch, 1968; Bannasch et al., 1972 and 1974).

There is some indication that the agranular membranes, which are often abundant and in close contact with glycogen in precancerous storage cells, are usually transformed into granular membranes by the addition of ribosomes during the reduction of glycogen (Bannasch, 1968). We consider unusual combinations of smooth and rough endoplasmic reticulum as intermediary stages of this membrane transformation. One may interpret them as more or less dense complexes consisting of multiple ergastoplasmic pockets and smooth membranes, the latter being associated with glycogen (fig. 4.13). Characteristically these pockets of ergastoplasm enclose glycogen-free islands within zones of accumulated glycogen. The islands contain large amounts of free ribosomes or, sometimes, mitochondria. Within small areas of the cytoplasm, the formation of such pockets thus leads to a state which is characteristic of the entire cytoplasm of the ultimate tumor cell. Here a lack of glycogen and of agranular membranes is usually observed together with an unusual abundance of free ribosomes or granular membranes (Bannasch, 1968 and 1975). We, therefore, consider the pockets of ergastoplasm to be an early indication of neoplastic cell transformation. This interpretation is in accordance with recent results of Hruban and coworkers (1972), who found very similar structures in slow-growing Morris hepatomas.

In the basophilic cells of rat livers poisoned with diethylnitrosamine or butter yellow, other investigators have demonstrated an increase in both cytoplasmic RNA content (Hobik and Grundmann, 1962) RNA

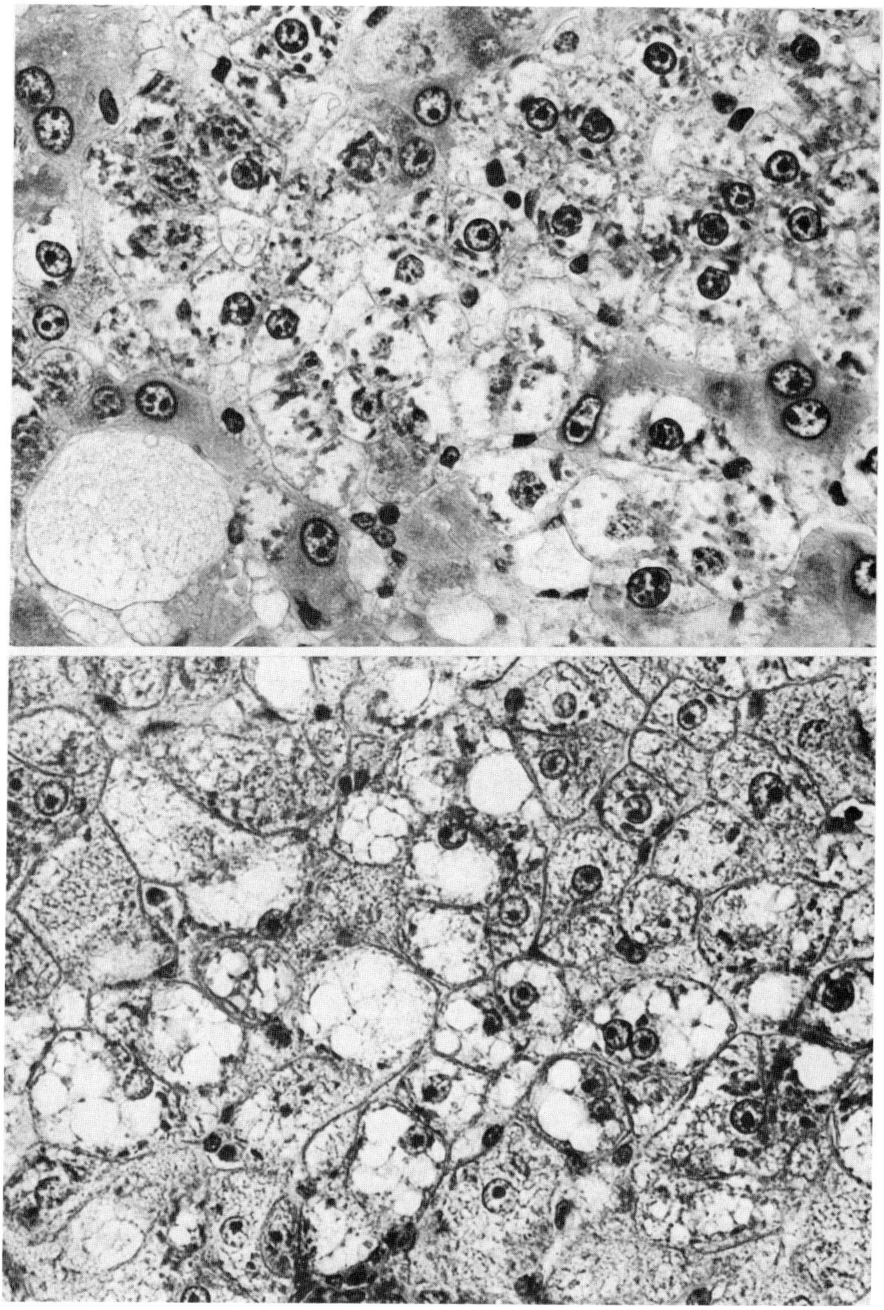

4.11 Light microscopic appearance of so-called hyperplastic liver nodules induced in the rat by TAA. (top) Some cells poor in glycogen and rich in basophilic material in the midst of a focus of glycogen storage. (bottom) Several fat-storing cells in addition to acidophilic and faintly basophilic cells. (Both micrographs: H and E, × 480.) (Bannasch et al., 1974, *Virch. Arch. Abt. B.* 17:29–50.)

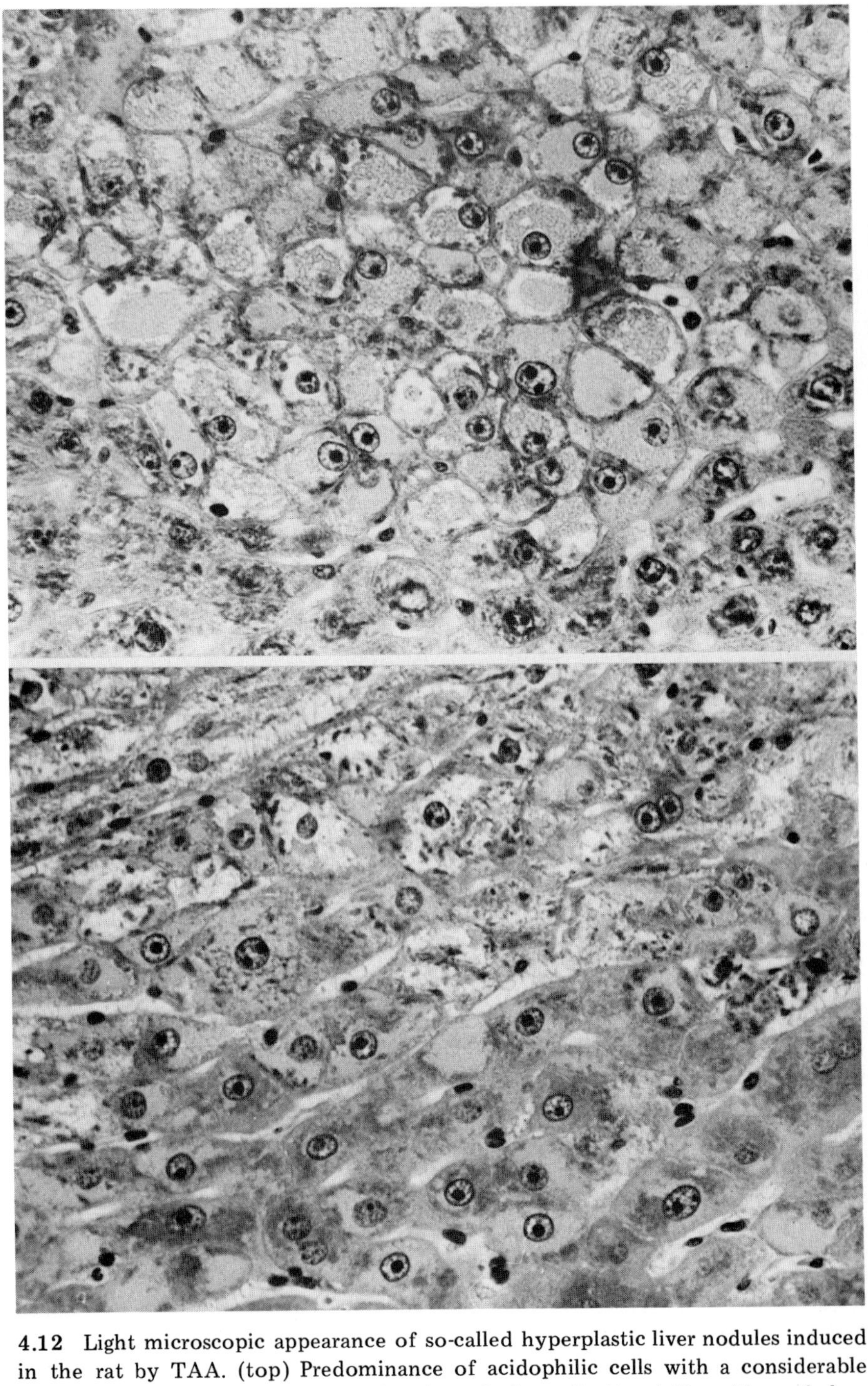

4.12 Light microscopic appearance of so-called hyperplastic liver nodules induced in the rat by TAA. (top) Predominance of acidophilic cells with a considerable hypertrophy of the smooth endoplasmic reticulum. (bottom) Transition of clear and acidophilic to basophilic cells, especially at the bottom right. (Both micrographs: H and E, × 390.) (Bannasch et al., 1974, *Virch. Arch. Abt. B.* 17:29–50.)

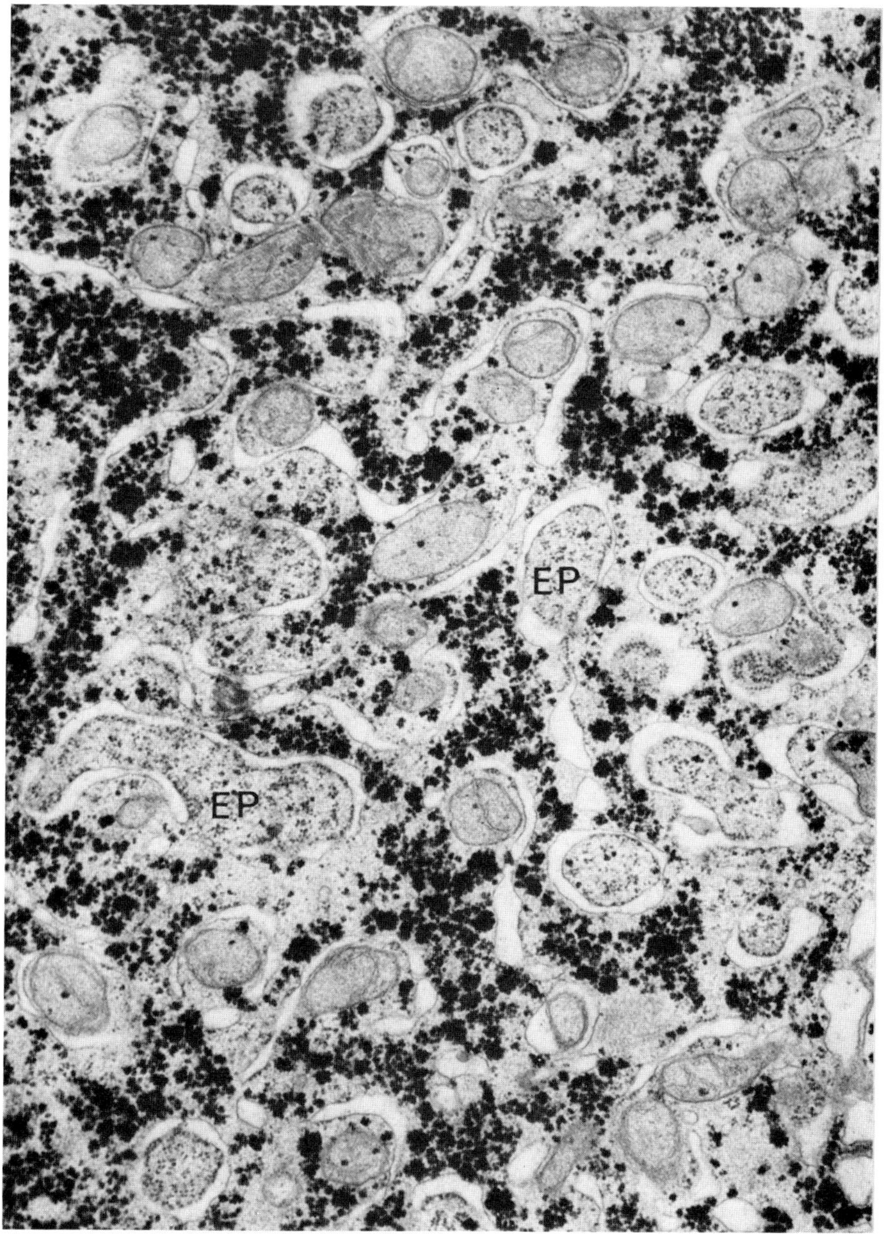

4.13 Electron micrograph of an NNM-induced "intermediary" cell of the rat liver: multiple ergastoplasm pockets (EP) enclose islands poor in glycogen but rich in free ribosomes. Between the pockets are many glycogen particles, most of which form rosettes. The glycogen zones are usually lined by smooth parts of the endoplasmic reticulum-cisternae. Lead hydroxide. (× 35,150.)

synthesis (Oehlert and Hartje, 1963; Daoust and Simard, 1968). Daoust (1972) and Taper and collaborators (1971) think that the histochemically demonstrable loss of activity of certain RNases in what they call "hyperbasophilic cells" is causally related to the increase in cytoplasmic basophilia.

At about the same time the glycogen content of the cytoplasm is gradually being reduced and the number of ribosomes is increasing, the fine structure of the nucleus is often observed to alter remarkably (Romen and Bannasch, 1973; see also Sugihara et al., 1972): the nuclei are enlarged, and the condensed chromatin is reduced except for small remnants (fig. 4.14). The nucleoli are also enlarged, but the relationship of granular and fibrillar components of this organelle remains almost unchanged. These alterations might indicate a so-called "functional nuclear swelling" related to the conversion of the intracellular metabolism during the transformation of glycogen storage cells into glycogen-poor tumor cells.

It has been suggested that the hyperplastic liver nodules represent a reasonably homogeneous cell population (Farber, 1973). Cytochemical and electron microscopical investigations reported so far, however, clearly show that the nodules are often composed of a rather heterogeneous cell population (Bannasch, 1968, 1974b, and 1975; Bannasch et al., 1974). At least four different cellular prototypes can be distinguished (fig. 4.9, 4.11, and 4.12): (1) "clear" glycogen storage cells with a dislocation and relative reduction of the granular endoplasmic reticulum; (2) "acidophilic" cells with a hypertrophy of the agranular endoplasmic reticulum; (3) fat-storing cells; and (4) tumor cells poor in glycogen and rich in ribosomes. In addition, there are all possible intermediary types. Thus, most hyperplastic nodules are probably a mixture of precancerous, cancerous, and diverse intermediary cells. If this suggestion proves true, one must question whether definite steps in the process of carcinogenesis can be adequately explored by biochemical investigations of hyperplastic nodules. Moreover, the cellular heterogeneity of the nodules does not totally support the suggestion that the nodules develop by clonal selection (see Farber, 1973).

Another important question is whether the hyperplastic nodule is an obligatory step in the development of liver cell cancer. Many morphological observations indicate that all cellular alterations characteristic of the nodules may also occur without a nodular arrangement in the midst of an otherwise normally structured liver parenchyma. If such foci are mainly composed of glycogen-poor basophilic cells, it seems justified to speak of a "microcarcinoma" (Grundmann and Sieburg, 1962) or a "carcinoma in situ." This interpretation is supported by the observation by various authors that basophilic foci of this type may already show prominent atypical cellular changes and a considerable increase in mitosis (e.g., Grundmann and Sieburg, 1962; Oehlert and

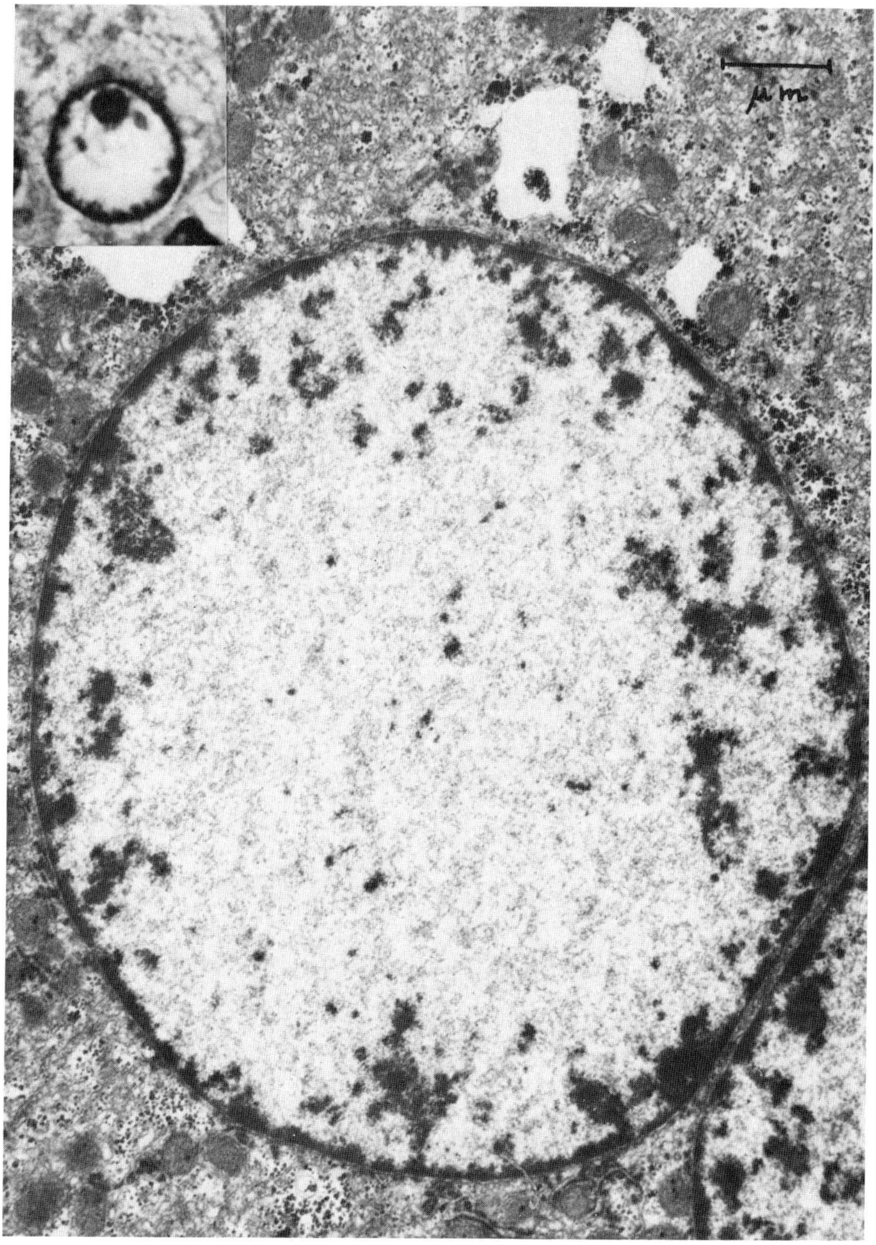

4.14 Electron micrograph of the nucleus of a glycogen-poor tumor cell induced in the rat by NNM; note reduction of the condensed chromatin. The remaining condensed chromatin in confined to the nuclear periphery. Lead hydroxide. (× 14,480.) Inset: "Vesicular" appearance of the nucleus of a tumor cell in the light microscope (H and E, × 190.) (Romen and Bannasch, 1973, *Virch. Arch. Abt. B.* 13: 267–296.)

Hartje, 1963; Daoust and Molnar, 1964; Bannasch et al., 1972). It is highly probable, therefore, that the hyperplastic nodule is only one of at least two different histological formations in which the neoplastic transformation can take place.

4.5 Quantitative Aspects of the Different Cytotoxic Patterns

In order to get more precise information concerning the cytoplasmic enlargement induced by carcinogens and the participation of the various cytotoxic patterns in cellular transformation, we have undertaken some quantitative studies of our qualitative observations (Bannasch et al., 1972). For an approximate determination of cell areas we chose a simple, indirect method that permits the registration of great numbers of cells: with a Zeiss measuring ocular we counted the number of liver cell nuclei within contiguous parenchymal areas not containing portal tracts of central veins.[1] In addition, we determined the percentage of the various cytotoxic patterns occurring in the total population of hepatocytes at the end of seven weeks of poisoning, and at three time intervals after withdrawal of the carcinogen. A complete cross-section from the middle of the central liver lobe of each animal was examined. For the final estimation, data from three animals of each group were pooled. The figures for each experimental stage are based on about 200,000 single counts.

We distinguished six types of altered hepatocytes in these experiments (fig. 4.15): (1) glycogen-deficient hepatocytes with a disorganization of the ergastoplasm; (2) "clear" glycogen storage cells; (3) "acidophilic" cells; (4) fat-storing cells; (5) basophilic tumor cells; and (6) the remaining altered cells, called X-cells. The X-cells appeared normal at first glance, but most of them were actually enlarged to twice their normal size. This anomaly is clearly revealed by a comparison with normal liver cells at the same magnification (fig. 4.16). The electron microscope shows that both cells are rich in glycogen and smooth endoplasmic reticulum, and perhaps also in some other cytoplasmic organelles. Small X-cells were therefore quite difficult to distinguish from normal cells.

Table 4.1 shows the percentage of the six types of cells in the different experimental stages. The cells deficient in glycogen comprise 3.6 percent of the cell population at the end of the period of poisoning. Four weeks later, their number has decreased to 0.2 percent; 48 weeks later, none can be found. The clear glycogen storage cells, on the other hand, make up more than half of the liver parenchyma not only during the period of poisoning but even one year after withdrawal of the carcinogen. It is only after 61–87 weeks that the cellular proportions

[1] Counting liver cell nuclei can result in misleading information unless the relatively large number of binucleate cells is taken into account. See chapter 2.—Ed.

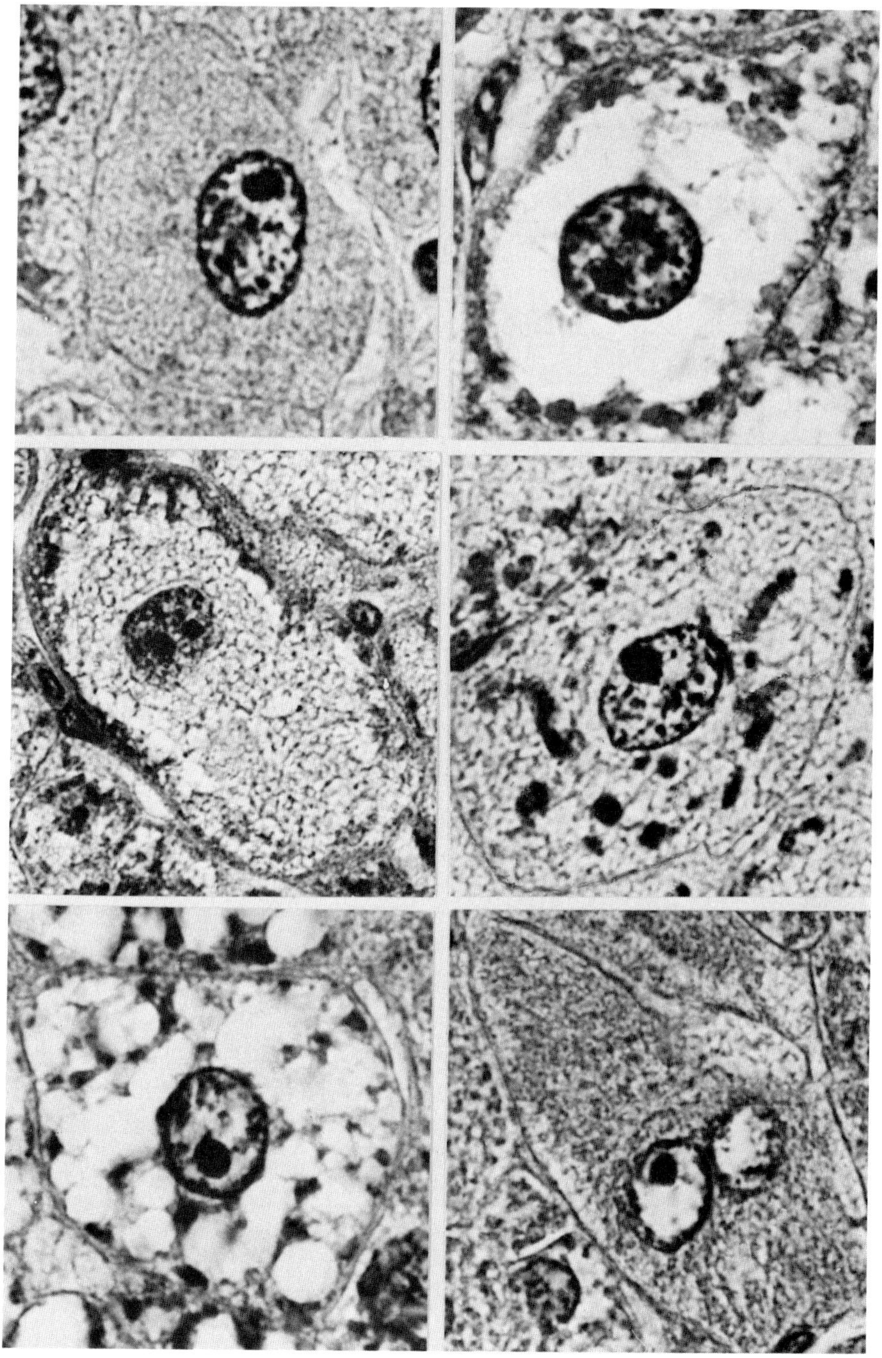

4.15 Light microscopic appearance of different types of pathologically altered hepatocytes in the NNM-intoxicated rat liver (see text for detail). (a) Glycogen-deficient cell (H and E, × 1830). (b) Clear glycogen storage cell (H and E, × 1740). (c) Acidophilic glycogen storage cell (H and E, × 970). (d) Large "X-cell" (H and E, × 1640). (e) Fat-storing cell (H and E, × 1740). (f) Tumor cell (H and E, × 1350). (Bannasch et al., 1972, *Z. Krebsforsch* 77:108–133.)

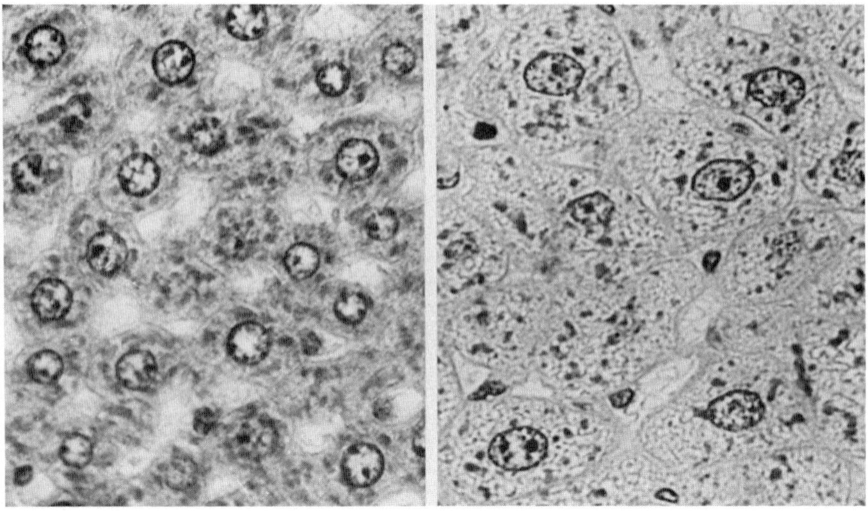

4.16 Light microscopic comparison of normal hepatocytes (a) of an untreated rat liver with enlarged "X-cells" and (b) of an NNM-intoxicated rat liver. Note dense aggregation of basophilic bodies in the normal hepatocytes and distinct acidophilic network and loosely distributed basophilic bodies in the cytoplasm of the X-cells. (Both micrographs: H and E, × 680.) (Bannasch et al., 1972, *Z. Krebsforsch* 77:108–133.)

Table 4.1 Percentage of different types of hepatocytes, mitoses, and necroses in nitrosomorpholine-treated rats (at the end of 7 weeks of poisoning, and 4, 52, and 61–87 weeks after cessation).[a]

	7 Weeks NNM-intoxication (%)	4 Weeks after stop (%)	52 Weeks after stop (%)	61–87 Weeks after stop (%)
Glycogen-deficient cells	3.55	0.19	—	—
"Clear" glycogen storage cells	58.03	53.70	55.70	15.28
"Acidophilic" cells	1.43	1.36	2.44	2.94
Fat storage cells	0.01	0.50	1.89	0.38
Hepatoma cells	—	0.09	5.44	62.30
X-cells (rest of population)	36.51	43.91	34.40	18.72
Mitoses	0.10	0.17	0.08	0.35
Necroses	0.37	0.08	0.05	0.03
Total	100.00	100.00	100.00	100.00

[a]See Bannasch et al., 1972, *Z. Krebsforsch.* 77:108–133.

change basically: clear storage cells are reduced to 15 percent, while the numbers of basophilic tumor cells and of mitosis are considerably increased. This observation agrees with the assumption that most tumor cells develop from the clear glycogen storage cells. Other possible precursors are the acidophilic cells and the large X-cells. Thus, all types of pathologically altered hepatocytes that persist after withdrawal of the carcinogen may be potentially precancerous.

The results of the indirect estimation of cell areas by counting the number of liver cell nuclei in a given field indicate that nearly the whole of the liver parenchyma undergoes irreversible cellular lesions. As demonstrated in table 4.2, the mean number of nuclei of the total population of hepatocytes is reduced from 36 to 24 already during the phase of intoxication. This change corresponds to an enlargement of the mean cell area of about 1.5 times. After withdrawal of the carcinogen, the cell areas at first continue to enlarge. X-Cells, as well as clear glycogen storage cells, are particularly involved in this cellular enlargement; their mean area is doubled 4 weeks after cessation of the carcinogenic treatment. Even one year after withdrawal of the carcinogen, enlargement factors of 1.38 for the whole population, 1.48 for the clear glycogen storage cells, and 1.65 for the X-cells are found. In the two first animal groups, 4 and 52 weeks after treatment the enlargement factor is partially due to a polyploidization, as shown by nuclear measurements (Romen et al., 1972). However, one year after withdrawal of the carcinogen, this factor should not matter greatly because by then the nuclear volume is again nearly normal. The cellular enlargement,

Table 4.2 Mean nuclear number per measuring field (MNN) and resulting enlargement factors of the mean cell areas (MCA) of the total population, of the "clear" glycogen storage cells, and of the X-cells in rat liver parenchyma poisoned by nitrosomorpholine.[a]

	Total population		"Clear" glycogen storage cells	
Experimental stages	MNN	Enlargement factor MCA	MNN	Enlargement factor MCA
7 Weeks NNM-intoxication	24.66	1.47	26.10	1.39
4 Weeks after stop	21.76	1.67	22.35	1.62
52 Weeks after stop	26.17	1.38	24.40	1.48
61–87 Weeks after stop	30.88	1.17	23.69	1.53

[a] See Bannasch et al., 1972, *Z. Krebsforsch.* 77:108–133.

therefore, appears to be due mainly to the excessive glycogen storage and/or a proliferation of membranes of the agranular endoplasmic reticulum. The significance of the glycogen accumulation for the enlargement of the hepatocytes becomes obvious again when, during the final cellular transformation, the mean areas of the cells are reduced along with the gradual reduction of glycogen: the enlargement factor for the total population diminishes to 1.17 by 61–87 weeks, although enlarged glycogen storage cells and X-cells are still present.

4.6 Conclusions

From the data presented here it seems that the glycogen-free basophilic tumor cell is the result of a long process that begins with an excessive storage of glycogen and perhaps passes through a stage of membrane proliferation or accumulation of fat. While poorly differentiated and fast-growing hepatomas are usually poor in glycogen, highly differentiated and slowly growing tumors may still contain considerable amounts of this polysaccharide (see Bannasch, 1968; Hruban et al., 1971; Bannash, 1975). This observation holds true not only for experimental but also for human tumors (fig. 4.17) (Garancis et al., 1969; Hamperl, 1970; Bannasch and Klinge, 1971; Schaff et al., 1971; Phillips et al., 1973). It is highly probable, therefore, that the results obtained in rats are valid also for man. In this context, it should be noted that in six instances of inborn hepato-renal glycogenosis (v. Gierke's disease) in children, hepatocellular tumors were found seven to twenty-three years after birth (see Bannasch and Klinge, 1971; Spycher and Gitzelmann, 1971).

X-Cells		Controls	
MNN	Enlargement factor MCA	MNN	Enlargement factor MCA
20.66	1.76		
17.80	2.04	36.25	1
21.91	1.65		
24.04	1.51		

The biochemical mechanism responsible for the reduction of the glycogen initially stored in excess is as unknown as the origin of the hepatocellular glycogenosis induced by carcinogens. Our suggestion that it might be related to the beginning of a Warburg type of glycolysis (Bannasch, 1968) seems to be in keeping with the results reported by several authors. These investigators found that quickly growing hepatomas with intense glycolytic activity are almost free of glycogen, whereas slowly growing tumors with low glycolytic activity are often relatively rich in glycogen (see Weber, 1968; Hruban et al., 1971).

Since the investigations of Aisenberg (1961), Potter (1964), and Weinhouse (1966) on transplantable hepatomas, a causal relationship between the aerobic glycolysis and tumor development has been rejected by most authors (for a differing opinion see Burk and Woods, 1967; Burk et al., 1967). Only the relationship between glycolysis and tumor growth is generally accepted, since many investigators have shown that the intensity of glycolysis is directly correlated with the growth rate of tumors. Some authors have reported that slowly growing hepatomas with minimal glycolysis may, after some passages, change into fast-growing tumors with marked glycolysis. This possibility suggests the need for a reinvestigation of the relationship between aerobic glycolysis and tumor development. At the present time the possibility cannot be excluded that the original transplants were only precancerous tissues or tumors in their early stages and were transformed into true hepatomas only after several passages. In any case, the question arises whether the anomaly of carbohydrate metabolism, which manifests itself in the early excessive glycogen storage, might be responsible for the late development of a Warburg type of glycolysis. Since there are some

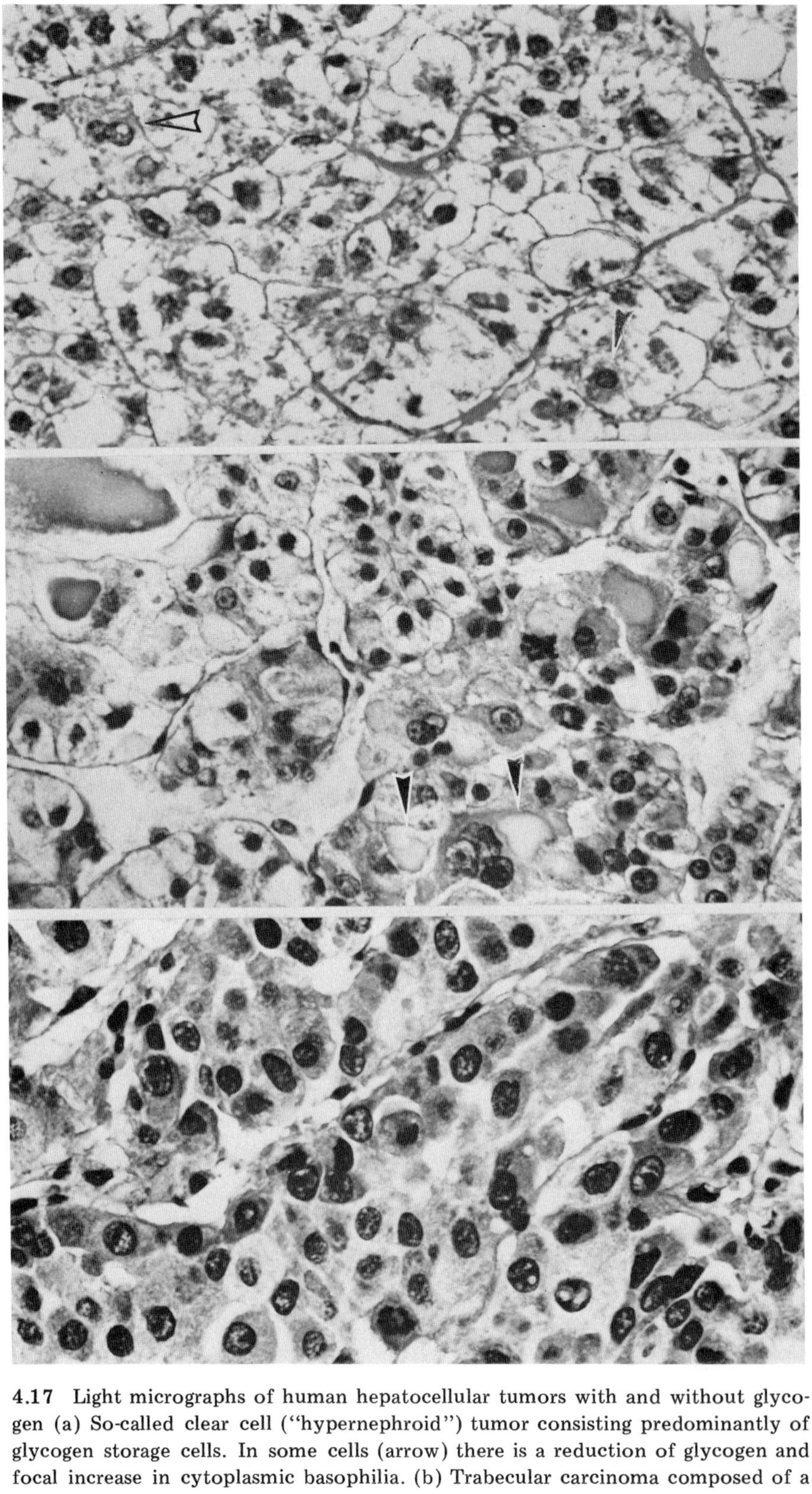

4.17 Light micrographs of human hepatocellular tumors with and without glycogen (a) So-called clear cell ("hypernephroid") tumor consisting predominantly of glycogen storage cells. In some cells (arrow) there is a reduction of glycogen and focal increase in cytoplasmic basophilia. (b) Trabecular carcinoma composed of a mixed population of clear, acidophilic (arrow), and basophilic cells. (c) Glycogen-free basophilic carcinoma. (All: H and E, × 460.) (Bannasch and Klinge, 1971, *Virch. Arch. Abt. A. Path. Anat.* 352:157–164.)

indications that excessive storage of carbohydrates during the preneoplastic phase is not a phenomenon occurring during carcinogenesis in the liver only (Bannasch, 1974b), the answer to that question could be of importance for a better understanding of carcinogenesis in general.

4.7 Acknowledgment

I heartily thank K. Aterman and F.-D. Dallenbach for helping me with the English translation.

4.8 References

Aisenberg, A. C. 1961. *The Glycolysis and Respiration of Tumors.* New York: Academic Press.

Bader, G., Stiller, D., and Ullrich, K. 1971. Ultrastructural changes in toxically injured and proliferating liver cells due to long-term application of diethylnitrosamine. *Arch. Geschwulstforsch.* 37:327–343.

Bannasch, P. 1968. *The Cytoplasm of Hepatocytes during Carcinogenesis. Recent Results in Cancer Research,* vol. 19. New York: Springer.

Bannasch, P. 1969. Grundsätzliche cytopathologische Unterschiede in der Genese von Lebercirrhose und Leberzellcarcinom. *Verh. Dtsch. Ges. Path.* 53:335–341.

Bannasch, P. 1974a. The cytologic heterogeneity of the so called hyperplastic liver nodules. *XI. Int. Cancer Congr.,* abstract of papers, p. 464, Florence.

Bannasch, P. 1974b. Carcinogen-induced cellular thesaurismoses and neoplastic cell transformation. In *Special Topics in Carcinogensis. Recent Results in Cancer Research,* vol. 44, ed. E. Grundmann, pp. 115–126. New York: Springer-Verlag.

Bannasch, P. 1975. Die Cytologie der Hepatocarcinogenese. In *Handbuch der Allg. Pathologie,* Bd. VI/7, Geschwülste. New York: Springer.

Bannasch, P., and Angerer, H. 1974. Glykogen und Glukose-6-Phosphatase während der Kanzerisierung der Rattenleber durch N-Nitrosomorpholin. *Arch. Geschwulstforsch.* 43:105–114.

Bannasch, P., and Klinge, O. 1971. Hepatozelluläre Glykogenose und Hepatombildung beim Menschen. *Virch. Arch. Abt. A., Path. Anat.* 352:157–164.

Bannasch, P., and Müller, H. A. 1964. Lichtmikroskopische Untersuchungen über die Wirkung von N-Nitrosomorpholin auf die Leber von Ratte und Maus. *Arzneim. Forsch. (Drug. Res.)* 14:805–814.

Bannasch, P., and Reiss, W. 1971. Histogenese und Cytogenese cholangiozellulärer Tumoren bei Nitrosomorpholin-vergifteten Ratten. Zugleich ein Beitrag zur Morphogenese der Cystenleber. *Z. Krebsforsch.* 76:193–215.

Bannasch, P., Hesse, I., and Angerer, H. 1974. Hepatozelluläre Glykogenose und die Genese sogenannter hyperplastischer Knoten in der Thioacetamid-vergifteten Rattenleber. *Virsch. Arch. Abt. B.* 17:29–50.

Bannasch, P., Papenburg, J., and Ross, W. 1972. Cytomorphologische und morphometrische Studien der Hepatocarcinogenese. I. Reversible und irreversible Veränderungen am Cytoplasma der Leberparenchymzellen bei Nitrosomorpholin-vergifteten Ratten. *Z. Krebsforsch.* 77:108–133.

Bruni, C. 1960. Hyaline degeneration of rat liver cells studied with the electron microscope. *Lab. Invest.* 9:209–215.

Bruni, C. 1973. Distinctive cells similar to fetal hepatocytes associated with liver carcinogenesis by diethylnitrosamine. Electron microscopic study. *J. Nat. Cancer Inst.* 50:1513–1528.

Burk, D., and Woods, M. 1967. Newer aspects of glucose fermentation in cancer growth and control. *Arch. Geschwulstforsch.* 28:305–319.

Burk, D., Woods, M., and Hunter, J. 1967. On the significance of glucolysis for cancer growth. *J. Nat. Cancer Inst.* 38:839–863.

Daoust, R. 1972. Focal loss of ribonuclease activity in preneoplastic rat liver. *Cancer Res.* 32:2502–2509.

Daoust, R., and Calami, R. 1971. Hyperbasophilic foci as sites of neoplastic transformation in hepatic parenchyma. *Cancer Res.* 31:1290–1296.

Daoust, R., and Molnar, F. 1964. Cellular populations and mitotic activity in rat liver parenchyma during azo dye carcinogenesis. *Cancer Res.* 24:898-1909.

Daoust, R., and Simard, A. 1968. Radioautographic study on RNA labeling in rat liver parenchyma during DAB carcinogenesis. *Cancer Res.* 28:874-880.

Drochmans, P., and Scherer, E. 1972. Enzyme defects in hepatocarcinogenesis. *J. Cell Biol.* 55:63a.

Epstein, S., Ito, N., Merkow, L., and Farber, E. 1967. Cellular analysis of liver carcinogenesis: the induction of large hyperplastic nodules in the liver with 2-fluorenylacetamide or ethionine and some aspects of their morphology and glycogen metabolism. *Cancer Res.* 27:1702-1711.

Epstein, S. M., McNary, J., Bartus, B., and Farber, E. 1968. Chemical carcinogensis: persistence of bound forms of 2-fluorenylacetamide. *Science* 162:907-908.

Ericsson, L. E., Orrenius, St., and Holm, J. 1966. Alterations in carnine liver cells induced by protein deficiency. Ultrastructural and biochemical observations. *Exp. Mol. Path.* 5:329-349.

Farber, E. 1963. Ethionine Carcinogenesis. *Advan. Cancer Res.* 7:383-474.

Farber, E. 1973. Hyperplastic liver nodules. In *Methods in Cancer Research*, ed. H. Busch, Vol. 7, pp. 345-375. New York: Academic Press.

Forget, A., and Daoust, R. 1970. Histochemical study on rat liver glycogen during DAB carcinogenesis. *Int. J. Cancer* 5:404-409.

Friedrich-Freksa, H., Gössner, W., and Börner, P. 1969. Histochemische Untersuchungen der Cancerogenese in der Rattenleber nach Dauergaben von Diäthylnitrosamin. *K. Krebsforsch.* 72:226-239.

Friedrich-Freksa, H., Papadopulu, G., and Gössner, W. 1969. Histochemische Untertsuchungen der Cancerogenese in der Rattenleber nach zeitlich begrenzter Verabfolgung von Diäthylnitrosamin. *Z. Krebsforsch.* 72:240-253.

Garancis, J. C., Tang, T., Panares, R., and Jurevics, I. 1969. Hepatic adenoma. Biochemical and electron microscopic study. *Cancer Res.* 24:560-568.

Grundmann, E., and Sieburg, H. 1962. Die Histogenese und Cytogenese des Lebercarcinoms der Ratte durch Diäthylnitrosamin im lichtmikroskopischen Bild. *Beitr. Path. Anat.* 126:57-90.

Hamperl, H. 1970. On the "adrenal rest-tumors" (hypernephromas) of the liver. *Z. Krebsforsch.* 74:310-317.

Heine, W.-D., Braun, A., and Bannasch, P. 1974. The kinetics of cell proliferation in the rat liver during the application of n-Nitrosomorpholine in different concentrations. *X. Int. Congr. Int. Acad. Pathol.*, abstract of papers, p. 41, Hamburg.

Hobik, H. P., and Grundmann, E. 1962. Quantitative Veränderungen der DNS und RNS in der Rattenleberzelle während der Carcinogenese durch Diäthylnitrosamin. *Beitr. Path. Anat.* 127:25-48.

Hruban, Z., Mochizuki, Y. Morris, H. P., and Slesers, A. 1972. Endoplasmic reticulum, lipid, and glycogen of Morris hepatomas. *Lab. Invest.* 26:86-99.

Hruban, Z., Morris, H. P., Mochizuki, Y., Meranze, D. R., and Slesers, A. 1971. Light microscopic observations of Morris hepatomas. *Cancer Res.* 31:752-762.

Karasaki, S. 1969. The fine structure of proliferating cells in preneoplastic rat livers during azo-dye carcinogenesis. *J. Cell Biol.* 40:322-335.

Langer, K. H. 1965. *Vergleichende histologische und funktionelle Untersuchungen an Niere, Leber und Herzmuskel nach experimentellen Blutverlusten und orthostatischen Kollapsen am Kaninchen.* Inang.-Dissertation, Würzburg.

Lesch, R., Bauer, Ch., and Reutter, W. 1973. The development of cholangiofibrosis and hepatomas in glactosamine induced cirrhotic rat livers. *Virch. Arch. Abt. B. Zellplath.* 12:285–289.

Moulin, M.-Ch., and Daoust, R. 1971. Glucose-6-phosphatase activity in rat liver parenchyma during azo dye carcinogenesis. *Int. J. Cancer* 8:81–85.

Newberne, P. M., and Wogan, G. N. 1968. Sequential morphologic changes in aflatoxin B_1 carcinogenesis in the rat. *Cancer Res.* 28:770–781.

Oehlert, W., and Hartje, J. 1963. Die Veränderungen des Eiweiss- und Ribonucleinsäurestoffwechsels während der experimentellen Cancerisierung durch Diäthylnitrosamin. *Beitr. Path. Anat.* 128:376–415.

Opie, E. L. 1946. Mobilization of basophile substance (ribonucleic acid) in the cytoplasm of liver cells with the production of tumors by butter yellow. *J. Exp. Med.* 84:91–106.

Phillips, M. J., Langer, B., Stone, R., Fisher, M., and Ritchie, S. 1973. Benign liver cell tumors. *Cancer* 32:463–470.

Porter, K. R., and Bruni, C. 1959. An electron microscope study of the early effects of 3′-methyl-DAB on rat liver cells. *Cancer Res.* 19:997–1009.

Potter, R. V. 1964. Biochemical studies on minimal deviation hepatomas. In *Cellular Control Mechanisms and Cancer,* eds. P. Emmelot and O. Mühlbock, pp. 190–210.

Reuber, M. D. 1965. Development of preneoplastic and neoplastic lesions of the liver in male rats given 0.025 percent N-2-fluorenyldiacetamide. *J. Nat. Cancer Inst.* 34:697–724.

Romen, W., and Bannasch, P. 1972. Karyokinese und Kernstruktur. I. Mitosen und Mitosestörungen in Hepatozyten und Hepatomzellen der Nitrosomorpholinvergifteten Rattenleber. *Virch. Arch. Abt. B.* 11:24–33.

Romen, W., and Bannasch, P. 1973. Karyokinese und Kernstruktur während der Hepatocarcinogenese. II. Die Feinstruktur des Zellkerns in Hepatozyten und Hepatomzellen der Nitrosomorpholinvergifteten Rattenleber. *Virch. Arch. Abt. B.* 13:267–296.

Romen, W., Ross, W., and Bannasch, P. 1972. Cytomorphologische und morphometrische Studien den Hepatocarcinogenese. II. Reversibilität von Kerngrössenänderungen in der Nitrosomorpholin-vergifteten Rattenleber. *Z. Krebsforsch.* 77: 134–140.

Schaff, Zs., Lapis, K., and Safrany, L. 1971. The ultrastructure of primary hepatocellular cancer in man. *Virch. Arch. Abt. A. Path. Anat.* 352:340–358.

Schauer, A., and Kunze, E. 1968. Enzymhistochemische und autoradiographische Untersuchungen während der Cancerisierung der Rattenleber mit Diäthilnitrosamin. *Z. Krebsforsch.* 70:252–266.

Scherer, E., Hoffmann, M., Emmelot, P., and Friedrich-Freksa, H. 1972. Quantitative study on foci of altered liver cells induced in the rat by a single dose of diethynitrosamine and partial hepatectomy. *J. Nat. Cancer Inst.* 49:93–106.

Schmitz-Moormann, P., Gedick, P., and Dharamandhach, A. 1972. Histologische und histochemische Frühveränderungen bei der experimentellen Erzeugung von Lebercarcinomen durch Diäthylnitrosamin. *Z. Krebsforsch.* 77:9–16.

Smuckler, E. A., and Arcasoy, M. 1969. Structural and functional changes of the endoplasmic reticulum of hepatic parenchymal cells. *Int. Rev. Path.* 7:305–418.

Soostmeyer, R. 1940. Glykogengehalt und Zellstrukturen der Leber während des anaplylaktischen Schocks. *Virch. Arch.* 306:554–569.

Spycher, M. A., and Gitzelmann R. 1971. Glycogenosis type I (glucose-6-phosphatase deficiency): ultrastructural alterations of hepatocytes in a tumor bearing liver. *Virch. Arch. Abt. B. Zellpath.* 8:133–142.

Stäubli, W., Hess, R., and Weibel, E. R. 1969. Correlated morphometric and biochemical studies on the liver cell. II. Effects of phenobarbital on rat hepatocytes. *J. Cell Biol.* 42:92–112.

Steiner, J. W., Miyai, K., and Phillips, M. J. 1964. Electron microscopy of membrane-particle arrays in liver cells of ethionine-intoxicated rats. *Amer. J. Path.* 44:169–213.

Sugihara, R., Hiasa, Y., and Ito, N. 1972. Ultrastructural changes in nuclei and nucleoli of rat liver cells treated with hepatocarcinogens. *Gann* 63:419–426.

Svoboda, D., Grady H., and Higginson, J. 1966. The effect of chronic protein deficiency in rats. II. Biochemical and ultrastructural changes. *Lab. Invest.* 15:731–749.

Taper, H. S., Fort, L., and Bruchler, J.-M. 1971. Histochemical activity of alkaline and acid nucleases in the rat liver parenchyma during N-nitrosomorpholine carcinogenesis. *Cancer Res.* 31:913–916.

Theodossiou, A., Bannasch, P., and Reuss, R. 1971. Glykogen und endoplasmatisches Reticulum der Leberzelle nach hohen Dosen des Carcinogens N-Nitrosomopholin. *Virch. Arch. Abt. B. Zellpath.* 7:126–146.

Thoenes, W., and Bannasch, P. 1962. Elektronen-und lichtmikroskopische Untersuchungen am Cytoplasma der Leberzellen nach akuter und chronischer Thioacetamid-Vergiftung. *Virch. Arch.* 335:556–583.

Weber, G. 1968. Carbohydrate metabolism in cancer cells and the molecular correlation concept. *Naturwissenschaften* 55:418–429.

Weinhouse, S. 1966. Glycolysis, respiration, and enzyme deletions in slow-growing hepatic tumors. In *U.S. Japan Joint Conference on Biological and Biochemical Evaluations of Malignancy in Experimental Hepatomas, Kyoto, 1965. Gann Monograph* no. 1, pp. 99–115. Tokyo: Japanese Cancer Association.

4.9 Discussion

P. Newberne In your studies with NNM-induced tumors, did you see any specific changes in the endothelium, and did you observe any angiosarcomas?

Bannasch Yes, we had many angiosarcomas, and I am afraid that we made many false diagnoses at first. We combined them with anaplastic carcinoma in the beginning, but we are now sure that we have many angiosarcomas and we may have early stage of the angiosarcomas, too, but we do not know in this case of any precancerous or preneoplastic condition.

P. Newberne We observe about 40 percent angiosarcomas in the liver and lung in rats fed NNM or nitrite plus morpholine.

Butler It is an advantage in studying mechanisms to separate different populations of cells and distinguish, if possible, early carcinoma from other cell changes. This obviously is an advantage if you are interested in testing, because so far, at this meeting, most have indicated that you must wait for the metastatic invasion before you can diagnose carcinoma with certainty. The lesions induced by aflatoxin in our system are in a sense similar to the lesions Dr. Bannasch has described. We induce these lesions by feeding aflatoxin at 5 ppm for six weeks, stop the diet, and then feed only the stock laboratory diet. Forty to fifty weeks later, all animals have carcinoma, diagnosed by classical, conventional methodology. We find basically the same early lesions with aflatoxin as those described by Dr. Bannasch in his studies with NNM and TAA, but we have different interpretations of significance. Both the basophilic foci and the vacuolated cells are present in the same liver following exposure to a carcinogenic dose of a number of chemical carcinogens. We feel that the vacuolated foci are degenerative and that it is probably the small basophilic foci, or perhaps the larger pink, rather washed-out cells, that contribute the important lesion.

Rogers It seems to me that it is useful to use the criterion of hyperplasia, that is, ^{3}H-thymidine uptake or mitosis, for this distinction. Like you, I think the small cell groups which are hyperplastic are the most likely precursors of cancer. When we look at the vacuolated cells, generally they are not hyperplastic. I think that is what you said also, is that correct?

Bannasch I perfectly agree, of course, but the vacuolated, glycogen-filled cells are the first pathological cell population which we see, and later, the basophilic cells appear, as a rule in very close connection to the vacuolated cells. I think we have a clear indication that the glycogen cell is at least one population of precancerous cells because, in one case, we have lung metastasis from the liver where metastases are primarily this type of cell.

Butler In our animals treated with either aflatoxin or nitrosamine we have always found the small foci of basophilic cells proliferating earlier, usually in the first few weeks, followed by the clear cell foci.

Bannasch The glycogen storage cells are not easy to find. Early stages are difficult to find. We are not the only ones to say that the clear glycogen storage cells are the first cell population that occurs.

Rogers Do you think it depends on the carcinogen?

Bannasch No. I do not think so.

P. Newberne Basically there does not seem to be that much disagreement among us. There are two types of lesions: the glycogen storage focus and the basophilic one. The times that we find them may vary, as we have pointed out (Newberne and Wogan, 1967), and you can modify time of appearance remarkably by dietary means, as Dr. Rogers has shown (Rogers and Newberne, 1969). If rats are given a carcinogenic dose of aflatoxin and fed a diet marginal in lipotropes, you will find the basophilic focal lesions as soon as you complete the three-week dosing schedule. In rats fed a complete stock diet, it takes about four or five months before the basophilic foci appear in significant numbers.

Bannasch Are you sure that these are not the alterations I demonstrated first, which, in our case at least, are indistinguishable from these hyperbasophilic foci; but they are reversible and the basophilic foci are not reversible.

P. Newberne We have labeled them with ^{3}H-thymidine, and we think we can follow them through from a very small focal area on to hepatocellular carcinoma. Furthermore, I agree with you that these vacuolated areas—the glycogen storage areas—are probably degenerative lesions, and often within them we see small foci areas of hyperplastic, basophilic cells growing. I think maybe that is one of the lesions that develops into cancer.

Bannasch I think you cannot say anything about a basophilic focus if the carcinogenic treatment goes on. The only way to distinguish between these two lesions is to make stop experiments, and if they persist I would say they are true basophilic foci, not if they are reversible.

Rogers These are stop experiments; a carcinogenic dose of aflatoxin is given during the initial three weeks of the experiment, and thereafter the animal gets the diet only.

Bannasch And, in spite of this, the first lesions you can see are the basophilic?

Rogers Yes, which have increased mitoses, increased thymidine uptake, and abnormalities of enzyme staining.

McCollum I have a question about these foci of glycogen storage: How often do you see them in your control animals?

Butler We have not observed the lesion in the liver of control rats except in very old rats, in which case we sometimes come across foci of glycogen storage, large, clear, vacuolated cells. I have also seen them in the liver of rats recovered from an acute single dose aflatoxin where, in our hands, such treatment never produces carcinoma.

Yoon In older rats, not that infrequently, you see focal areas of glycogen accumulation. On the whole, I think I have seen them more in females than males.

Leffert We see in tissue culture that all of these steps are part of the same sequence and that what you are seeing is the variation due to the fact that there are different populations of cells in the nodule. Therefore the responses are different and ultimately the kinds of changes that cause the cell to replicate are all the same. You are just starting from different points, which could be determined nutritionally, hormonally, or otherwise.

Bannasch We are trying other enzymatic reactions; one is a cytochemical reactant called adenylcyclase.

Laqueur The question arises about histochemical identification and the qualitative accuracy of some methods. From some reported work it is doubtful whether some precipitates even represent enzymatic activity; the lead precipitate which one sees may or may not be the result of enzymatic activity. This is basically the criticism of many techniques, and I do not think any better reaction has been worked out since the earlier studies proposed such identification.

Weisburger Which strain of rat did you use in your study?

Bannasch In our first experiment we used BD strains 1 and 2, but afterwards we used only the Sprague-Dawley because it was too difficult to get these other animals.

Weisburger What kind of spontaneous tumor incidence does the BD strain have?

Bannasch We have never seen a spontaneous hepatocellular carcinoma.

Butler I have had a colony for four years and I have never seen a Fischer rat with a spontaneous hepatocarcinoma. In our old rats, which are a Wistar strain, we have never seen a spontaneous liver tumor.

Laqueur Has anybody seen hepatomas or large nodule lesions in the liver composed of basophilic cells? I am not talking about these isolated lesions that you showed us which are free of glycogen, but a tumor.

Bannasch Yes. I would say most liver tumors are composed of the basophilic cells.

Laqueur I do not get that impression from having seen the pictures so far, that is, tumor nodules composed of the small cells with very intense cytoplasmic basophilia.

Bannasch I think that there might be one additional argument for the sequence we propose. If you take all of these experiments on transplantable tumors, recent studies of Reuber and Morris by light and electron microscopy in slow-growing tumors, nearly the same morphology was observed as described in the glycogen storage group: lots of glycogen, hypertrophy of smooth endoplasmic reticulum. The medium growth rate is characterized by some tumors which have these signs still, but most of them are already free of glycogen and basophilic cells. The fast-growing tumors are all relatively free of glycogen and intensely basophilic, suggesting a clear correlation.

Butler How much do you think all these stages in glycogen or phosphatase are secondary to vascular effects? We have been trying to perfuse these livers, and even when you have miniscule little nodules you cannot perfuse them through the portal vein, which works beautifully for a normal liver. You perfuse them through the hepatic artery, and they do not perfuse well. They seem to have different blood supply. How much will this affect glycogen?

Bannasch I cannot answer that question, but you can perfuse them quite well if you increase the pressure.

Butler Yes, well, we do not want to do that because it will destroy our tissue.

Knook I have some questions about your cell counting. First of all, at which age of the animals did you start your experiments?

Bannasch I cannot say the age because we always take animals with a given weight; they are about 200 grams, probably three months old.

Knook You mentioned that after one-half year you found an enlargement factor of about 1.2, and considering this, I should think that you have already had their main change to tetraploid cells.

Bannasch Oh, yes. That would seem so because they are adult rats.

Knook In combining normal cells with abnormal and talking about the size of the cells or the enlargement, we have to take into account the localization of the cells in the lobules since there are large changes or large differences in the cell size depending on the localization. Did you consider this in your calculation?

Bannasch No. We counted all over the lobule. It was not possible to distinguish in this experiment between peripheral and central lesions.

Knook Did you count any binucleate cells?

Bannasch No, we have not. But I think one can propose that the binucleate cell has double the amount of cytoplasm, and we count the nuclei so it will not matter.

Knook If there are changes in the percentage of binucleate cells during experiments, you will find changes in your cell volume.

Bannasch I think that, at any rate, we will have to repeat the morphometric studies because three animals from one group are not enough for statistical evaluation.

Yoon Our experiment was about the same as Dr. Bannasch's; in other words, the glycogen accumulation came first and then the basophilic foci. Probably the carcinogen makes a difference.

P. Newberne I am not convinced that the type of carcinogen makes that much difference. We see the same lesions with several different liver cell carcinogens.

Squires It is my impression that you consider the basophilic focus an irreversible state.

Bannasch If I find such foci in an experiment, in a stop experiment, that is not under the influence of carcinogen, I would consider it irreversible.

Squires So then you interpretation is really not based on any morphologic or functional features of the focus lesion. I mean, if you consider that focus irrefutably committed to cancer, then, stop or no stop, it makes no difference.

Bannasch There is a difference because you have basophilic focal alternations, under the influence of the carcinogen, which are reversible.

Squires You are not identifying that focus, per se, as an irreversible lesion?

Bannasch No, I could say this only within the context of the experiment, so I think there are reversible alterations indistinguishable from the hyperbasophilic focus which will probably never progress to cancer.

Squires If you look at an animal that had a prolonged exposure to carcinogens followed by a six-month observation period, and then, at the end of that six-month period, you found hyperbasophilic foci—small ones and big ones—are these irreversible?

Bannasch In some cases we give the carcinogens at 50 mg percent for three weeks and then stop and look at the animals for one to two years, or 12 mg percent for seven weeks and stop, or one or two years at 12 mg percent continuously, or 6 mg percent continuously up to about two years. At the end of the treatment, we have practically no basophilic foci after stopping the experiment. In the later stages we have only glycogen storage foci with clear cells or eosinophilic cells, and then one year after carcinogens, we have some small basophilic foci. In the last group we have a few glycogen storage foci and very many basophilic foci; this is the way we came to our sequence.

Leffert To my understanding, the basophilia is really looking at the quantity of polysomes. No cell can grow until a certain portion of its polysomes have been formed. So what you are really asking is, "When a cell is irrefutably committed to cancer, what genetic changes have taken place so that the polysomes do not break down?" In other words, when a cell is finished dividing, it has two choices: its polysomal material can disaggregate, which would be correlated with loss of basophilia, or it may not; it can continuously stay in the cycle—that is, the signals to which the cell is responding, the internal concentrations of nutrients, etc., are such that the polysomes do not break down. That cell will then be irrefutably committed to continue another round of growth, so that the fact that you see basophilia in certain early times would mean it probably had been stimulated to cell division and polysome reaggregation; but it may be that the genetic changes that require continued exposure to the carcinogens have yet to take place. For example, you may break DNA; it would be the time at which those breaks become reintegrated in a mutated way that the polysomal apparatus is constitutively stabilized so that the cell continuously grows, and that is what you call a carcinoma.

Squires But none of these precancerous changes that you just hypothesized is determined by the morphology.

Leffert No. I would say that you cannot definitely conclude that from current morphological evidence.

Squires But if the experiment is stopped, and even a year later you still see basophilic nodules, that still does not tell you that that cell is genetically transformed—just irreversibly changed.

I wanted to make sure that I understood the irreversible changes within the framework of your procedure and that they cannot be applied as a morphological criteria in general.

Bannasch No, only under these circumstances. But may I make one other comment? This regimen with nearly total loss of glycogen during the application of the carcinogen and intense cytoplasmic basophilia—you will not find later reversibility of glycogen loss, and you can find, instead of the basophilic areas, glycogen storage areas.

Harris When we see these binucleated cells, of course, we are in a section that goes through the cell; if it does not include a portion of the nucleus you do not know whether it is a binucleate cell or a mononucleate cell. I was wondering what percentage of the cells might be sectioned in that manner without your realizing they were binucleate cells. Do you have any comment on this point?

Knook That depends on the strain and the age of the animals. There are large changes in numbers of binucleate cells during the enlargement of the liver.

Goyings Dr. Bannasch, you also repeated these types of studies with the mouse, and you found them to be similar?

Bannasch We did some investigations ten years ago with NNM in mice. The tumors seemed to grow very slowly, but the cytologic alterations were quite similar to those of the rat.

Goyings Have you had an opportunity to label the carcinogen itself and use some of the techniques that you are utilizing in determining where these reactions take place?

Bannasch We have not had time, but we shall do it.

Butler Dr. Neil is preparing some labeled aflatoxin B_1 for autoradiography to see what occurs. Another thing about these basophilic nodules, whatever they are. I have seen them under two situations where you do not get carcinomas: one is single-dose aflatoxin; another is the pyrrolizidine alkaloids. We have followed such lesions and, under the conditions of the experiment, the liver returns eventually to normal. About irreversible nodules, we always see some of them. Every time you take a section, if you have the time right, they are there. And yet, as Becker said yesterday, he always lands up with, at the most, two or three carcinomas. Well, we only land up with a few carcinomas; so either they coalesce, or most of them disappear. I would suspect most of them disappear.

J. Newberne Do you know anything about the half-life of the compound?

Bannasch No.

Leffert Generally, in carcinogenesis in the liver, is it known whether the carcinogen selects the cell that is already in the liver that is destined to become carcinoma, or does the carcinogen induce the change? In other words, it is back to the old question, "Are you raising the spontaneous rate in producing tumors or is the carcinogen having a direct effect?" This is an open question to the speakers.

Butler In our hands, our rats do not get liver tumors unless some agent is applied, so this spontaneous incidence is very low.

Leffert Is there a relationship in the nontoxic range between the quantity of foci and dose?

Butler There is one paper by Scherer et al. (1972) which reflects the correlation between dose, hepatectomy, and glucose phosphatase deficiency. They found an increase in numbers of glucose-phosphatase-deficient areas after hepatectomy.

Squires In the strain A mouse, which has the genetic predisposition to develop lung tumors, it appears that the carcinogen is enhancing the spontaneous incidence.

Rogers In the rat, when we alter the final tumor incidence by changing the diet, we have more of these early lesions; so in that sense there is a parallel.

Baldwin I think I would answer Dr. Leffert's comment about the carcinogen giving a selective advantage to a cell that is already there by saying you can show that this is not the case by immunological methods. The population that develops following an exposure to the carcinogens is not one with a selective advantage but a population susceptible to the transformation change.

P. Newberne I would like to make one comment relative to dose response. We have done studies in rats with nitrite and morpholine using concentrations up to 1000 ppm of both chemicals. When both nitrite and morpholine is included at 1000 ppm, we get 100 percent liver cancer in a very short period of time. As the morpholine level is decreased, you still get a very high incidence and in a relatively short period of time. With most ratios and with NNM, we find a dose response and, in all cases, lesions in the livers of the rats seem to go through this same series of changes we are discussing here. The difference is the rate at which such lesions develop. When you drop the nitrite level down lower, somewhere around 50 or 100, they go through the same pattern, but much more slowly. The end result is the same, but you speed up the process when you raise the level of both of these. But the nitrite concentration has the greatest influence.

Butler Is there a difference in the relation of hepatocarcinomas and angiosarcomas with different regimens?

P. Newberne There does not appear to be.

5 Microsomal Enzyme Systems and Drug Toxicities

James R. Gillette
Chemical Pharmacology Laboratory
National Heart, Lung and Blood Institute
Bethesda, Maryland 20014

5.1 Introduction

Many drugs are converted in the body to various metabolites that elicit therapeutical and toxicological responses. In most cases these metabolites bring about their effects by combining reversibly at various action sites in the tissues. Many toxicities, however, including tissue necrosis, mutagenesis, and carcinogenesis, are considered by a number of authorities to be mediated by chemically reactive metabolites of compounds rather than by the foreign compounds themselves. It is suspected that drugs, as well as environmental toxicants, may bring about serious kinds of toxicities (including cellular and tissue necrosis, blood dyscrasias, toxicity and damage to the fetus, and hypersensitivity reaction), as well as carcinogenesis and mutagenesis, through the formation of reactive metabolites. There has been considerable research in recent years on drugs and other foreign compounds that are converted in the body to chemically reactive metabolites, which then combine irreversibly with macromolecules such as DNA, RNA, protein, and glycogen.

The laboratory at the National Heart, Lung and Blood Institute of the National Institutes of Health has been engaged in this area of research for many years. We have generally approached the problem by identifying toxicities caused by various drugs and other foreign compounds and determining whether the toxicities are mediated by reactive metabolites or through some other mechanism. Many reactive metabolites are formed by mixed-function oxidases, which are preferentially located in the endoplasmic reticulum of the liver, although the enzymes are also present in some extrahepatic tissues. The effectiveness of reactive metabolites in evoking toxicities depends on many factors, most of which are not yet clearly understood. Among these factors, however, are kinetics for the formation and elimination of the reactive metabolites.

In attempting to relate the formation of chemically reactive metabolites to specific toxicities, our laboratory has conceived the idea that the incidence and severity of any given toxicity could ideally be determined from the product mathematical functions; that is, the incidence and severity is equal to ABCD. In this equation, A is the proportion of the dose of the toxicant that is converted to a chemically reactive metabolits; B is the proportion of the chemically reactive metabolite that becomes covalently bound to various cellular components; C is the proportion of the covalently bound metabolite that is attached to vitally important cellular components; and D is the proportion of C that cannot be replaced or repaired or that leads to the formation of genetic abnormalities. According to this concept, the values of C and D vary with the toxicant, the tissue in which the covalent binding occurs, and the mechanism of toxicity. When the product of C times D is small, toxicities may not occur even when considerable amounts of covalently bound metabolites are found in tissues. If the product of C times D is high, for a given kind of reaction, the toxicity may occur even when only small amounts of covalently bound metabolites are found in the tissues.

Since very little is known about the mechanism of most toxicities, the numerical values of C and D are virtually impossible to estimate with the present state of knowledge, and therefore measurements of covalent binding alone cannot be used to predict whether a given compound will cause a given kind of toxicity. However, when it has been established that the drug causes a toxicity, we can determine whether the toxicity is mediated through the formation of a chemically reactive metabolite by correlating changes in the amount of covalently bound metabolite (the product of A times B) with changes in the incidence and severity of the toxicity. Furthermore, we can determine whether treatments that alter the rates of drug metabolism also after either A or B.

This paper presents a few examples of these kinds of reactions and interactions.

5.2 Microsomal Enzymes, Chemical Metabolites, and Cell Injury

Chemically inert substances can be converted to reactive metabolites by a variety of reactions. For example, secondary amines such as N-methyl-4-aminoazobenzene, primary amines such as aminobiphenyl, and acetylated primary amines such as 2-acetylaminofluorene are N-hydroxylated by enzymes systems including the cytochrome P-450 or amine N-oxidase. Dialkylnitrosamines are N-dimethylated by cytochrome P-450 enzymes to monoalkylnitrosamines, which appear to spontaneously rearrange to other less well-understood or identified metabolites believed by some to be alkyl carbonium ions (Magee and Barnes, 1969). Intestinal microflora produce an enzyme, β-glucosidase, which hydrolyzes cycasin to methylazoxymethanol, which in turn acts as a methylating agent (Laqueur, 1964). Pyrrolizidine alkaloids are thought to be dehydrogenated to chemically reactive derivatives (Mattocks, 1973); polycyclic hydrocarbons undergo epoxidation by the cytochrome P-450 enzyme system to form highly potent arylating agents (Daly et al., 1972); and urethane may undergo an N-hydroxylation before it is converted to an ethylating agent (Mirvish, 1968). These types of reactions have been implicated in the formation of chemically reactive metabolites that result in toxicity, carcinogenesis, or mutagenesis (Gillette et al., 1974).

In the context of this workshop, discussion will be limited to examples of reactive metabolites associated with hepatic necrosis; examples are now available for other tissues, including the lung, kidney, testis, bone marrow, and the form elements of the blood (Gillette et al 1974).

For some time investigators have realized that the effects of an active metabolite that acts by combining reversibly with action sites within or on a cell can often be related to its plasma concentration. However, if there is tissue damage caused by covalent binding of reactive metabolites to tissue macromolecules, it is not logical to expect a relationship between the plasma level of the metabolite and the severity of the induced lesion. Nevertheless, there should be some indication of a relationship between the extent and severity of the lesion and the amount of covalently bound metabolite.

Since the liver is a major site of enzymatic transformation of foreign chemicals, it is the site at which one can expect often to find evidence of tissue damage. The lesion may be severe or it may be minimal. The induction of liver necrosis by halobenzenes illustrates many of the ways by which the severity of drug-induced tissue damage can be affected by changes in the formation or the fate of chemically reactive metabolites. These compounds are converted to their epoxide forms by the cytochrome P-450 enzyme system in the microsomal fraction of the liver, and although the parent compounds are chemically inert, some of the metabolites become covalently bound to macromolecules and effect liver injury. It has been observed recently that the magnitude of the covalent binding of labeled metabolites of the halobenzenes parallels their toxic effects. For example, chlorobenzene, bromobenzene, iodobenzene, and *o*-dichlorobenzene all lead to liver necrosis associated with covalent binding of reactive metabolites to the protein of the liver cell. On the other hand, fluorobenzene and *p*-dichlorobenzene do not cause liver necrosis and are not bound to any significant degree (Brodie at al., 1971; Reid et al; 1971b; Reid, 1973). Of considerable interest, particularly to those concerned with tissue lesions, is the finding that ^{14}C-bromobenzene covalently binds preferentially in the centrilobular area of the liver, the same site where necrosis occurs. Equally interesting is the observation that pretreatment of rats with phenobarbital, to stimulate the cytochrome P-450 drug-metabolizing enzymes in the liver, results in an increased rate of disappearance of bromobenzene from the animal tissues, covalent binding of labeled bromobenzene in the liver, and severity of liver necrosis (Reid, 1973; Reid et al., 1971a).

metabolizing enzymes in the liver decrease also the rate of disappearance of bromobenzene, the covalent binding of labeled compound, and the severity of liver necrosis (Reid et al., 1971a; Reid et al., 1971b; Reid, 1973; Mitchell et al., 1971; Jollow et al., 1974). These observations indicate that necrosis is caused by the covalent binding of a metabolite of bromobenzene not by the parent compound. There is reason, then, to believe that the severity of the necrosis depends on the rate of formation of the active metabolite.

Further investigations of the formation of active intermediates from bromobenzene has shown that, in the presence of reduced triphosphopyridine nucleotide and oxygen, bromobenzene is activated by microsomes to a substance that binds covalently with glutathione (GSH), which traps the active intermediate (Brodie et al., 1971; Jollow et al., 1972 and 1974). If rats are pretreated with phenobarbital, which increases the activity of the drug-metabolizing enzymes, the amount of the complex is increased. Furthermore, carbon monoxide, which lowers the activity of cytochrome P-450 enzymes, also decreases the formation of the complex. These observations clearly indicate that the reactive metabolite is formed by a cytochrome P-450 enzyme.

In the bromobenzene system in rats, the size of the dose is clearly related to the proportion of the dose that is bound to liver microsomes; in addition, the effects of phenobarbital pretreatment also appear to depend on the dose of bromobenzene. For example, when doses of bromobenzene are below the toxic level, pretreatment with phenobarbital decreases the covalent binding; when doses of the chemical are high, pretreatment with phenobarbital increases covalent binding.

Studies to determine the reason for the threshold dose of bromobenzene and the variable effects of phenobarbital treatment revealed that liver cells have several ways of converting bromobenzene to 3,4-bromobenzene epoxide and then to chemically inert metabolites (Daly et al., 1972; Jollow et al., 1974; Azouz et al., 1953). An enzyme in the soluble fraction of liver can catalyze a reaction between the epoxide and GSH to form a conjugate (Brodie et al., 1971) that is ultimately excreted in the urine as a mercapturic acid (Knight and Young, 1958; Baumann and Preusse, 1879). Another enzyme, which is present in the endoplasmic reticulum of the liver cell, catalyzes the addition of water to the epoxide to form a dihydrodial, which may be oxided to a catechol (Azouz et al., 1953). Thus, the steady-state concentration of the epoxide depends not only on the rate of formation of the epoxide but also on the rate at which the epoxide is converted to the phenol, the GSH conjugate, and the dihydriol. In the case of a high concentration of the epoxide, there would presumably be a reaction with macromolecules in the liver followed by liver necrosis. Furthermore, experimentation has shown that the degree of covalent binding of bromobenzene and the severity of necrosis of the liver depend on the concentrations of GSH in the liver, which in turn depend on the relative rates at which GSH is being synthesized and consumed in the formation of the GSH bromobenzene conjugate (Jollow et al., 1974; Gillette, 1973).

Phenobarbital may increase the activity of GSH transferase, as well as the epoxide hydrase, in rats (Daly et al., 1972; Jollow et al., 1974), since it decreases the covalent binding of subtoxic doses of bromobenzene in this species (Reid, 1973). By increasing the rate of bromobenzene epoxide formation, phenobarbital hastens the depletion of liver GSH. When bromobenzene is given at toxic doses, pretreatment with phenobarbital increases the amount of covalent binding of bromobenzene metabolites as well as the severity of liver necrosis probably because the increase in rate of bromobenzene epoxide formation is in excess of the increase in hydrase activity, and the rate of GSH conjugation formation is limited by GSH synthesis (Jollow et al., 1974). The covalent binding is markedly decreased and toxicity largely prevented by SKF 525-A, which inhibits the formation of bromobenzene epoxide and thus slows depletion of GSH in the liver.

An additional area of interest in bromobenzene research concerns 3-methylcholanthrene. If rats are pretreated with 3-methylcholanthrene, the severity of liver necrosis from bromobenzene is appreciably diminished (Zampaglione et al., 1973; Reid et al., 1971b). The reason is that there is a change in the relative proportions of the kinds of epoxides formed. For example, very little 2,3-bromobenzene epoxide is formed in untreated rats, but large amounts of it are formed in rats pretreated with 3-methylcholanthrene (Zampaglione et al., 1973; Jollow et al., 1972).

In recent years a great deal of information about activation of chemicals in biological systems has derived from studies with carbon tetrachloride. Before its toxicity became evident, carbon tetrachloride was used in man as an anthelmintic. Only during the past few years has the mechanism by which carbon tetrachloride induces liver damage been clarified (Judah et al., 1970; Rechnagel, 1967; Slater, 1966; Plaa and Larson, 1964). The toxic action of carbon tetrachloride is mediated through an active metabolite. The lethal effects of this chemical are increased by pretreatment with phenobarbital (McLean and McLean, 1969; Traiger and Plaa, 1971), DDT, or isopropyl alcohol, all of which induce drug-metabolizing enzymes. Treatments that decrease the activity of drug-metabolizing enzymes diminish the toxicity of carbon tetrachloride. It is generally believed that the active metabolite causes centrilobular necrosis by promoting lipid peroxidation within the cell. The lipid peroxidation may be initiated by trichloromethyl free radical either by combining covalently with phospholipid or by reacting with phospholipid to form chloroform and phospholipid free radicals. Even in the latter case, however, the covalent binding of the metabolite to proteins (Cessi et al., 1966; Reynolds, 1967; Rao and Recknagel, 1968) and lipids (Gordis, 1969) should parallel the severity of the necrosis because they all arise from a common intermediate. Thus pretreatment of rats with dibenamine, which decreases drug-metabolizing enzyme activity, prevents the toxic effects of carbon tetrachloride and decreases the covalent binding of carbon tetrachloride metabolites to lipids (Maling, 1972). Possibly, dibenamine impairs the liver enzyme system that catalyzes the formation of the reactive metabolite. The increase in toxic effects of carbon tetrachloride associated with pretreatment of rats with phenobarbital is also associated with an increase in covalent binding of carbon tetrachloride (Reynolds et al., 1972). Although it was suggested earlier that carbon tetrachloride was homolytically split to form free radicals in the body (Butler, 1961), it now appears that the formation of the chemically reactive metabolite occurs by reductive cleavage of carbon tetrachloride to a free radical and a chloride ion. Furthermore, an enzyme of the cytochrome P-450 system appears to be involved in catalyzing the formation of a free radical; this conclusion is based on the inhibition of binding by carbon monoxide (Corsini et al., 1972; Sipes et al., 1972; Uehleke et al., 1973) and by an antibody against liver microsomal NADPH reduced nicotineamide-adenine dinucleotide phosphate) cytochrome *c* reductase (Krishna et al., 1973).

In view of this above data, it is interesting that, in addition to causing centrilobular necrosis and fatty infiltration of the liver, carbon tetrachloride destroys the liver microsomal enzyme P-450. The toxic effects of carbon tetrachloride may, therefore, be self-limiting if the initial dose was not overwhelming (Glende, 1972). Mechanisms by which carbon tetrachloride destroys the enzyme system are not clear, but studies in vitro have shown that carbon-tetrachloride-induced destruction of cytochrome P-450 may be a result of lipid peroxidation (Reiner et al., 1972). Earlier suggestions (based on studies in vivo) that antioxidants such as α-tocopherol or diphenyl *p*-phenylenediamine did not prevent the destruction of cytochrome P-450 (Castro et al., 1968) conflicted with the earlier observation that addition of ethylenediaminotetraacetate to liver microsomal systems blocked both the NADPH-dependent lipid peroxidation and the cytochrome P-450 destruction caused by carbon tetrachloride. The results in vivo may not have been correct, since it has been recently observed that antioxidants have little effect on lipid peroxidation induced by carbon tetrachloride (Gillette et al., 1974).

GSH markedly inhibits the covalent binding of carbon tetrachloride to rat liver microsomes (Corsini et al., 1972) but does not appreciably decrease in vitro covalent binding of carbon tetrachloride metabolite to liver microsomes of rabbits (Uehleke et al., 1973). Although liver levels of GSH ar not depleted after the administration of carbon tetrachloride to rats, it seems likely that GSH may protect the liver against the toxic effects of carbon tetrachloride; this probability is based on the observation that pretreatment with diethylmaleate, which depletes liver GSH without accompanying toxicity, increases both the toxicity and the covalent binding of carbon tetrachloride and its metabolites in vivo (Boyland and Chasseaud, 1970; Gillette, 1973; Gillette et al., 1974; Maling et al., 1974). The mechanism, however, is obscure.

Of the many remaining examples of liver damage mediated by chemically reactive metabolites of parent compounds, two may be briefly summarized as follows: there is considerable evidence that the hepatic necrosis caused by pyrrolizidine alkaloids is mediated by a chemically reactive intermediate that covalently binds to liver macromolecules (Mattocks, 1972; White et al., 1973; Mattocks and White, 1971). The toxic effect of amanita toadstools is also mediated through toxic, active metabolites (Floersheim, 1966a, b) because compounds that stimulate drug metabolism increase toxicity, while inhibitors of drug metabolism prevent the toxicities.

5.3 Conclusions

Returning to the idea of a mathematical function, ABCD, one can see that the concept has helped to elucidate several important principles in the formation of chemically reactive metabolites. Inducers and inhibitors of drug metabolism after the magnitude of covalent binding only to the extent that they change A or B or both. In some instances, inducers of drug metabolism cannot increase A or B as long as the drug and the chemically reactive metabolites are eliminated solely by first-order processes. Thus, the effects of inducers on the covalent binding of reactive metabolites may differ with the dose of the toxicant and the animal species. Moreover, an inducer may increase the toxicity of a drug even though it does not markedly alter the biological half-life of the drug, or it may have little effect on the toxic effects of the drug even though it markedly shortens the drug's biological half-life. In addition, a treatment may increase the toxicity of a drug in some tissues of an animal and decrease it in others.

It is clear that many drugs can be converted to chemically reactive metabolites and covalently bound to microsomal protein. It is not clear, however, to what extent the covalent binding can be used to predict the incidence of various kinds of toxicity in animals or man. However, studies of liver microsomes in vitro should be extremely useful in revealing which in a series of analogues have similar pharmacological activity and which should be developed or discarded as thereapeutic agents. As the mechanisms by which various kinds of toxicities and carcinogenicities are mediated by chemically reactive metabolites come clear, people concerned with the development of new drugs will be in a better position to make appropriate choices.

5.4 References

Azouz, W.M., Parke, D. V., and Williams, R. T. 1953. Studies in detoxication; determination of catechols in urine, and formation of catechols in rabbits receiving halogenobenzenes and other compounds. Dihydroxylation in vivo. *Biochem. J.* 55:146–151.

Baumann, E., and Preusse, C. 1879. Uber bromphenylmercapgursaure. *Ber. Deut. Chem. Ges.* 12:806–810.

Boyland, E. and Chasseaud, L. F. 1970. The effect of some carbonyl compounds on rat liver glutathione levels. *Biochem. Pharmacol.* 19:1526–1528.

Brodie, B. B., Reid, W. D., and Chu, A. K. 1971. Possible mechanism of liver necrosis caused by aromatic organic compounds. *Proc. Nat. Acad. Sci. USA* 68:160–164.

Butler, T. C. 1961. Reduction of carbon tetrachloride in vivo and reduction of carbon tetrachloride and chloroform in vitro by tissues and tissue constituents. *J. Pharmacol. Exp. Ther.* 134:311–319.

Castro, J., Sasame, H., Sussman, H. and Gillette, J. R. 1968. Diverse effects of SKF 525-A and antioxidants on carbon tetrachloride-induced changes in liver microsomal P-450 content and ethylmorphine metabolism. *Life Sci.* 7:129–136.

Cessi, C., Colombini, C., and Mameli, L. 1966. The reaction of liver proteins with a metabolite of carbon tetrachloride. *Biochem. J.* 101:46C–47C.

Corsini, G., Sipes, I. G., Krishna, G., and Brodie, B.B. 1972. Studies on covalent binding of drugs to liver microsomes. *Fed. Proc.* 31:1882.

Daly, J. W., Jerina, D. M., and Witkop, B. 1972. Arene oxides and the NIH shift; the metabolism, toxicity and carcinogenicity of aromatic compounds. *Experientia* 28:1129–1149.

Floersheim, G. L. 1966a. Schutzwirkung hepatotoxischer Stoffe gegen letale Dosen eines Toxins aus Amanita phalloides (Phalloidin). *Biochem. Pharmacol.* 15:1589–1593.

Floersheim, G. L. 1966b. Protektion gegen Amanitatoxine. *Helvet. Physiol.* 24:219–228.

Gillette, J. R. 1973. Factors that affect the covalent binding and toxicity of drugs. In *Pharmacology and the Future of Man*, vol. 2, *Toxicological Problems*, ed. T. A. Loomis, pp. 187--202. Basel: S. Karger

Gillette, J. R., Mitchell, J. R., and Brodie, B. B. 1974. Biochemical mechanisms of drug toxicity. *Ann. Rev. Pharmacol.* 14:271–288.

Glende, E. A. 1972. Carbon tetrachloride-induced protection against carbon tetrachloride toxicity. The role of the liver microsomal drug-metabolizing system. *Biochem. Pharmacol.* 21:1697–1702.

Gordis, E. 1969. Lipid metabolites of carbon tetrachloride. *J. Clin. Invest.* 48:203–209.

Jollow, D. J., Mitchell, J. R., Zampaglione, N., and Gillette, J. R., 1972. *5th Int. Congr. Pharmacol.* Abstr. Vol. Papers, 117.

Jollow, D. J., Mitchell, J. R., Zampaglione, N., and Gillette, J. R. 1974. Bromobenzene-induced liver necrosis. Protective role of glutathione and evidence for 3,4-bromobenzene oxide as the hepatotoxic metabolite. *Pharmacology* 11:151–169.

Judah, J. D., McLean, A. E. M., and McLean, E. K. 1970. Biochemical mechanisms of liver injury. *Am. J. Med.* 49:609–616.

Knight, R. J., and Young, L. 1958. Biochemical studies of toxic agents. II. The occurrence of premercapturic acids. *Biochem. J.* 70:111-119.

Krishna, G., Sipes, I. G., and Gillette, J. R. 1973. Trichloromethane free radical formation from carbon tetrachloride by liver microsomes: a reductive or homolytic cleavage? *Pharmacologist* 15:260.

Laqueur, G. L. 1964. Carcinogenic effects of cycad meal and cycasin, methylazoxymethanol glycoside, in rats and effects of cycasin in germ free rats. *Fed. Proc.* 23:1386-1388.

McLean, A. E. M., and McLean, E. K. 1966. The effect of diet and 1,1,1-trichloro-2,2-bis(*p*-chlorophenyl)ethane (DDT) on microsomal hydroxylating enzyme and on sensitivity of rats to carbon tetrachloride poisoning. *Biochem. J.* 100:564-571.

Magee, P. N., and Barnes, J. M. 1969. Diet and toxicity. *Brit. Med. Bull.* 25:278-281.

Maling, H. M., Eichelbaum, F. M., Saul, W., Sipes, I. G., Brown, E. A. B., and Gillette, J. R. 1974. The nature of the protection against CCl_4-induced hepatotoxicity produced by pretreatment with Dibenamine [N-(2-chloroethyl-)dibenzylamine]. *Biochem. Pharmacol.* 23:1479-1491.

Mattocks, A. R. 1972. Acute hepatotoxicity and pyrrotic metabolites in rats dosed with pyrrolizidine alkaloids. *Chem.-Biol. Interactions* 5:227-242.

Mattocks, A. R., and White, I. N. 1971. The conversion of pyrrolizidine alkaloids to N-oxides and to dihydropyrrolizidine derivatives by rat liver microsomes in vitro. *Chem.-Biol. Interactions* 3:383-396.

Mattocks, A. R. 1973. Mechanisms of pyrrolizidine alkaloid toxicity. In *Pharmacology and the Future of Man*, vol. 2, *Toxicological Problems*, ed. T. A. Loomis, pp. 114-123. Basel: S. Karger.

Mirvish, S. S. 1968. The carcinogenic action and metabolism of urethan and N-hydroxyurethan. *Advan. Cancer Res.* 11:1-42.

Mitchell, J. R., Reid, W. D., and Christie, B. 1971. Bromobenzene-induced hepatic necrosis: species differences and protection by SKF 525-A. *Res. Comm. Chem. Path. Pharmacol.* 2:877-888.

Plaa, G. L., and Larson, R. E. 1964. CCl_4-Induced liver damage. Current concepts regarding mechanisms of action. *Arch. Environ. Health.* 9:536-543.

Rao, K. S., and Rechnagel, R. N. 1968. Early onset of lipoperoxidation in rat liver after carbon tetrachloride administration. *Exp. Mol. Path.* 9:271-278.

Rechnagel, R. O. 1967. Carbon tetrachloride hepatotoxicity. *Pharmacol. Rev.* 19:145-208.

Reid, W. D. 1973. Relationship between tissue necrosis and covalent binding of toxic metabolites of halogenated aromatic hydrocarbons. In *Pharmacology and the Future of Man*, vol. 2, *Toxicological Problems*, ed. T. A. Loomis, pp. 62-74. Basel: S. Karger.

Reis, W. D., Christie, B., Krishna, G., Mitchell, J. R., Moskowitz, J., and Brodie, B. B. 1971a. Bromobenzene metabolism and hepatic necrosis. *Pharmacology* 6:41-55.

Reid, W. D., Christie, B., Eichelbaum, M., and Krishna, G. 1971b. 3-Methylcholanthrene blocks hepatic necrosis induced by administration of bromobenzene or carbon tetrachloride. *Exp. Mol. Path,* 15:363-372.

Reiner, O., Athanassopoulos, S., Hellmer, K. H., Murray, R. E., and Uehleke, H. 1972. Tetrachlorkohlenstoff in Lebermikrosomen, Liperperoxidation und Zerstürung von Cytochrom P-450. *Arch. Toxicol.* 29:219-233.

Reynolds, E. S. 1967. Liver parenchymal cell injury. IV. Pattern of incorporation of carbon and chlorine from carbon tetrachloride into chemical constituents of liver in vivo. *J. Pharmacol. Exp. Ther.* 155:117–126.

Reynolds, E. S., Ree, H. J., and Moslen, M. T. 1972. Liver parenchymal cell injury. IX. Phenobarbital potentiation of endoplasmic reticulum denaturation following tetrachloride. *Lab. Invest.* 26:290–299.

Sipes, I. G., Corsini, G., Krishna, G., and Gillette, J. R. 1972. *Proc. 5th Int. Congr. Pharmacol.* Abstr. Vol. Papers, 215.

Slater, T. F. 1966. Necrogenic action of carbon tetrachloride in the rat: a speculative mechanism based on activation. *Nature* 209:36–40.

Traiger, G. J., and Plaa, G. L. 1971. Differences in the potentiation of carbon tetrachloride in rats by ethanol and isopropanol pretreatment. *Toxicol. Appl. Pharmacol.* 20:105–112.

Uehleke, H., Hellmer, K. H., and Tabarelli, S. 1973. Binding of ^{14}C-carbon tetra-Chloride to microsomal proteins in vitro and formation of $CHCL_3$ by reduced liver microsomes. *Xenobiotica* 3:1–11.

White, I. N. H., Mattocks, A. R., and Butler, W. H. 1973. The conversion of the pyrrolizidine alkaloid retrorsine to pyrrolic derivatives in vivo and in vitro and its acute toxicity to various animal species. *Chem. Biol. Interact.* 6:207–218.

Zampaglione, N., Jollow, D. J., Mitchell, J. R., Stripp, B. Hamrick, M., and Gillette, J. R. 1973. Role of detoxifying enzymes in bromobenzene-induced liver necrosis. *J. Pharmacol. Exp. Ther.* 187:218–227.

5.5 Discussion

Weisburger It has been suggested that any drug which is bound to proteins or other cellular constituents is a very dangerous drug as far as being a carcinogen; I would like to have your comments on that.

Gillette I do not agree. If we applied that to a large number of common drugs we have on the market today, we would find that we would have to throw away 50 percent of them. Many noncarcinogens can be covalently bonded. We have a number of examples which, measured in vitro, are covalently bonded. I just simply cannot agree that all of them are carcinogens.

Weisburger Would you think that perhaps covalent binding is necessary for the pharmacological action of the drug?

Gillette No. The answer is definitely not. With barbiturates you will not find any covalent binding in the brain.

Leffert Have any studies been done to localize the P-450 activity in the liver lobules? Why do you see toxicity in the centrilobular region?

Gillette Why there is a zonal distribution of toxicity isn't clear. We frequently suggest that zonal distribution reflects the distribution of the enzyme, but we have not really precluded other possibilities. For example, it is also possible that the flow of reactive metabolites would be from the portal region. If part of the reactive metabolite can get out, then there would be a gradient of the reactive metabolite from the periportal region to the centrilobular region.

Leffert Where is the GSH-dependent transferase system located, and is it possible to inhibit toxin binding by administering GSH?

Gillette I am not sure exactly where the GSH transferase system is located. Histochemical evidence, however, indicates that GSH is distributed rather evenly throughout the liver. Since GSH does not readily penetrate the cell membrane, it would not be a good antidote, but administration of cysteinamine or cysteine blocks bromobenzene toxicity.

Goyings Looking at some of the data you presented, I wondered: If you assume that covalent bonding is important in a carcinogenic response, would not that create an artificial situation with the type of dosage regimens which we use currently? Are we moving our metabolism in such a way that we are allowing to take place that which under normal physiological doses would not occur?

Gillette This is true, especially when you go to very high concentrations. In fact, we have frequently found dose thresholds below which there is little covalent binding that is, AB increases with the dose. This is one of the reasons I have some reservations about the Delaney amendment. The question is: Is there any dose that is completely safe, where there would be no covalent bonding at all, especially with the DNA? It is an unanswerable question, because if the drug combined with only one molecule of DNA we would not detect covalent bonding.

Tate Your answer suggested that you did not feel that all compounds that produced covalent binding were carcinogenic. Does the converse of that apply?

Gillette The issue has been raised that if covalent binding to DNA is a requisite to the carcinogenic activity, any time you did not see such binding to DNA, that would be some presumptive evidence that the compound was safe. The problem is: How can you prove that it is not bound at low levels?

Weisburger We have found that both aniline and acetanilide bind to proteins and are not carcinogenic, as far as we know.

Baldwin It seems to be almost a naive argument to say that a compound must covalently bind to something to act as a carcinogen, because you are really saying that you have at one end of the spectrum of reactivity a compound which gives covalent binding and that is the only reaction that will give you this change. But a multiple weak ionic binding will probably give you the same degree of interaction that the covalent function will give.

Grice In most of our carcinogenesis tests we use a relatively pure system. The animals are of one line, the diets specified, we have them contained in cages with no external stress, and we try to insult them with one specific compound. Do you think we can get anything more than a rough dose effect that is of value in extrapolation? Do you think there is any value in trying to find out whether four or five molecules are going to cause an effect, when you consider that you will not be able to extrapolate this to man?

Gillette Well, I would question whether or not we have really as good control in tests as we think we have. Do you use wood chips as bedding, as soft wood chips are prime inducers? I do not know that you can really isolate all of the factors.

Grice What I am really asking is: Can you get anything more than a rough estimate?

Gillette Can you have any better rough estimate than we have for any pharmacological effect? I think one of the things that we can do is to look carefully at cases in which we have a number of different strains of species of animals in which we have diametrically opposite results. Possibly a compound is perfectly safe in one strain and carcinogenic in another strain. I am suggesting that there are a number of factors by which this could occur, some of which would be in the mechanism of development of the cancer. But a lot of it may just be due to metabolism. By measuring covalent bonding, we can actually determine whether the difference in effect is really dependent on differences in metabolism or on other factors that can affect the incidence rate of cancer.

P. Newberne Can we come back to the zonal localization of enzymes? There is a class of compounds, namely, pesticides such as Mirex and DDT, characteristically affecting centrilobular cells which causes these cells to hypertrophy. What is your impression of what is going on here? These cells are certainly not exposed to it first; it enters vis the portal area.

Gillette Well, most of these compounds are inducers so that actually what you are looking at is essentially equivalent to the effects we find with phenobarbital. You are raising the point, I think, of: Are these reversible or not. I do not know.

The mechanism is independent of hormonal control, although the ultimate effect, that is, how much activity you can get with a given inducer, will be altered by the presence of various hormones.

J. Newberne Those of us who are testing carcinogens are obligated to overdose animals; what you show would indicate that the final change in a system is not even closely related to real usage of drugs. I think we probably need to reconsider this whole matter of dose as it becomes a very critical part of this.

Gillette There are a number of compounds and toxicants that we have studied that do not affect the GSH levels, and yet we have dose thresholds because of the way certain compounds are eliminated by the kidney. The fact that some compounds are actively secreted by the kidney means we can saturate the transport system as we go up in dose, and under those conditions we can then increase the proportion of the dose that has to be metabolized by the body.

Butler The problems of testing have been mentioned. Some people have suggested maximum tolerated dose for carcinogenesis testing. Presumably from what you are saying you think this is not the best course.

Gillette I think we should know where we are on the dose-response map, and I think one of the ways of getting at that kind of information is measurement of covalent binding. I would also suggest that it might be useful in determining whether the pattern of metabolism changes during the dosage for over one or two years. I have never been sure whether, as one continues the testing of the compound, how much we change metabolism. I do not think it has ever been studied in this way, but here we have a nice simple tool to decide whether or not we do get important changes in metabolism, as measured by the formation of reactive metabolites.

Butler There is some evidence that if you keep on feeding inducers a plateau is reached, then, after a time, this appears to break down?

Squires In some instances it has been reported that the activity of the enzyme system goes down even though the proliferation of smooth endoplasmic reticulum is retained with the continuation of the chronic treatment. Is there any example in which the toxicity would be decreased by giving a compound at a high level?

Gillette We can saturate the formation of the reactive metabolite without affecting the rate constant for the elimination. Under these conditions you should have a limiting effect as far as toxicity is concerned.

Squires Just let me return to the subject of testing. In empirical testing with maximum tolerated dose, for example, there could be circumstances, perhaps, where you would be masking the toxic effect by excessive dosage. Is this possible?

Gillette Several years ago, it was indicated that the chlorocyclizine, which is an antihistamine, was a teratogen in rats, but only under certain circumstances; that is, if it was given on the tenth or eleventh or twelfth day of pregnancy you would get it. If you started with treatment at day zero and gave it to the animals, the abnormality did not occur because you got altered metabolism in the mother and, consequently, did not have the reactive metabolite or the pharmacologically active compound at the critical time.

I am quite concerned about these false positives, because obviously if I am going to say that half of the drugs that we are taking can become covalently bonded, then what kind of situation are we walking into? False negatives are equally of concern, the point being that ideally we would like to have a test that would predict everything that is positive in man and nothing that is negative, but here I am pointing out errors on both sides of the fence and saying that we have to resolve both issues before we can really be satisfied with the testing methods that we have.

6 Light Microscopy of Rat Hepatic Neoplasia

Glenys Jones
Shell Research Laboratories
Sittingbourne, Kent
England

W. H. Butler
Imperial Chemical Industries, Ltd.
Macclesfield, Cheshire
England

6.1 Introduction

Interest in the morphology and biological behavior of malignant tumors of the rat liver was initiated by the discovery of the carcinogenic aminoazo dyes. D-aminoazotoluene, the active principle of scarlet red, *p*-dimethylaminoazobenzene, and 3′-methyl-4-dimethylaminoazobenzene are highly specific for the liver; they produce malignant tumors of similar morphology in all rats maintained on a suitable feeding regime (Kinosita, 1955; Orr, 1940; Opie, 1944a,b). Other agents producing a high incidence of metastasizing liver cell tumors include 2-acetylaminofluorene (Bielschowsky, 1944), ethionine (Farber, 1956), cycasin (Laqueur et al., 1963), aflatoxin (Butler and Barnes, 1963; Newberne et al., 1967; Newberne and Butler, 1969), and Ponceau 3R (Grice et al., 1961). The lesions are predominantly of parenchymal cell origin and consist of two main histological types, trabecular and glandular, irrespective of the inducing compound. Tumors of bile duct origin, true cholangiocarcinomas, do occur but are uncommon. Angiosarcomas and reticuloendothelial tumors may also arise in the liver (Maltoni and Lefemine, 1974; Newberne and Shank, 1973; Gillman and Hallowes, 1972a,b). Metastatic lesions from extrahepatic primary tumors will not be discussed in this paper.

The reported incidence of spontaneous malignant tumors in rat liver is low—less than 1 percent (McCoy, 1909; Crain, 1958; Gilbert and Gillman, 1958). A few spontaneous sarcomas and reticuloendothelial tumors were reported by McCoy (1909), and biliary cystadenomas are seen in aged rats. There is no good account of the morphology of these spontaneous tumors in the recent literature.

6.2 Morphology

Primary malignant tumors of the liver may be single or multiple and are found in all lobes, though in the absence of nodular cirrhosis, the right upper lobe is the commonest site (Opie, 1944b). They vary from a few millimeters to many centimeters in diameter and contain solid, cystic, necrotic, and hemorrhagic areas. Invasion of the surrounding liver, liver capsule, and adjacent structures such as the diaphragm, pancreas, and lymph nodes occurs. Pedunculated forms which may undergo torsions and infarction are sometimes found. Lymphomas, lymphosarcoma, and leukemia may involve the liver, where they produce both focal lesions and a diffuse infiltration of the organ. Two benign lesions, parenchymal cell adenomas and biliary cystadenomas, are included in this review.

Primary tumors can be divided into four main groups:

1. *Parenchymal cell tumors.* Trabecular and glandular hepatocarcinomas. Benign hepatic tumors (adenoma).
2. *Tumors of bile duct origin.* Cholangiocarcinoma. Biliary cystadenomas.
3. *Anaplastic tumors.*
4. *Sarcomas.* Angiosarcomas. Kupffer cell sarcomas. Fibrosarcoma.

Other tumors, such as pseudolipomas, have been reported (Hemm, 1973). The lymphomas will be considered as a separate group. These classification should not be considered rigid, as there is considerable overlap between the poorly differentiated tumors of groups 1 and 2 and little agreement on the cell of origin of biliary cystadenomas and the anaplastic tumors. There have been reports of mixed tumors (Stewart and Snell, 1959), described as mixed biliary and parenchymal cell tumors and as "carcinosarcoma."

6.2.1 Parenchymal Cell Tumors

The malignant parenchymal cell lesions occur in two histological forms, trabecular and glandular.

Trabecular Carcinoma

Well-differentiated tumors are composed of cords of regular parenchymal cells, many cells thick, in which sinusoids are lined by normal endothelium (fig. 6.1). The poorly differentiated lesions contain large, blood-filled spaces lined by atypical cells (fig. 6.2). Necrosis and inflammation of the center of large papillary structures forming cystic spaces is common. Solid tumors composed of confluent sheets of parenchymal cells are frequently seen (fig. 6.3). Areas of infarction and hemorrhage scarring are characteristic of the large lesions. Many tumors contain foci of fatty change (fig. 6.4) and large, swollen cells (fig. 6.5). The staining characteristics of the tumor cells are variable. Eosinophilic cells with pale vesicular nuclei are characteristic of the well-differentiated tumors (fig. 6.6). Some tumors consist of small basophilic cells, and others contain a variety of forms with variation in nuclear size, shape, and cytoplasmic staining (fig. 6.7). Within the same lesion, both the basophilic and eosinophilic cells may be present (fig. 6.8). In some poorly differentiated tumors, multiple abnormal nuclei with inclusions and a high mitotic rate with abnormal mitosis are seen.

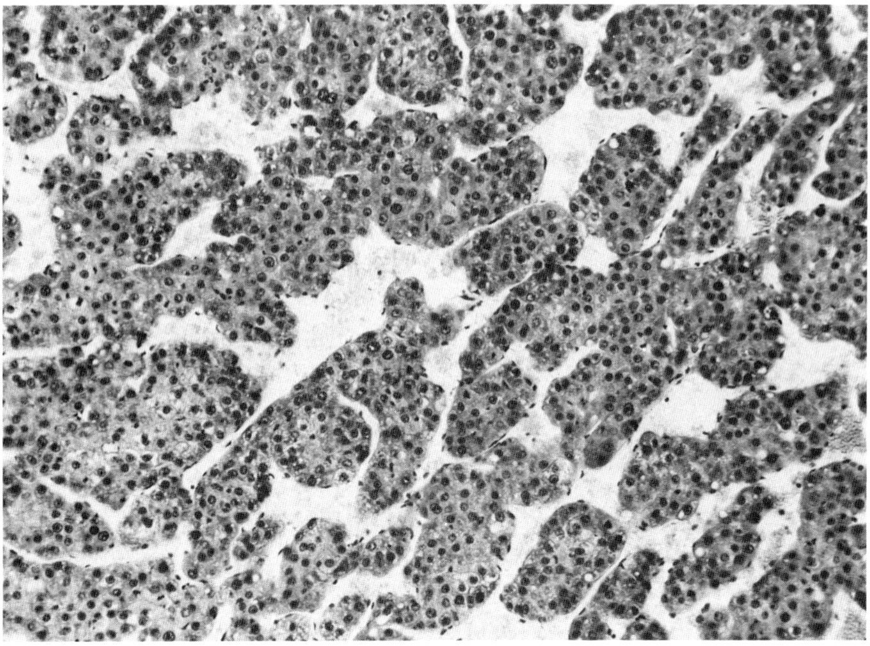

6.1 Histologic appearance of well-differentiated trabecular hepatocellular carcinoma with multicellular cords and sinusoids lined by normal endothelium. H and E (× 130).

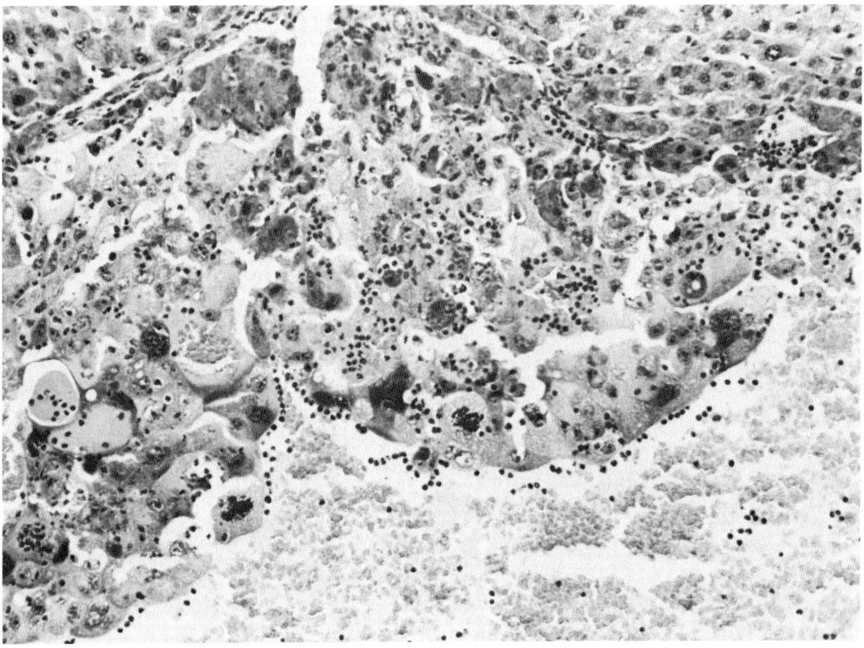

6.2 Poorly differentiated tumor; blood-filled spaces lined by atypical cells. H and E (× 130).

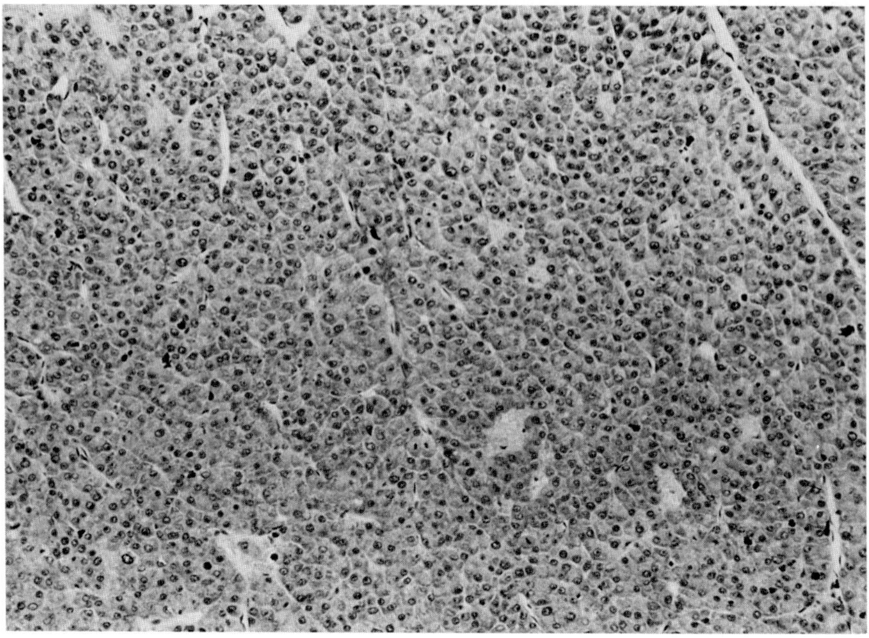

6.3 Solid parenchymal cell tumor composed of sheets of hepatic neoplastic cells. H and E (× 100).

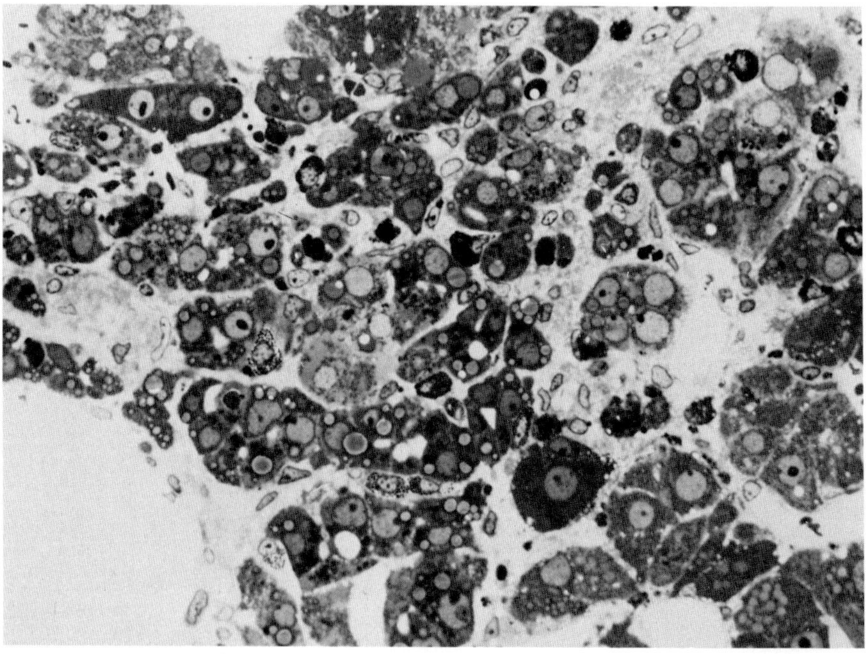

6.4 Area of tumor with fatty change. Epoxy resin 1 μ section stained with toluidine blue (× 160).

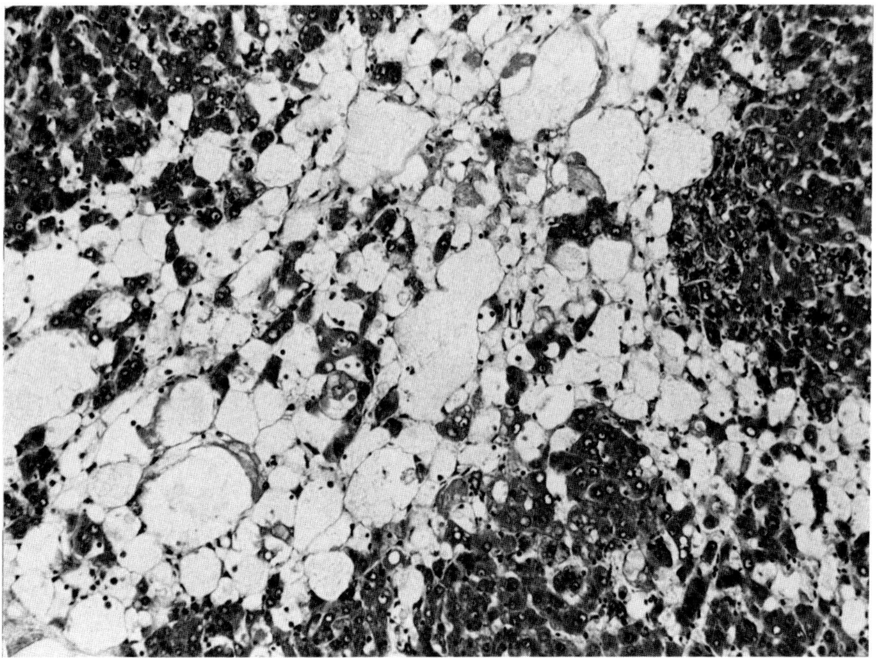

6.5 Large, swollen cells and lysis in one area of tumor. H and E (× 110).

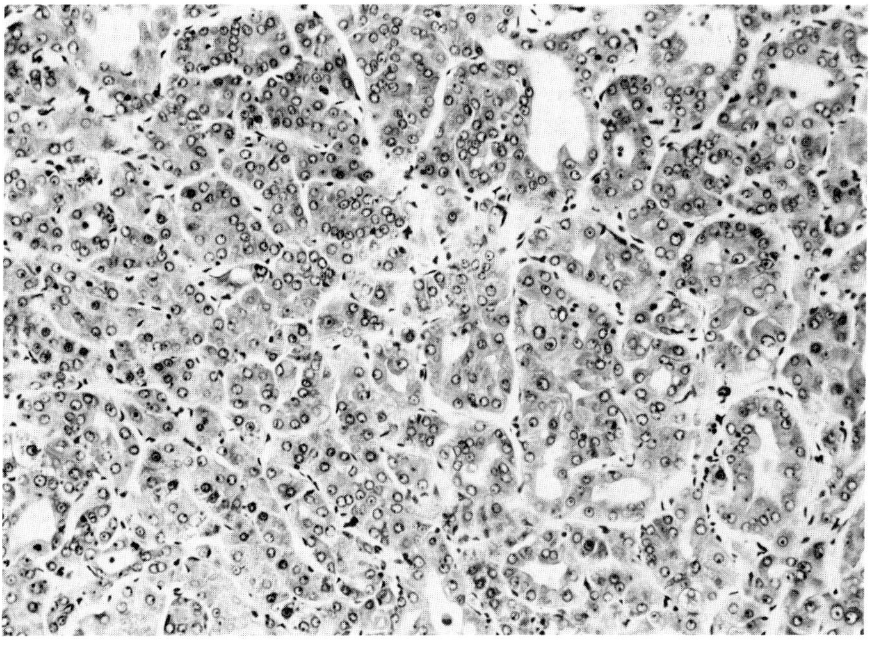

6.6 Eosinophilic cells with pale vesicular nuclei typical of well-differentiated tumor. H and E (× 130).

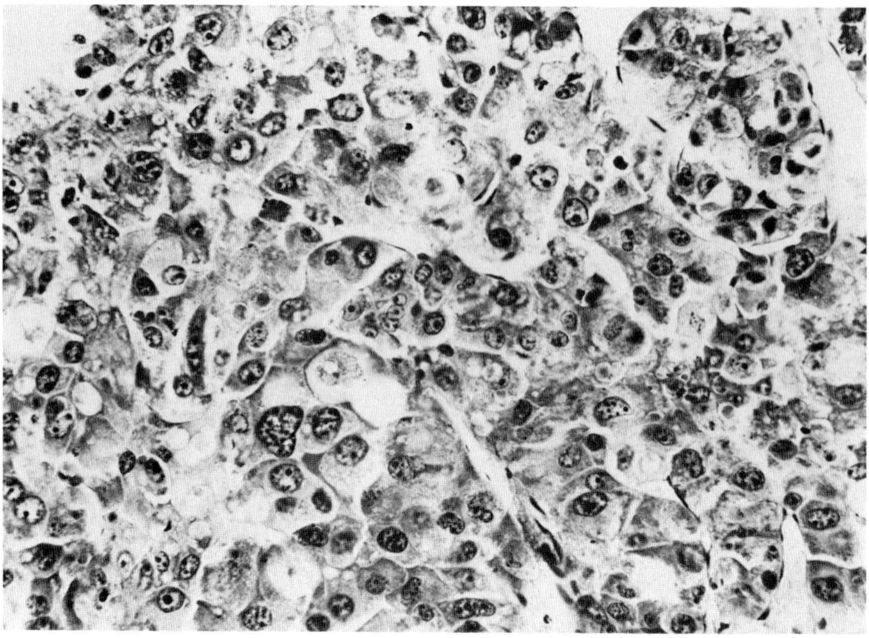

6.7 Variation in size, shape, and staining of cells that is characteristic of some tumors. H and E (240).

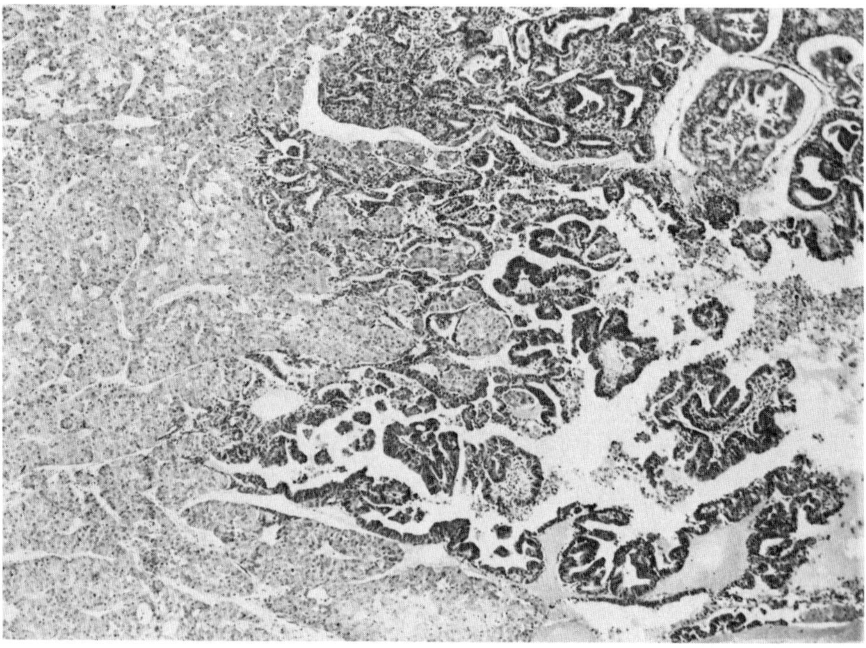

6.8 Two types of cells in same tumor. H and E (× 41).

Glandular Carcinoma

This carcinoma is characterized by acinar formation of the parenchymal cells; four or five cells with a basement membrane are arranged around a central lumen (fig. 6.9), a pattern found as the focus of more typical trabecular lesions. In some areas, a significant amount of stroma is present between the glandular elements (fig. 6.10). Cystic spaces, lined by flattened epithelium, contain amorphous material (fig. 6.11); necrotic parenchymal cells and inflammatory cells are found adjacent to small glandular structures and typical trabecular areas. The individual hepatocytes comprising the glandular structures vary considerably in form. Some are typical eosinophilic parenchymal cells with vesicular nuclei and prominent nucleoli in which the lumen consists of a common dilated duct (fig. 6.12). Other acini are composed of more basophilic, cuboidal cells. Some poorly differentiated lesions contain foci of hyperchromatic cells, distorted glandular structures (fig. 6.13) abundant stroma (fig. 6.14), and a dense reticulin network that resembles true cholangiocarcinoma; in such cases, accurate differentiation by light microscope is impossible. Mucus can be demonstrated in both the glandular and trabecular carcinomas. Islands of cartilage (fig. 6.15) and foci of calcification may occur in necrotic tumors.

"Pseudotubular" formations may be seen in the normal liver adjacent to parenchymal cell tumors. There is confusion in the literature concerning the definition and description of "pseudotubules" and their relevance to the neoplastic process. This subject is discussed fully by Stewart and Snell (1959), who describe pseudotubules as a double row of rectangular cells ensheathed by reticulin or collagen fibers. They are seen radiating out from the portal tracts after diverse forms of liver injury. There is no direct evidence, however, that these structures are relevant to tumor development (Farber, 1956). Tubular formations in cirrhotic nodules are similar to the glandular structures present in parenchymal cell tumors. They are considered to be intermediate between parenchymal and bilary epithelial cells and have been termed "biliary hepatocytes." Reactive reticuloendothelial cell proliferation may be seen in the normal liver adjacent to the neoplasm (fig. 6.16). Lacqueur et al, (1963) reported such intrasinusoidal proliferation at the margin of liver tumors induced by cycasin. We have also seen this phenomenon in our series of tumors induced by aflatoxin. These lesions contain foci of plump, active, lining cells and may closely resemble foci of angiosarcoma; they are usually considered to be a reactive phenomenon, however. Most cells are frequently seen in the fibrous areas of parenchymal cell neoplasms. Their significance is unknown.

6.2.2 Tumors of Bile Duct Origin

Cholangiocarcinomas of liver are uncommon and are morphologically similar to certain poorly differentiated trabecular carcinomas (fig. 6.17). Cholangiocarcinomas characteristically contain abundant

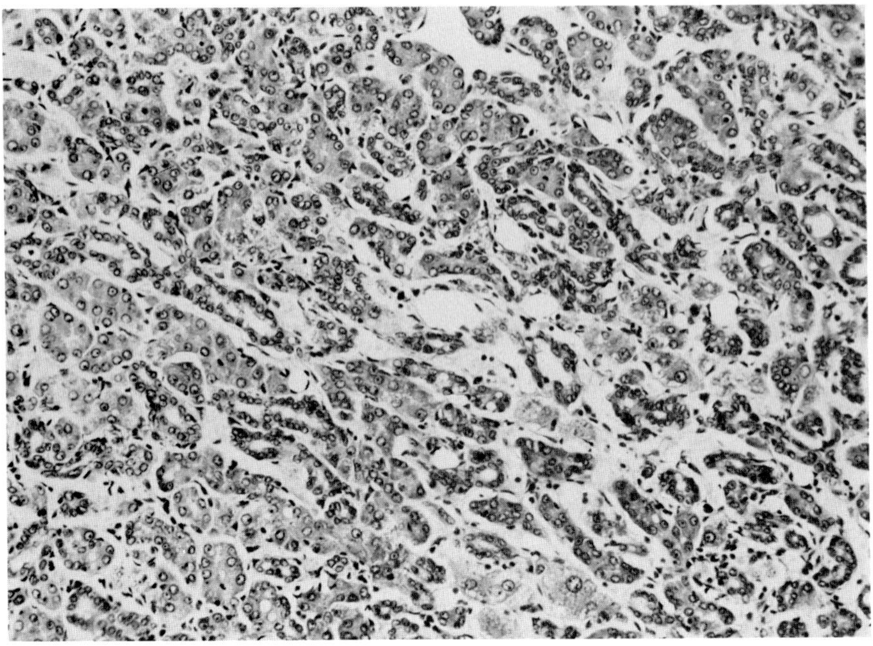

6.9 Acinar formations of tumor cells about a central lumen typical of glandular carcinoma. Basement membranes are present. H and E (× 100).

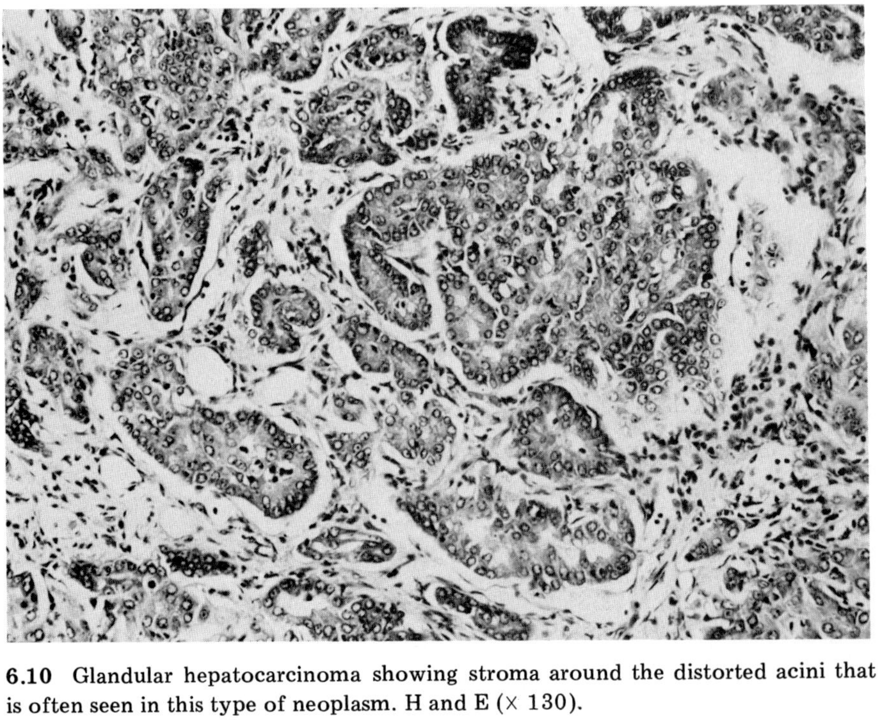

6.10 Glandular hepatocarcinoma showing stroma around the distorted acini that is often seen in this type of neoplasm. H and E (× 130).

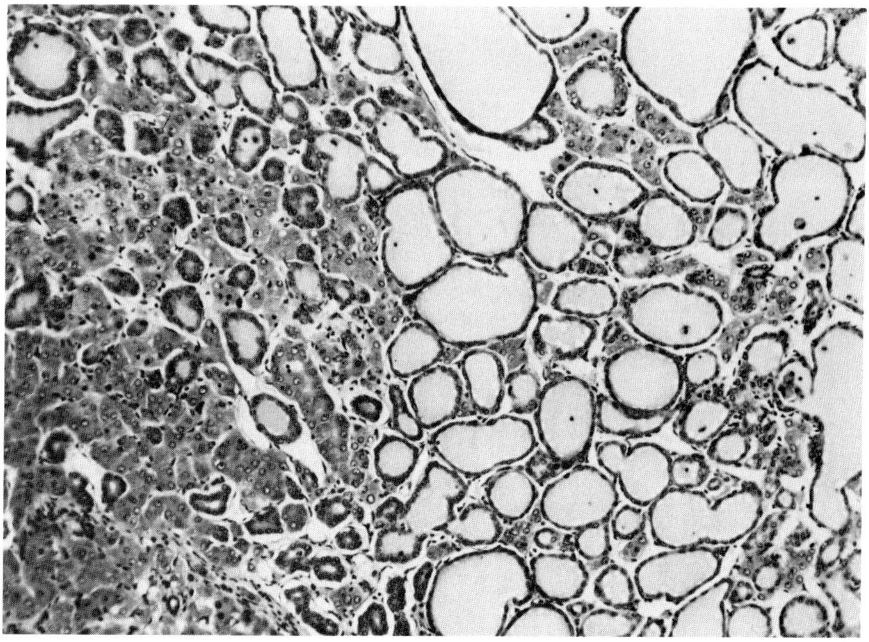

6.11 Cystic spaces lined by flattened epithelium, as seen in some areas of glandular carcinoma. Some spaces contain amorphous material, while others (not shown) contain necrotic material. H and E (× 100).

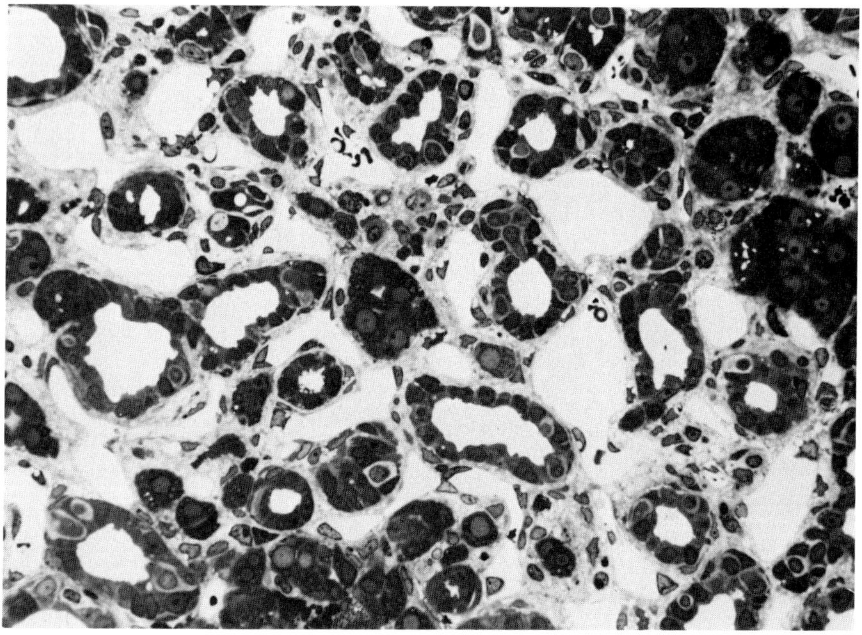

6.12 Variation in cells in glandular structures; vesicular nuclei, prominent nucleoli, and common dilated ducts often appear in various parts of the tumor. Epoxy resin 1-μ section stained with toluidine blue (× 160).

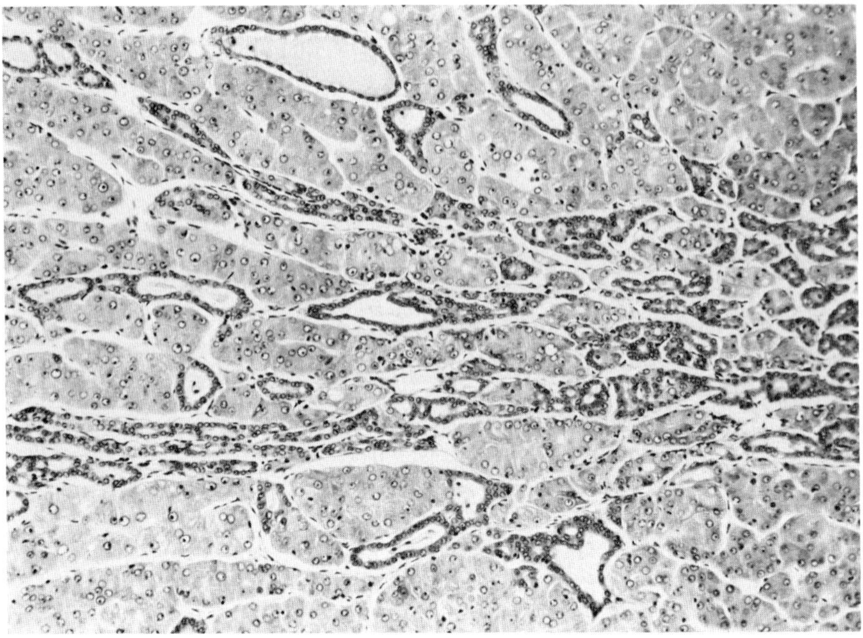

6.13 Poorly differentiated lesion. Note foci of hyperchromatic cells and distorted glandular structures. H and E (× 100).

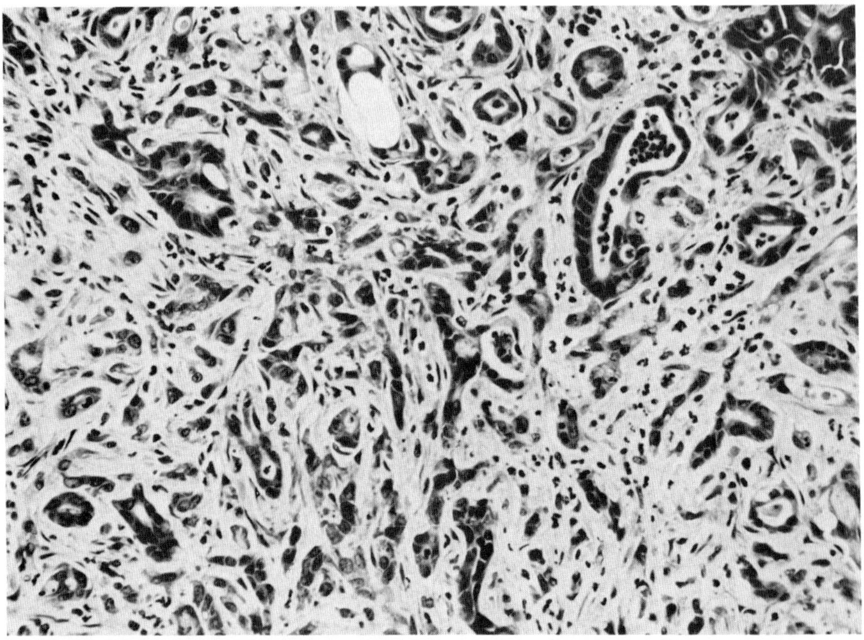

6.14 Glandular hepatocarcinoma with abundant stroma distorting the acinar pattern. This lesion resembles cholangeocarcinoma. H and E (× 130).

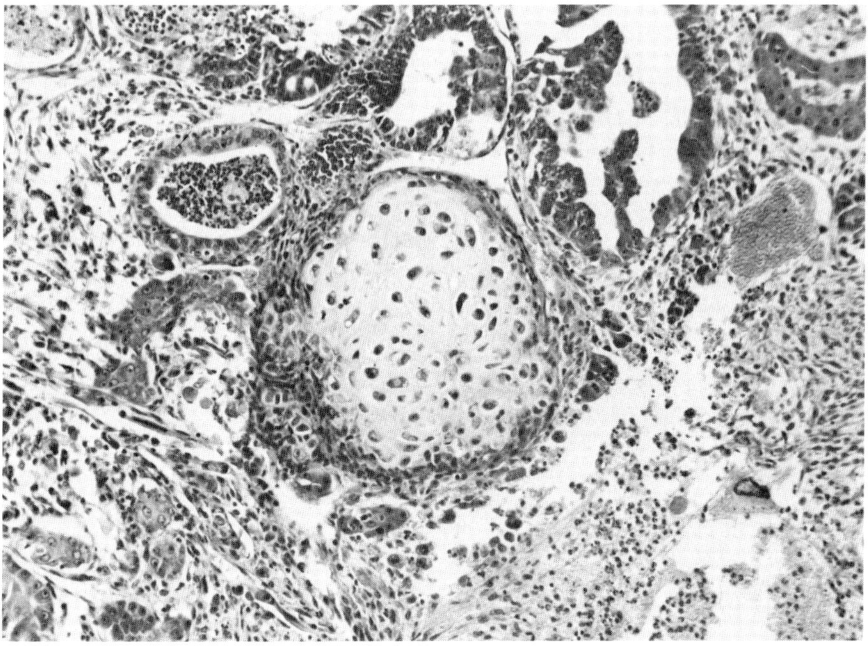

6.15 Cartilagenous metaplasia, sometimes seen in glandular tumors. H and E (× 100).

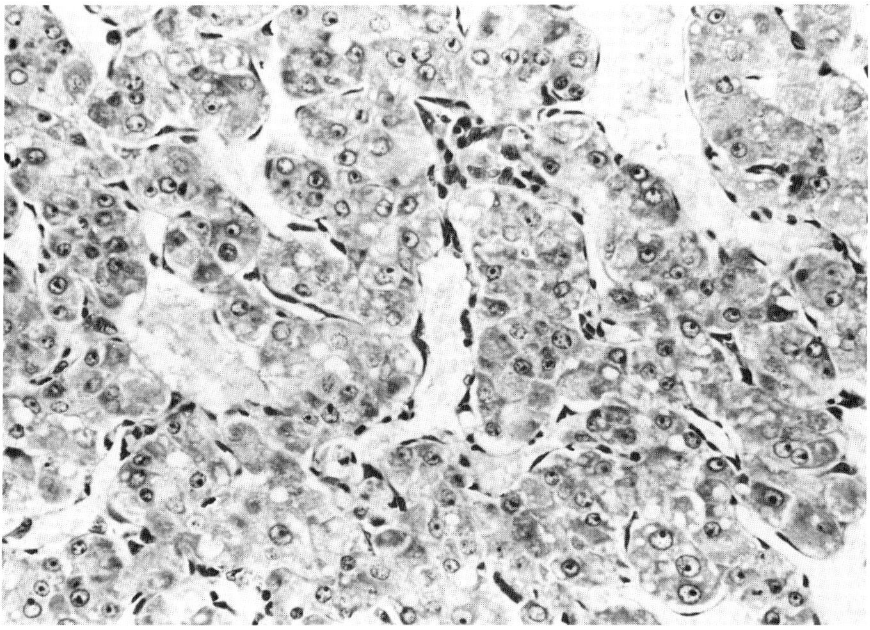

6.16 Reactive endothelial cell proliferation in more normal liver adjacent to the neoplasm. H and E (× 150).

scirrhus stroma; glandular structures composed of basophilic cells infiltrate this stroma and the adjacent liver tissue. No readily identifiable parenchymal cell element is present. These tumors have been reported in rats treated with aflatoxin (Butler, 1965) and butter yellow (Orr, 1940).

A wide range of proliferative lesions of biliary epithelium are seen after chemical injury, and, of these, biliary cystadenomas may be considered neoplastic (fig. 6.18). Many authors, however, consider them to be a degenerative phenomenon. These lesions are seen in diverse forms of liver injury and in aged rats; they consist of multilocular cysts lined by flattened or cuboidal cells separated by loose fibrous stroma; they occasionally contain cords of normal liver cells (Laqueur and Matsumoto, 1966). They vary in size but can reach several centimeters in diameter and may be found in the absence of diffuse bile duct hyperplasia. On the other hand, these lesions are frequently seen in livers containing hepatocellular carcinoma. The cells lining the cysts may be eosinophilic and resemble hepatocytes (fig. 6.19), but the larger multilocular lesions are lined by cuboidal, thin, compressed cells resembling neither hepatocytes nor biliary epithelium (fig. 6.20). The cell of origin of these lesions and their neoplastic nature is undecided and of considerable interest.

6.2.3 Anaplastic Tumors

Tumors consisting of sheets of pleomorphic cells with high mitotic rate, bizarre nuclei, and large central areas of necrosis and hemorrhage are difficult to classify and are best defined as anaplastic lesions (fig. 6.2).

6.2.4 Sarcomas

The endothelial cells, Kupffer cells (Wisse, 1974), and fibrocytes in the liver give rise to malignant neoplasms. Angiosarcomas have been reported after treatment with urethane and aflatoxin (Newberne et al., 1967), aflatoxin alone (Butler, unpublished observations), dimethylnitrosamine (Hadjiolov and Markow, 1973) and nitrosomorphaline. These tumors form flood cysts and hemorrhagic solid tumors, and they are usually multiple. The blood-filled spaces are lined in some areas by multiple layers of plump spindle cells having pale nuclei and in other areas by confluent sheets of poorly differentiated hyperchromatic cells, which invade the adjacent liver. A mesenchymal tumor, morphologically similar to the renal neoplasm reported by Hard and Butler (1970), may arise in the stump remaining after partial hepatectomy in rats treated with a single dose of dimethylnitrosamine. This lesion presumably arises from the granulation tissue at the ligature site (Craddock, 1971) (fig. 6.21).

Kupffer cells sarcomas can be induced by trypan blue (Gillman and Hallowes, 1972a,b). These are pleomorphic, solid, mesenchymal tumors with a high mitotic rate. Finally, fibrosarcomas arise from the reactive

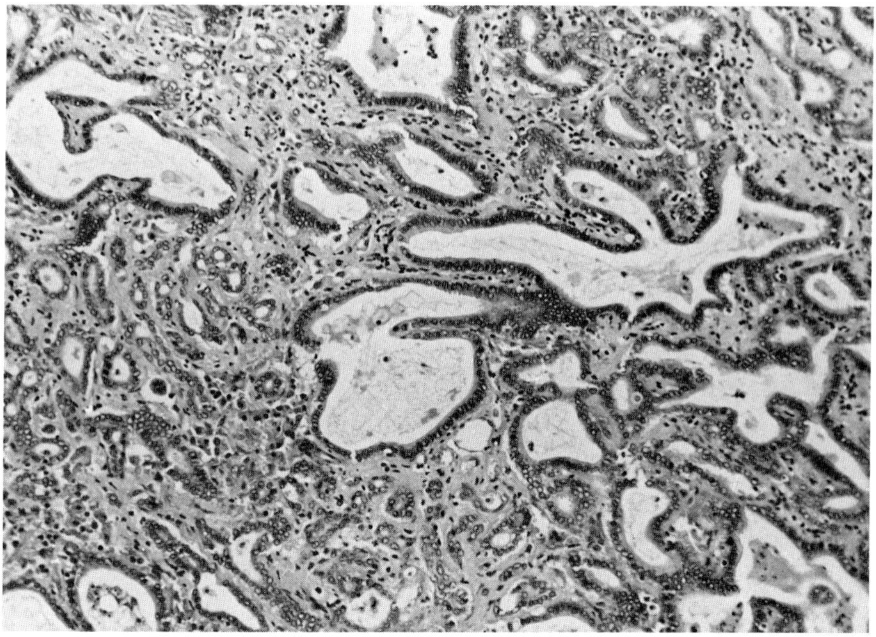

6.17 Cholangiocarcinoma, which is rare in liver of rats. Such tumors contain scirrhus stroma and neoplastic glandular structures. H and E (× 110).

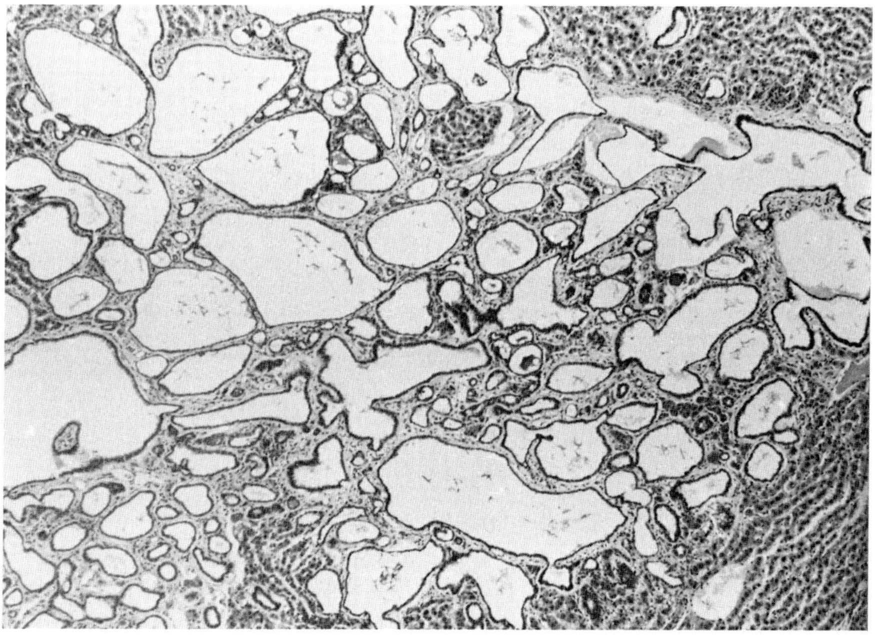

6.18 Biliary cystadenomas seen in the liver of rats exposed to any one of numerous chemical agents with hepatotoxic properties. H and E (× 100).

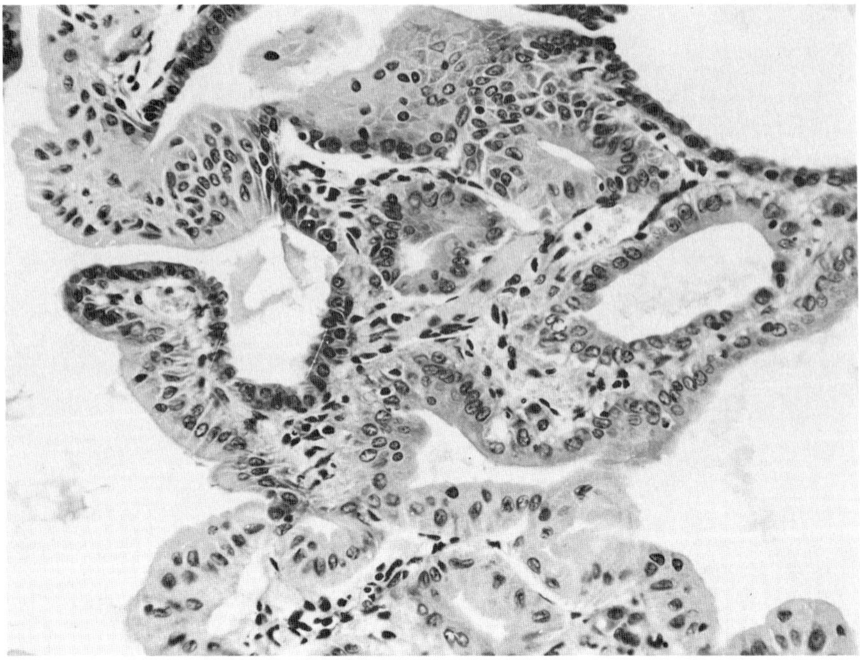

6.19 Area of bile duct tumor with living cells that resemble hepatocytes. H and E (× 210).

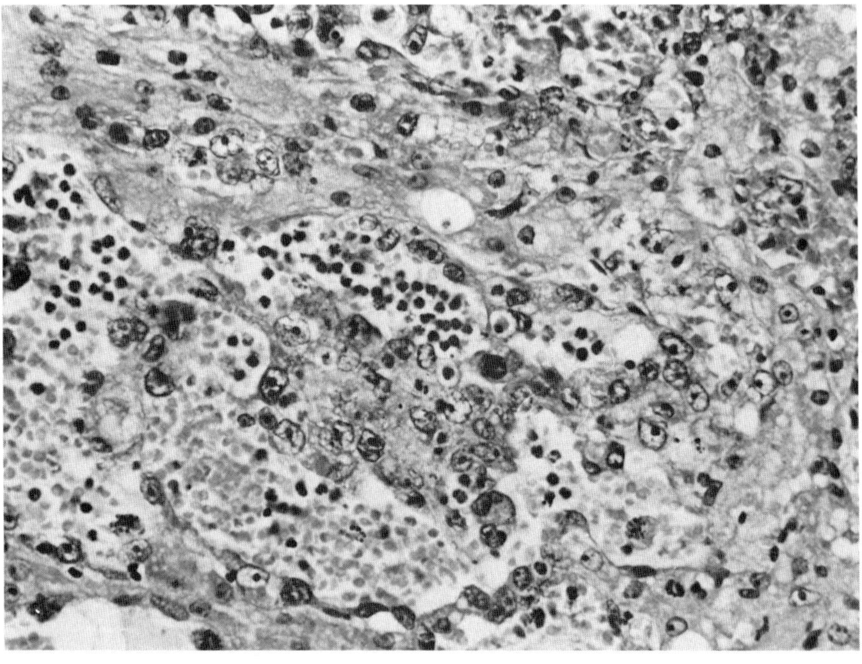

6.20 Glandular lesions lined by irregular parenchymal cells with areas of necrosis and hemorrhage; they resemble neither hepatocytes nor biliary epithelium. H and E (× 160).

fibrosis induced in the liver by the tapeworm *Taennia teaniaformis.* The tumors consist of sheets of atypical spindle cells with abnormal nuclei; they contain many mitotic figures.

Lymphomas

Lymphosarcomas characteristically produce foci of proliferating abnormal lymphocytes around portal tracts (fig. 6.22), central veins, and in the midzonal region; a diffuse infiltration of the sinusoids appears throughout the organ. When parenchymal cell tumors are present, they may also contain focal and/or diffuse infiltration by neoplastic lymphocytes. Neoplasms of the hemopoetic and the lymphoid system can be induced by 7,12-dimethylbenz(a)anthracene (Pollard and Kajima, 1967) and by derivatives of aromatic amines (Morris et al., 1960).

6.3 Biological Behavior

In all the groups of malignant tumors described, local invasion and distant metastases are common. Lung metastases from parenchymal cell tumors occur in 75 percent of rats fed 5 ppm dietary aflatoxin for six weeks (fig. 6.23). Invasion of adjacent liver tissue, vessels (fig. 6.24), liver capsule, and organs (e.g., diaphraghm, pancreas, and omentum) occurs. Cholangiocarcinomas metastasize as adenocarcinomas, but frequently the stromal component is lost (fig. 6.25). In addition, peritoneal spread with formation of ascites, lymphatic spread to abdominal nodes, and vascular dissemination to lungs, adrenal glands, and other organs are reported.

6.4 Interpretation of Morphology

Several problems arise in the interpretation of the morphology of these tumors. Accurate definition of a lesion's cell of origin is important in the study of pathogenesis. Typical parenchymal cell carcinomas contain two populations of cells; one is anaplastic, and the other is trabecular and often has foci of glandular structure. These lesions have been defined as mixed hepatocholangiocarcinomas (Svoboda, 1964). In several examples of trabecular lesions, we have demonstrated the presence of variable morphology within a single cord of liver cells, and many foci of glandular structure are composed of cells with the ultrastructural characteristics of parenchymal cells. Epoxy resin sections of 1μ clearly demonstrate that, in a significant number of such lesions, the cells of these foci contain little stroma, reticulin, or mucin, and are structurally similar to parenchymal cells. The presence of PAS-positive material in tumors has been reported as proof of their biliary cell origin (Firminger and Mulay, 1952). However, mucin and PAS-positive material can be found in typical trabecular carcinomas (Butler, 1971). Parenchymal cell tumors may develop the cystic forms by dilation of the central canaliculus or necrosis of the central portion of a papillary lesion. Dalton and

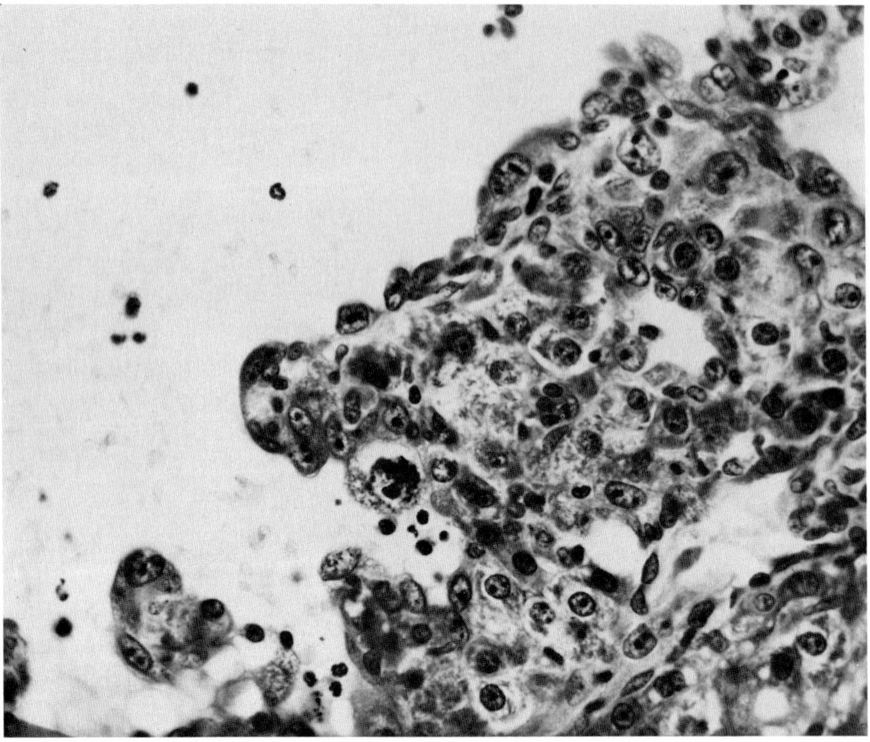

6.21 Angiosarcoma observed in liver of rats exposed to a variety of chemical agents. Plump, spindle-shaped cells line vascular spaces in some areas, whereas other sections contain confluent sheets of neoplastic cells. H and E (× 250).

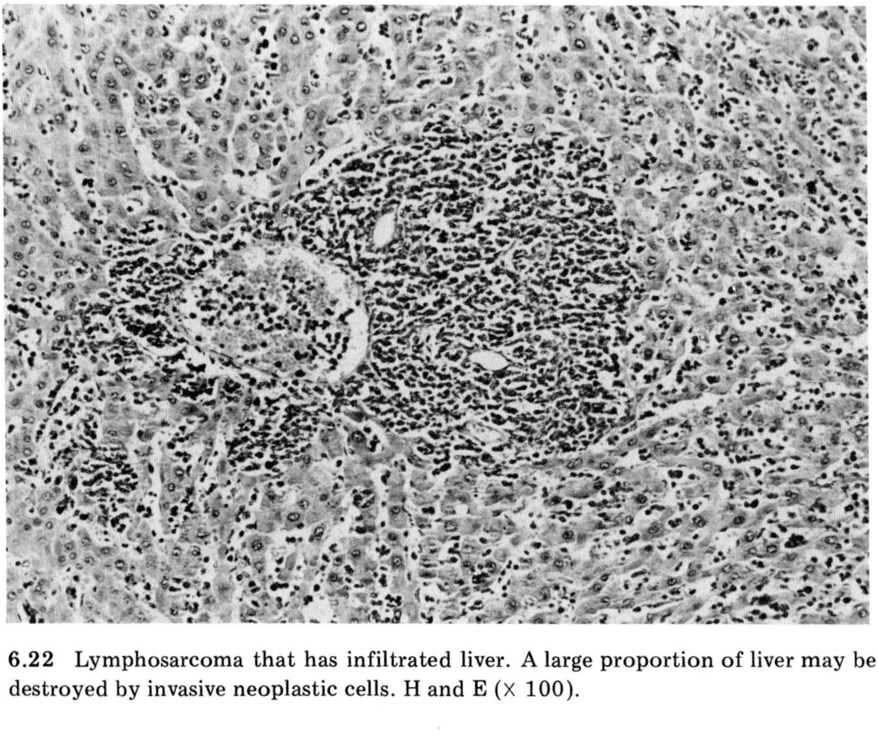

6.22 Lymphosarcoma that has infiltrated liver. A large proportion of liver may be destroyed by invasive neoplastic cells. H and E (× 100).

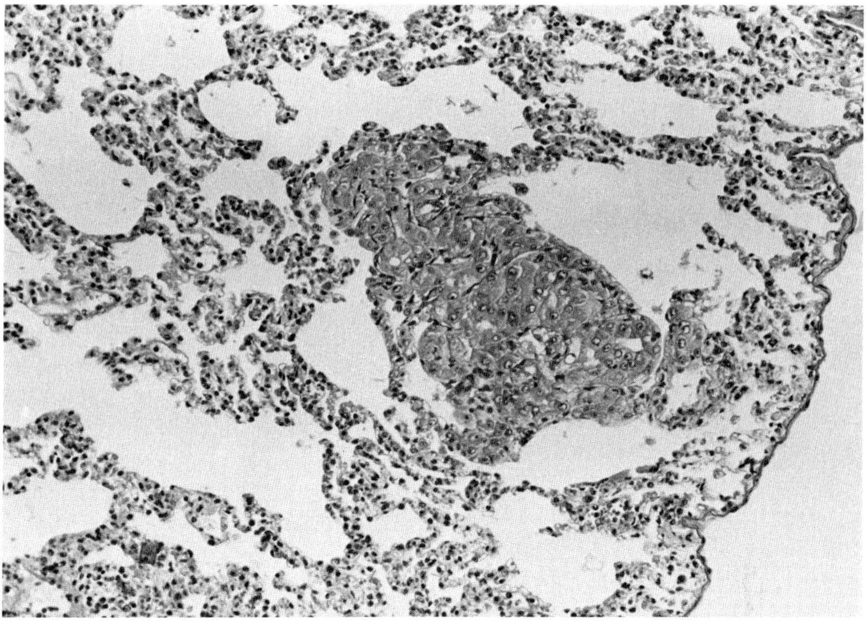

6.23 Lung with metastatic lesion of liver carcinoma. Some chemical agents result in virtually 100 percent of rats with metastases. H and E (× 70).

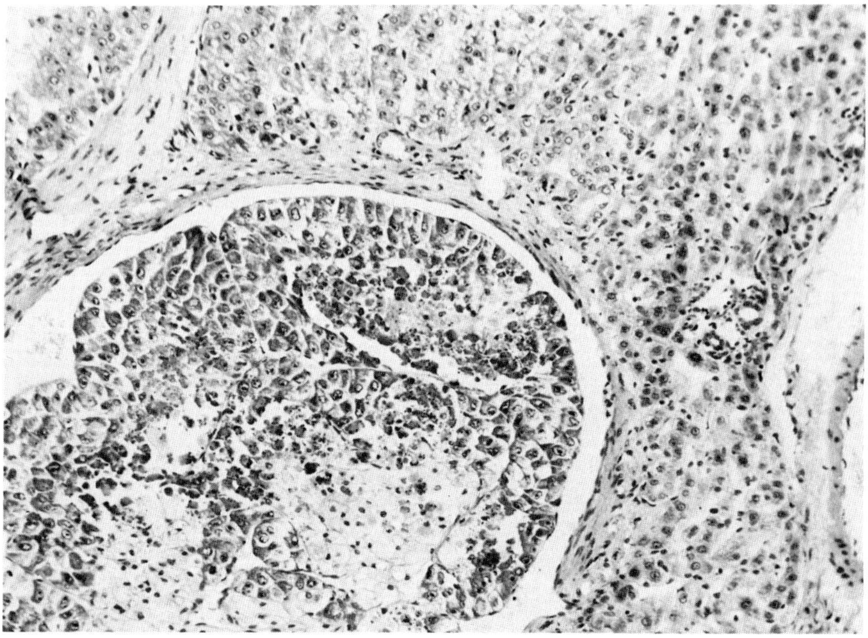

6.24 Embolus of liver cell tumor in vessel. The tumor may grow at site of invasion or spread to other organs and tissues. H and E (× 110).

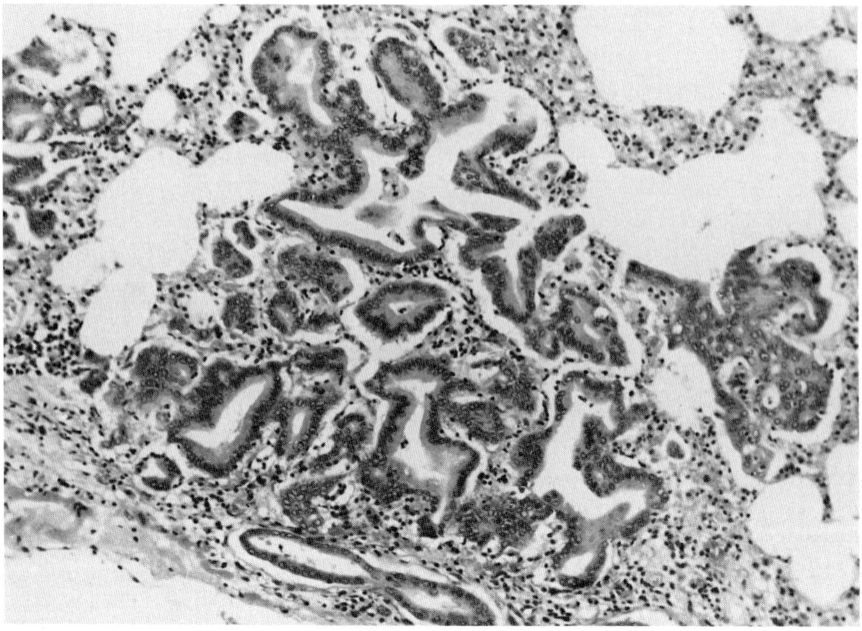

6.25 Lung with metastatic lesion from liver cholangiocarcinoma. H and E (× 120).

Edwards (1942), Stewart and Snell (1959), and Butler (1971) all conclude that the glandular form of "cystic adenocarcinoma" usually derives from parenchymal cells and that true cholangiocarcinoma of the rat liver is uncommon.

Some poorly differentiated anaplastic parenchymal cell lesions are easily confused with angiosarcomas. Parenchymal cell lesions consisting of large, vascular spaces lined by pleomorphic basophilic cells resemble angiosarcomas, and they present a problem in differential diagnosis. Areas of reticuloendothelial cell hyperplasia, found at the periphery of the tumors, may cause further confusion. Nuclear morphology and cytology of a metastasis can be helpful in making the diagnosis.

The definition and significance of parenchymal cell adenomas and biliary cystadenomas are subjects of much discussion. In the literature, "hepatoma" is used to describe lesions ranging from small hyperplastic foci to obvious, invasive carcinoma. Fitzhugh and Nelson (1947) reported adenomas, or low-grade hepatic cell carcinomas, induced by DDT, and Grasso et al. (1969) reported nodular hyperplasia induced by the dye Ponceau MX. Reuber and Glover (1967) described small hyperplastic nodules and "hepatomas" induced by carbon tetrachloride. All these lesions are composed of uniform trabeculae of hepatocytes with occasional cystic or fatty areas. The cells closely resemble normal hepatocytes, but portal tracts are absent and few mitoses are present. There is no evidence of local invasion of vessels, and the cells grow by compression, not invasion of adjacent liver tissues. These lesions do not metastisize. Whether these tumors are "premalignant" and will, if present for sufficient time, develop into carcinomas, is uncertain. It is possible that some are benign parenchymal cell adenomas. Their relationship to nodules found in the development of true carcinomas induced by aflatoxin (Newberne and Wogan, 1968) and ethionine (Farber, 1956) is unknown. The problem of distinguishing hepatomas from nodular hyperplasia is discussed elsewhere.

Biliary cystadenomas are frequently dismissed as being of minor importance in toxicity studies. Little interest has been shown in their pathogenesis and possible neoplastic nature (Stewart and Snell, 1959). They are induced by a wide variety of carcinogens including dimethylnitrosamine (Terracini et al., 1967), aflatoxin (Butler and Barnes, 1963), and methylazoxymethanol (Laqueur and Matsumoto, 1966). It is accepted that they are of biliary epithelial origin, but in many examples the lining cells are morphologically similar to hepatic parenchymal cells. If these lesions are true benign neoplasms of either biliary or parenchymal cell origin, reevaluation of their significance in chemical carcinogenesis is necessary.

6.5 Comparison of Tumors in Rats and Humans

It is useful to compare the morphology of these tumors in the rat with that of malignant liver tumors of humans. Many of these lesions are reported in the literature (Linder et al., 1974; Ohlsson and Norden, 1965). They include parenchymal cell tumors (Domeiri et al., 1971; Creemers and Jadin, 1968; Ma and Blackburn, 1973; O'Conor et al., 1972), angiosarcomas (Creech and Johnson, 1974), and cholangiocarcinomas (Hou, 1956). Specific agents have been associated with several of these lesions in man: aflatoxin and parenchymal cell neoplasms, vinyl chloride and angiosarcomas, thorotrast with endothelial sarcomas and carcinomas in the liver (Hassler et al., 1964; Suckow et al., 1961), and infestation with *Clonorchis sinensis* and cholangiosarcomas (Hou, 1956). The histological patterns of these tumors resemble those found in the rat. There is renewed interest in angiosarcomas because reports that vinyl chloride may induce angiosarcoma in the rat (Maltoni and Lefemine, 1974) were followed by the realization that similar lesions in human liver are related to exposure to vinyl chloride.

The coexistence of malignant parenchymal cell tumors and cirrhosis in human liver has been discussed at length; Linder et al. (1974) found cirrhosis to be associated with hepatocellular tumors in 23 percent to 70 percent of cases, Domeiri et al. (1971) found cirrhosis coexisting with carcinoma in 17 to 78 cases. Domeiri et al. (1971) also state that when cirrhosis was present, frequently both lobes of the liver were involved by tumor.

The biological behavior of human tumors parallels that in the rat. Distant metastases to the lung, lymph nodes, bone, and other organs are found in 30 percent to 85 percent of cases (Linder et al., 1974; Domeiri et al., 1971). Local invasion and spread within the abdomen was reported in both series. Linder et al. (1974) reported the incidence of tumors to be five times higher in males than in females, and a similar trend was reported in rats by Newberne and Butler (1969). Benign tumors in man and the definition of adenoma and hamartoma of the liver are discussed by Phillips et al. (1973); the morphology of these lesions is comparable to that found in certain benign parenchymal cell tumors of rats referred to previously. Interest in these lesions in man has been increased by their association with the oral contraceptives (Baum et al., 1973).

6.6 Conclusions

The majority of malignant neoplasms of rat liver originate in parenchymal cells. Poorly differentiated glandular forms of these lesions can be confused histologically with cholangiocarcinomas, and the similarities between angiosarcomas and some parenchymal cell lesions make diagnosis difficult. Reticuloendothelial lesions can be induced by a variety of compounds. Biliary cystadenomas and benign parenchymal cell adenomas are worthy of further consideration, but the term adenoma should not be used indiscriminately to describe hyperplastic nodules, "precancerous lesions," and lesions of unknown biological behavior. The morphology and biological behavior of malignant liver tumors in the rat is similar to that reported in man.

6.7 References

Baum, J. K., Holtz, F., Bookstein, J. J., and Klein, E. W. 1973. Possible association between benign hepatomas and oral contraceptives. *Lancet* 2:926.

Bielschowsky, F. 1944. Distant tumors produced by 2-amino- and 2-acetyl-fluorene. *Brit. J. Exp. Path.* 25:1-4.

Butler, W. H. 1965. Liver injury and aflatoxin. in *Mycotoxins in Foodstuffs*, ed. G. N. Wogan, Cambridge, Mass.: MIT Press.

Butler, W. H. 1971. Pathology of liver cancer in experimental animals. In *IARC Scientific Publications #1. Liver Cancer*, pp. 30–41. Lyons: International Agency for Research on Cancer.

Butler, W. H., and Barnes, J. M. 1963. Toxic effects of groundnut meal containing aflatoxin to rats and guinea pigs. *Brit. J. Cancer* 17:669–710.

Craddock, V. M. 1971. Liver carcinomas induced in rats by single administration of dimethylnitrosamine after partial hepatectomy. *J. Nat. Cancer Inst.* 47:889-907.

Crain, R. C. 1958. Spontaneous tumors in the Rochester strain of Wistar rat. *Amer. J. Path.* 34:311. 1974.

Creech, J. L., and Johnson, M. N. 1974. Angiosarcoma of liver in the manufacture of polyvinyl chloride. *J. Occup. Med.* 16:150–151.

Creemers, J., and Jadin, J. M. 1968. Ultrastructure of human hepatocellular carcinoma. *J. Microscop.* 7:257–264.

Dalton, A. J., and Edwards, J. E. 1942. Cytology of hepatic tumors and proliferating bile duct epithelium in the rat induced with *p*-dimethylaminoazobenzene. *J. Nat. Cancer Inst.* 3:319–329.

Domeiri, A., Huvos, A. G., Goldsmith, H. S., and Foote, F. W. 1971. Primary malignant tumors of the liver. *Cancer* 27:7–11.

Farber, E. 1956. Similarities in the sequence of early histological changes induced in the liver of the rat by ethionine, 2-acetylaminofluorene and 3-methyl-4-dimethylaminoazobenzene. *Cancer Res.* 16:142–148.

Firminger, H. K., and Mulay, A. S. 1952. Histochemical and morphologic differentiation of induced tumors of the liver in rats. *J. Nat. Cancer Inst.* 13:19–33.

Fitzhugh, O. G., and Nelson, A. A. 1947. The chronic oral toxicity of DDT. (2,2-bis(*p*-chlorophenyl-1,1,1-trichloroethane). *J. Pharmacol. Exp. Ther.* 89:18–30.

Gilbert, G., and Gillman, J. 1958. Spontaneous neoplasms in the albino rat. *S. Afr. J. Med. Sci.* 23:257.

Gillman, T., and Hallowes, R. C. 1972a. Ultrastructural changes in rat livers induced by repeated injections of trypan blue. *Cancer Res.* 32:2393–2399.

Gillman, T., and Hallowes, R. C. 1972b. Ultrastructural and histochemical observations on a transplantable reticuloendothelial tumor in rats. *Cancer Res.* 32: 2383–2392.

Grasso, P., Lansdown, A. B. G., Kiss, I. S., Gaumt, I. F., and Gangolli, S. D. 1969. Nodular hyperplasia in the rat liver following prolonged feeding on Ponceau MX. *Fd. Cosmet. Toxicol.* 7:422–425.

Grice, H. C., Mannell, W. A., and Allmark, M. G. 1961. Liver tumors in rats fed Ponceau 3R. *Toxicol. Appl. Pharmacol.* 3:509.

Hadjiolov, D., and Markow, D. 1973. Fine structure of heamangioendothelial sarcomas induced by N-nitrosodimethylamine. *Arch. Geschwulsforsch*, 42/2:120–126.

Hard, G. C., and Butler, W. H. 1970. Cellular analysis of renal neoplasia: Induction of renal tumors in dietary conditioned rats by dimethylnitrosamine, with a reappraisal of morphological characteristics. *Cancer Res.* 30:2796-2805.

Hassler, O., Bostrom, K., and Dahlback, O. L. 1964. Thorotrast tumors. *Acta Path. Microbiol. Scand.* 61:13-20.

Hemm, R. D. 1973. Hepatic capsular pseudolipoma in the rat. *J. Comp. Path.* 83:83-86.

Hou, P.-C. 1956. Relationship between primary carcinoma of the liver and infestation with *Chlonorchis sinensis. J. Path. Bacteriol.* 72:239.

Kinosita, R. 1955. Some recent findings concerning hepatomas induced with *p*-dimethylaminoazobenzene. *J. Nat. Cancer Inst.* 15:1443.

Laqueur, G. L., and Matsumoto, H. 1966. Neoplasms in female Fischer rats following intraperitoneal injection of methylazoxymethanol. *J. Nat. Cancer Inst.* 37: 217-232.

Laqueur, G. L., Mickelson, O., Whiting, M. G., and Kurland, L. T. 1963. Carcinogenic properties of nuts from *Cycas Cirunalis L.* indigenous to Guam. *J. Nat. Cancer Inst.* 31:919.

Linder, G. T., Crook, J. N., and Cohn, I. 1974. Primary liver carcinoma. *Cancer* 33:1624-1629. 1973.

Ma, M. H., and Blackburn, C. R. B. 1973. Fine structure of primary liver tumors and tumor-bearing livers in man. *Cancer Res.* 33:1766-1774.

McCoy, C. W., 1909. A preliminary report of tumors found in wild rats. *J. Med. Res.* 21:285.

Maltoni, C., and Lefemine, G. 1974. Le potenzialita dei saggi sperimentali nella predizione dei rischi oncogeni ambientali. Un esempio: il chloruro di vinile. *Rend. Sci. fis. mat. nat. (Lincei)* 66:1-11.

Morris, H. P., Velat, C. A., Wagner, B. P., Dahlgard, M., and Ray, F. E. 1960. Studies on the carcinogenicity in the rat of derivatives of aromatic amines related to N-2-fluorenylacetamide. *J. Nat. Cancer Inst.* 24:149-180.

Newberne, P. M., and Butler, W. H. 1969. Acute and chronic effects of aflatoxin on the liver of domestic and laboratory animals: a review. *Cancer Res.* 29:236-250.

Newberne, P. M., and Shank, R. C. 1973. Induction of liver and lung tumors in rats by the simultaneous administration of sodium nitrite and morpholine. *Fd. Cosmet. Toxicol.* 11:819-825.

Newberne, P. M., and Wogan, G. N. 1968. Sequential morphologic changes in aflatoxin B_1 carcinogenesis in the rat. *Cancer Res.* 28:770-781.

Newberne, P. M., Hunt, C. E., and Wogan, G. N. 1967. Neoplasms in the rat associated with the administration of urethan and aflatoxin. *Exp. Mol. Path.* 6:285-299.

O'Conor, G. T., Tralka, T. S., Hensen, E., and Vogel, C. L. 1972. Ultrastructural survey of primary liver cell carcinomas from Uganda. *J. Nat. Cancer Inst.* 48: 587-603.

Ohlsson, E. G., and Nordén, J. G. 1965. Primary carcinoma of the liver. *Acta Path. Microbiol. Scand.* 64:430-440.

Opie, E. L. 1944a. The influence of diet on the production of tumors of the liver by butter yellow. *J. Exp. Med.* 80:219.

Opie, E. L. 1944b. The pathogenesis of tumors of the liver produced by butter yellow. *J. Exp. Med.* 80:231.

Orr, J. W. 1940. The histology of the rat's liver during the course of carcinogenesis by butter-yellow (*p*-dimethylaminoazobenzene). *J. Path. Bacteriol.* 50:939.

Phillips, M. J., Langer, B., Stone, R., Fisher, M. M., and Ritchie, S. 1973. Benign liver cell tumors. *Cancer.* 32:463.

Pollard, M., and Kajima, M. 1967. Leukemia induced by 7,12-dimethylbenz(a)anthracene in germ free rats. *J. Nat. Cancer Inst.* 39:135–141.

Reuber, M. D., and Glover, E. L. 1967. Hyperplastic and early neoplastic lesions of the liver in Buffalo strain rats of various ages given subcutaneous carbon tetrachloride. *J. Nat. Cancer Inst.* 38:891–899.

Stewart, H., and Snell, K. 1959. The histopathology of experimental tumors of the liver of the rat. In *The Physiopathology of Cancer,* ed. F. Homburger. 2nd ed. New York: Paul B. Hoeber.

Suckow, E. E., Henegar, G. C., and Baserga, R. 1961. Tumors of the liver following administration of thorotrast. *Am. J. Clin. Path.* 38:663.

Svoboda, D. J. 1964. Fine structure of hepatomas induced rats with *p*-dimethylaminoazobenzene. *J. Nat. Cancer Inst.* 33:315–339.

Terracini, B., Magee, P. N., and Barnes, J. M. 1967. Hepatic pathology in rats on low dietary levels of dimethylnitrosamine. *Brit. J. Cancer* 21:559.

Wisse, E. 1974. Kupffer cell reactions in rat liver under various conditions as observed in the electron microscope. *J. Ultrastruct. Res.* 46:499–520.

6.8 Discussion

Bannasch The glandular lesions you showed have in many cases a basement membrane, and this might be an indication that they derived from bile ducts; but as I understand it, you believe they derived at least partly from hepatocytes. You would assume that hepatocytes in this condition produce basement membrane?

Butler Every carcinoma of any histological type we have looked at in the liver has a basement membrane. Those that you would accept as basophilic, trabecular carcinoma have basement membranes between the obvious hepatic cells and the endothelium. I think the basement membrane is associated with the endothelium. You can see overlapping layers of endothelium, and a basement membrane is present between each layer of endothelium. I think it is independent of the neoplastic cell and I do not think it can be used for differential diagnosis. Also one can see central necrosis of the thick trabeculae; that is the way I think the glandular appearance develops.

Bannasch Have you any criteria to discriminate between glandular lesions which are derived from hepatocytes and those which are derived from bile ducts?

Butler At one time I did, but not now. I thought the way to do it was by the presence of fibrous stroma, but when one looks at the range of lesions and the transition between bile duct and liver cell types, I don't think that this is correct. The case which Dr. Jones showed you was originally reported in the literature as a glandular carcinoma—I am now beginning to have my doubts about it. The only thing is that it metastasizes as an adenocarcinoma, and all the other ones I have seen metastasized as solid trabecular carcinoma.

Bannasch When glandular tumors metastasize, are they derived from hepatocytes or bile ducts?

Jones We showed that some metastasized in glandular form; others metastasized as typical trabecular lesions.

Bannasch We had the same difficulty in interpreting the lesion; but I would like to know why you call this tumor cholangiocarcinoma and not glandular hepatoma.

Jones Cholangiocarcinomas have been reported elsewhere, and of our series the one illustrated is most marked. Also the primary lesion was almost entirely that structure.

Laqueur I think we can diagnose only things we see—not what we want to see or what we think we ought to see, but what we can see—so I am perfectly satisfied in recognizing the adenomatous or cholangiocarcinoma structure if the cells resemble those that arise from bile ducts. I am also perfectly willing to recognize a pseudoglandular hepatocarcinoma which has holes in the center, which, I agree, are due to cellular necrosis.

Butler There may be two separate lesions. One is a glandular acinar growth; the other one is a pseudo gland formation developing in a larger mass of cells, probably because of nutritional factors. Necrosis occurs in the center of these lesions, and so spaces are formed containing debris. Histogenetically, bile duct epithelium and hepatic epithelium derive from the same area of endoderm. You can consider a hepatocarcinoma and a cholangiocarcinoma as different expressions of the same histogenetic background. I do not want to go into this argument in detail, but, leaving malignant lesions aside, in other proliferative situations following liver injury I have seen apparent transitional stages between the hepatocyte and the bile duct cell.

In considering angiosarcomas, Dr. Laqueur said that one of the striking things about these lesions was the eosinophilic material in the space of Disse. Do you think this is a product directly related to neoplasm, or is it membrane material, and is it seen in the vinyl chloride sarcomas?

Squires It appears as a granular material, and many people have commented upon it.

Tate A diagnosis of angiosarcoma seems likely for the lesions with blood-filled spaces, described by Dr. Jones. What's the gross appearance of the lesions? Are they nodular?

Butler They are nodular, hemorrhagic masses. I am not suggesting that angiosarcomas do not exist; following nitrosomorpholine a high incidence is reported. I think the incidence varies with the carcinogen. I have seen reasonably convincing angiosarcomas, but I do not think that, just because one sees these hemorrhagic nodules with blood-filled spaces, a diagnosis of vascular tumors is justified. Vascular lesions have been reported in old rats. We see in our old rats areas of telangiectasis as I am sure most people do.

7 Ultrastructure of Hepatic Neoplasia

W. H. Butler
Imperial Chemical Industries, Ltd.
Macclesfield, Cheshire
England

Glenys Jones
Shell Research Laboratories
Sittingbourne, Kent
England

7.1 Introduction

At a light microscope level, there is an apparent pleomorphism of tumor type in most studies of hepatic neoplasia. This has inevitably led to the problem of determining the histogenesis of the lesions. In any study of mechanism in a heterogenous population of cells, it is of obvious importance to determine which cell type is reacting to the administered compound. Do the trabecular, cholangio, and hemorrhagic anaplastic carcinomas represent the response of one cell type to a carcinogenic stimulus and the local environment, or are they three distinct neoplasms arising from different populations of cells within the same organ?

The electron microscope allows conclusions different from those reached through use of the light microscope. By way of illustration, this paper deals with carcinomas induced by aflatoxin; only malignant lesions, verified by the presence of invasion or metastases, are studied. It is important, however, to compare the ultrastructure of these tumors with those induced by ethionine (Merkow et al., 1972), 2-acetylaminofluorine 2-FAA (Merkow et al., 1969), and *p*-dimethylaminoazobenzene (DAB) (Svoboda, 1964).

7.2 Materials

The material presented in this report was derived from hepatic carcinomas induced by feeding 5 ppm aflatoxin to male Porton/Wistar or inbred Fischer rats. The compound was fed for six, nine, twelve, or fifteen weeks, after which a normal diet was resumed. The animals were killed when in poor condition, usually after the return to normal diet. Multiple, large, hepatic tumors were present, and metastasis found in 75 percent of the animals. In the absence of metastasis, lesions were considered to be malignant when the morphology was similar to the lesions that had metastasized and also where local invasion was present. Seventeen carcinomas satisfying these criteria were examined. The carcinomas were either diced into 2 percent veranol-buffered osmium tetroxide or perfused via the aorta with 2 percent phosphate-buffered glutaraldehyde and then fixed in 2 percent veranol-buffered osmium tetroxide. Large area slices of the perfused material (Jones and Butler, 1974) and the dices were embedded in Epon 812 and sectioned with either glass or diamond knives.

7.3 Carcinomas Induced by Aflatoxin

This group of carcinomas illustrates many of the features described in chapter 6 on morphology as revealed through the light microscope. When 1 μ plastic sections are examined, it is possible to recognize the trabecular carcinomas with vascular spaces lined by endothelium (fig. 7.1). The individual cells of the trabeculae show considerable variations in shape and size when cut either transversely or longitudinally. In many instances, small spaces between individual cells give the appearance of a glandular pattern (fig. 7.2). The anaplastic hemorrhagic tumors exhibit large, blood-filled spaces that are partially lined by elongated cells overlying an acellular amorphous material (fig. 7.3). Dark and light cells are seen under both light and electron microscopes (figs. 7.4 and 7.5). They are seldom noted in perfused fixed material, however, and may represent an artifact of fixation.

In the carcinomas having the most developed glandular pattern, the acini were lined by columnar cells lying on a fibrous stroma (fig. 7.6). PAS-positive, dense granules are present within the columnar cells, and PAS-positive material is seen within the lumen (fig. 7.7).

7.3.1 Trabecular Carcinomas

Trabecular carcinomas are formed from branching plates of hepatic parenchymal cells, which are often many cells thick (fig. 7.8). The vascular pole, which has sparse microvilli, abuts onto the basement membrane of the endothelium. The basement membrane is usually continuous but may be multiple and thickened (fig. 7.9). The endothelium may be of single-cell thickness but in some areas is overlapping. When the cells overlap, basement membrane can be seen between the layers. Pseudopods of the parenchyma can be seen extending between the endothelium, a phenomenon that may represent microinvasion (fig. 7.10). Clusters of macrophages with large phagocytic vacuoles underlie the endothelium (fig. 7.11). In some areas within this modified space of Disse, are collagen fibers and also amorphous granular material (fig. 7.12).

The neoplastic parenchymal cells show considerable variation both between and within the individual neoplasms. In the well-differentiated trabecular carcinomas, single-cell plates of hepatic parenchymal cells with well-organized stacks of rough endoplasmic reticulum are a com-

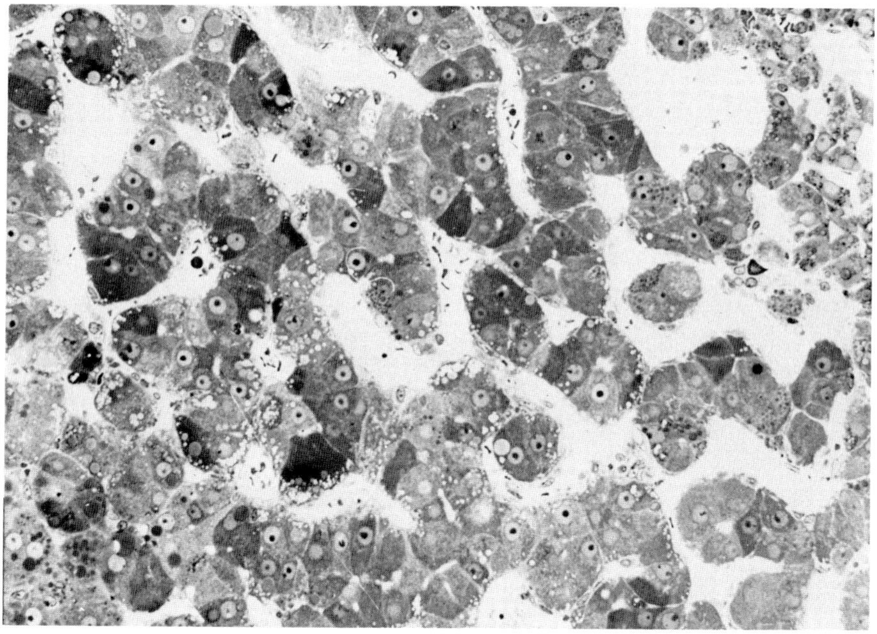

7.1 Trabecular hepatocellular carcinoma. The trabeculae are branched and may be two or more cells thick. Vascular spaces are lined by endothelium. Epoxy resin 1 μ section stained with toluidine blue (× 660).

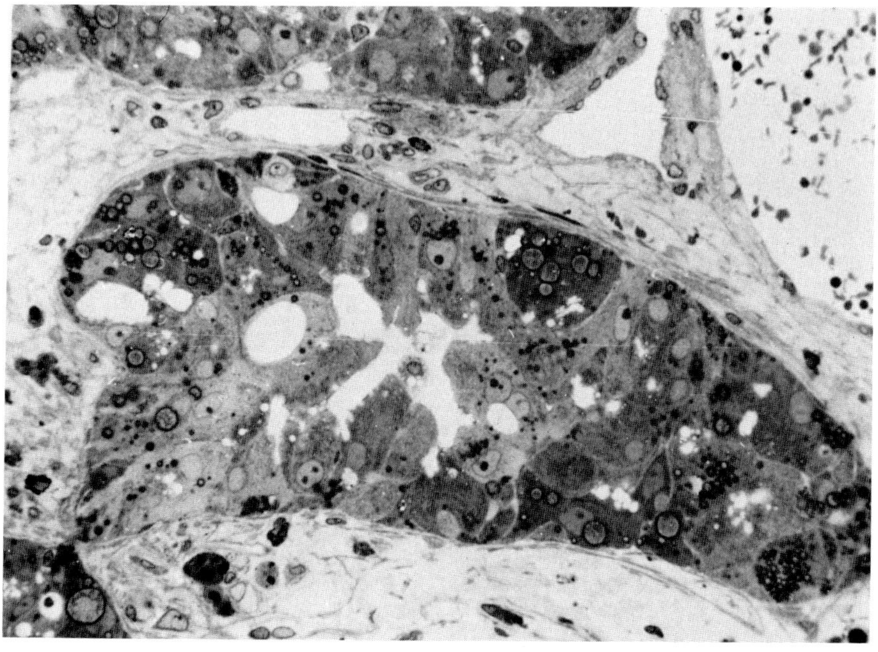

7.2 Trabecular hepatocellular carcinoma with dilated canalicular space giving an appearance of a glandular pattern. Epoxy resin 1 μ section stained with toluidine blue (× 680).

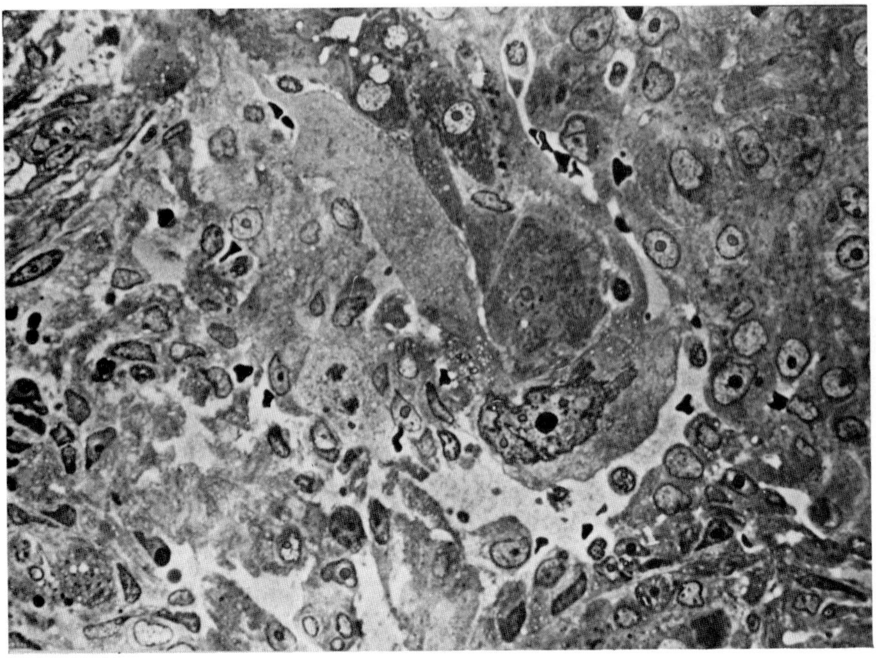

7.3 Anaplastic hepatocellular carcinoma showing disorderly arrangement of cells and amorphous material. Epoxy resin 1 μ section stained with basic fuchsin (× 510).

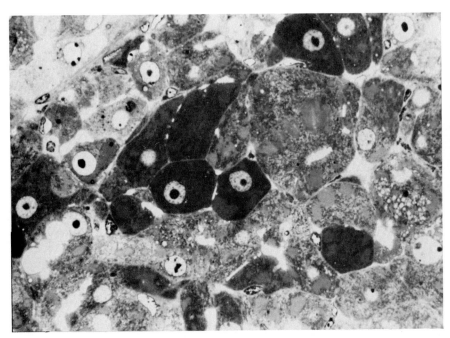

7.4 Trabecular carcinoma showing variation in staining with dark and light cells. Epoxy resin 1 μ section stained with toluidine blue (× 850).

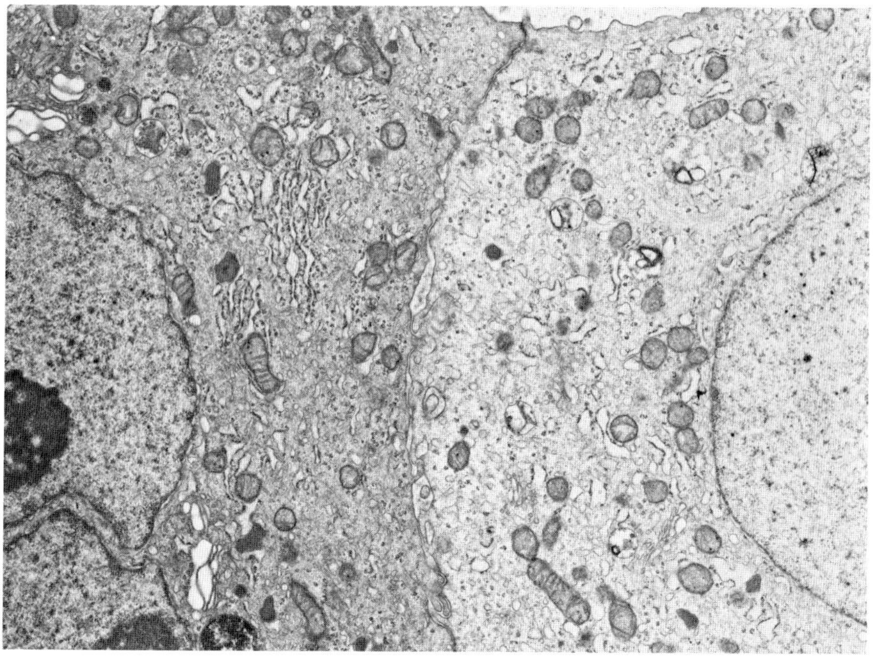

7.5 Electron micrograph of trabecular carcinoma showing dark and light cells (× 5,040).

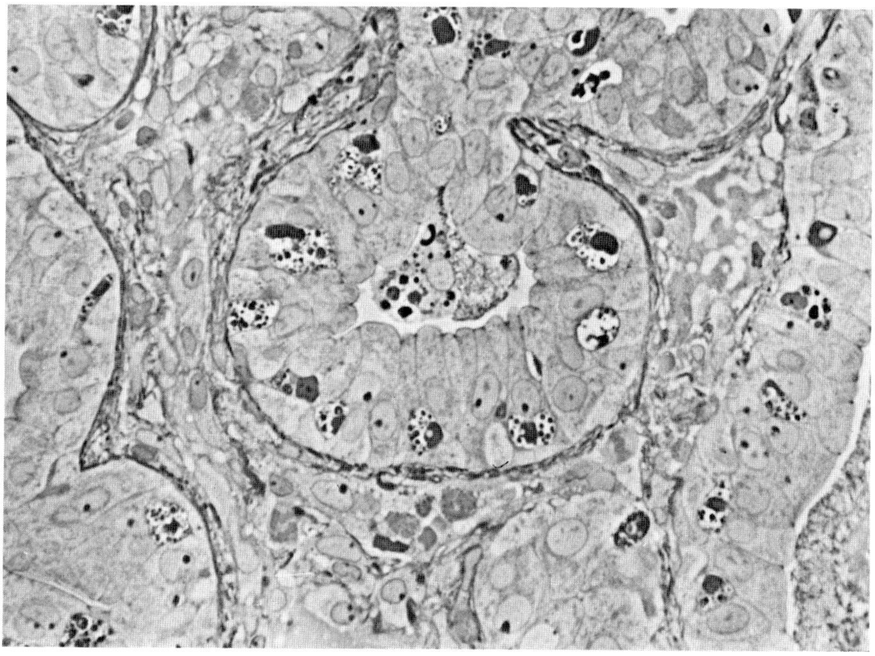

7.6 Light micrograph of a hepatocarcinoma with well-developed glandular pattern lined by columnar epithelium. PAS-positive material is present within the columnar cells. Epoxy resin 1 μ section stained with PAS (× 510).

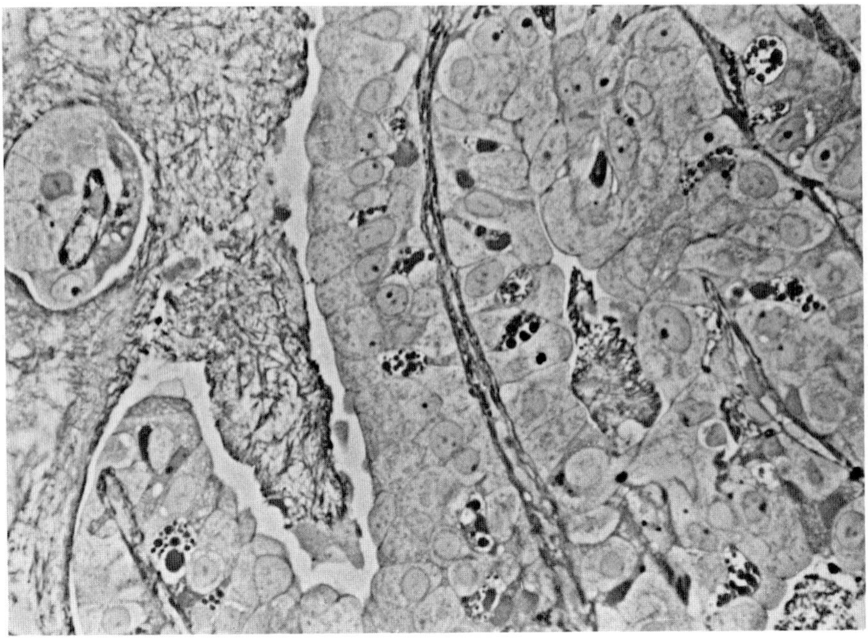

7.7 Glandular hepatocarcinoma with acinar spaces lined by columnar epithelium. PAS-positive material is present within the lumen. Epoxy resin 1 μ section stained with PAS ($\times$ 510).

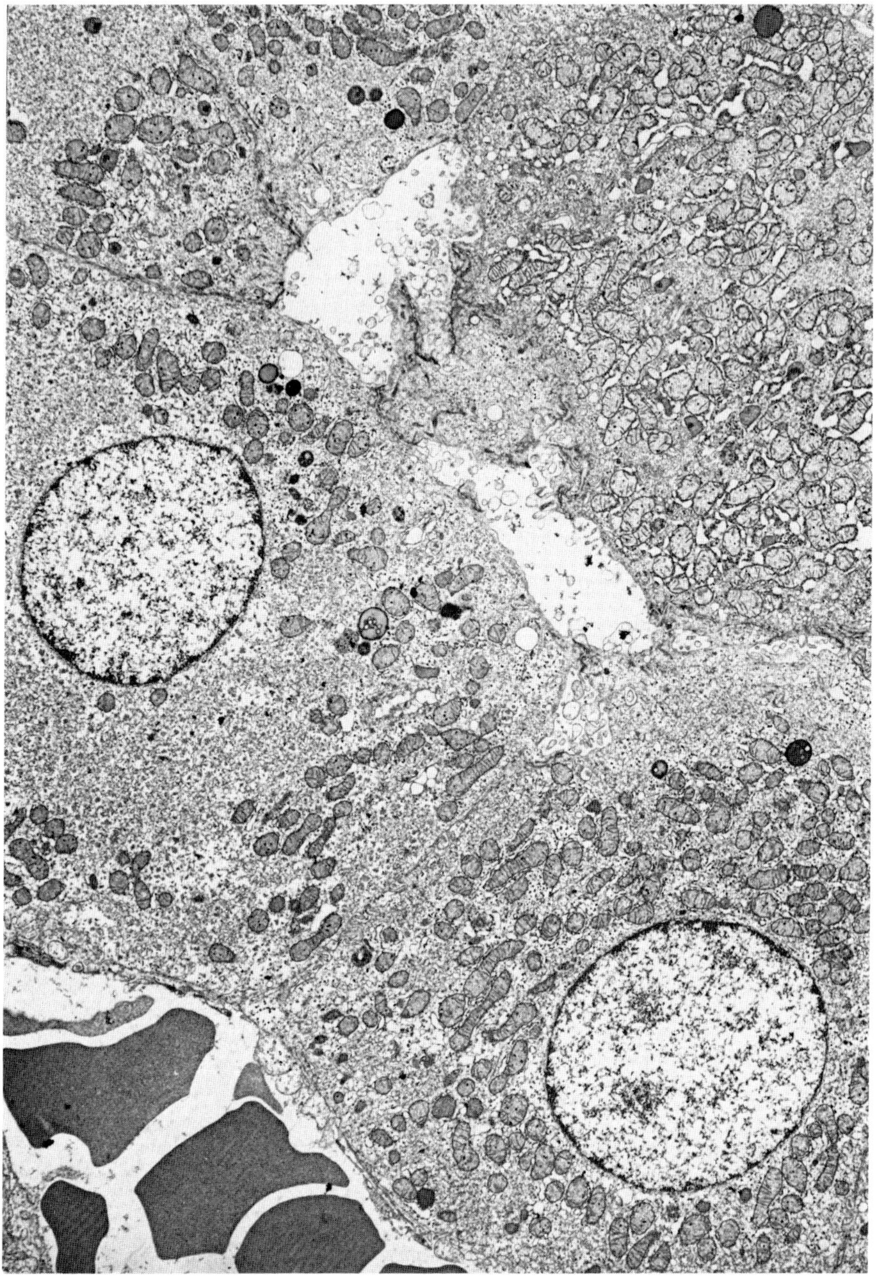

7.8 Low-power electron micrograph of trabecular carcinoma. In the vascular pole the microvilli are sparse (× 3,720).

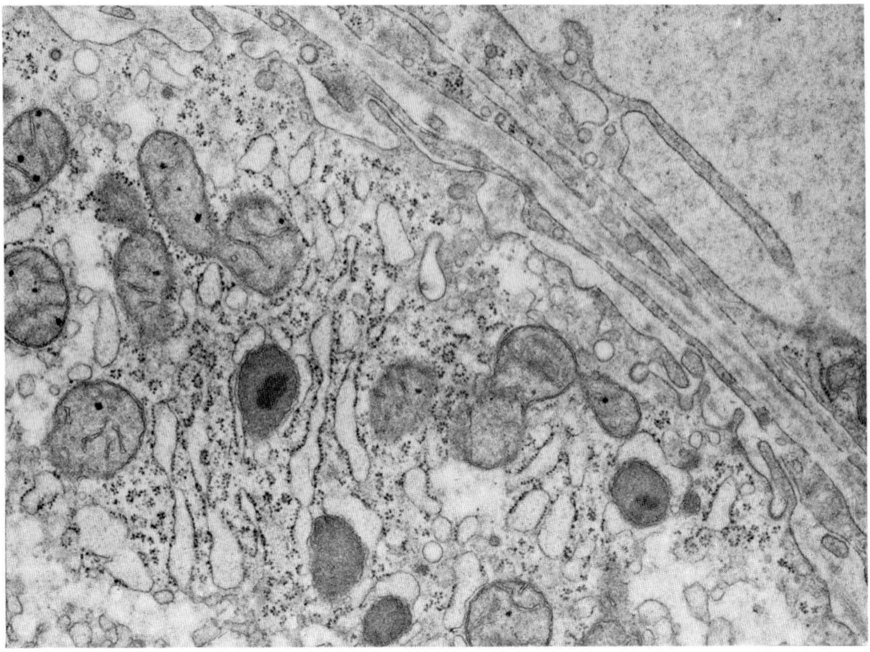

7.9 Vascular pole of trabecular carcinoma showing multiple basement membrane-like structures (× 16,900).

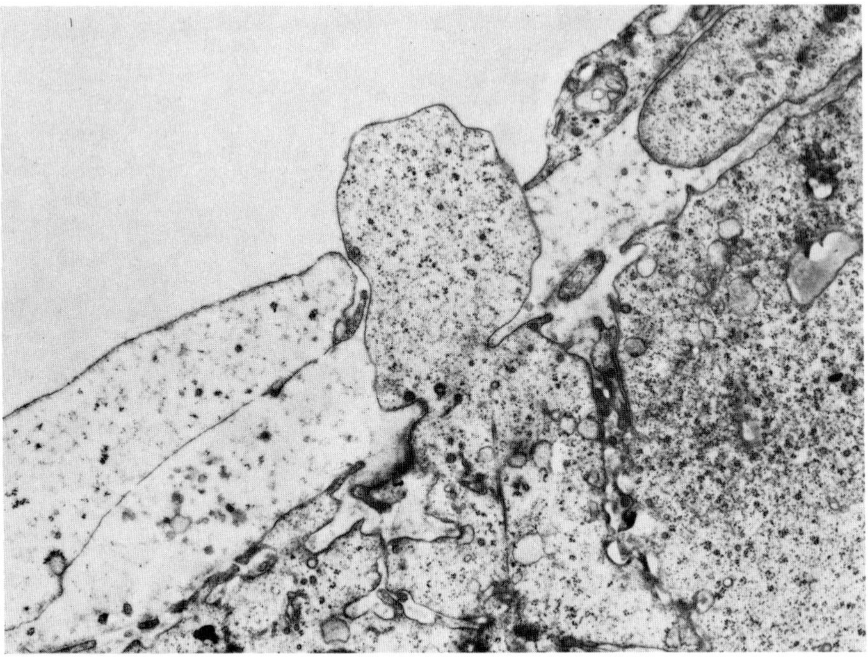

7.10 Vascular pole of trabecular carcinoma showing protrusion of cytoplasm between endothelium (× 16,900).

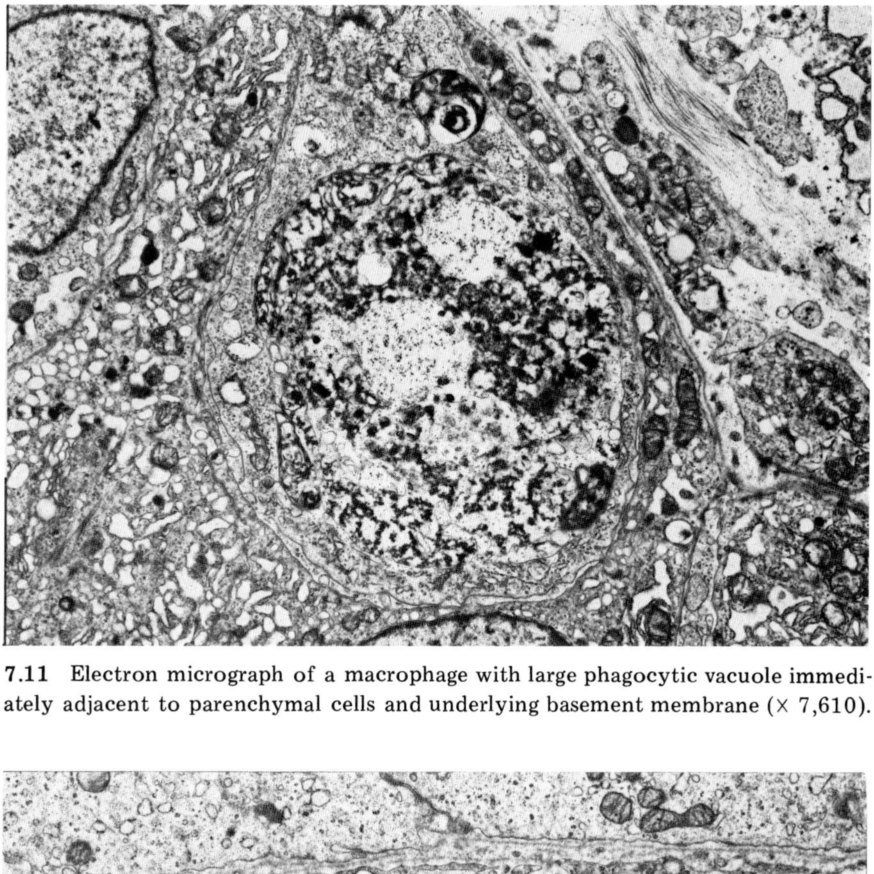

7.11 Electron micrograph of a macrophage with large phagocytic vacuole immediately adjacent to parenchymal cells and underlying basement membrane (× 7,610).

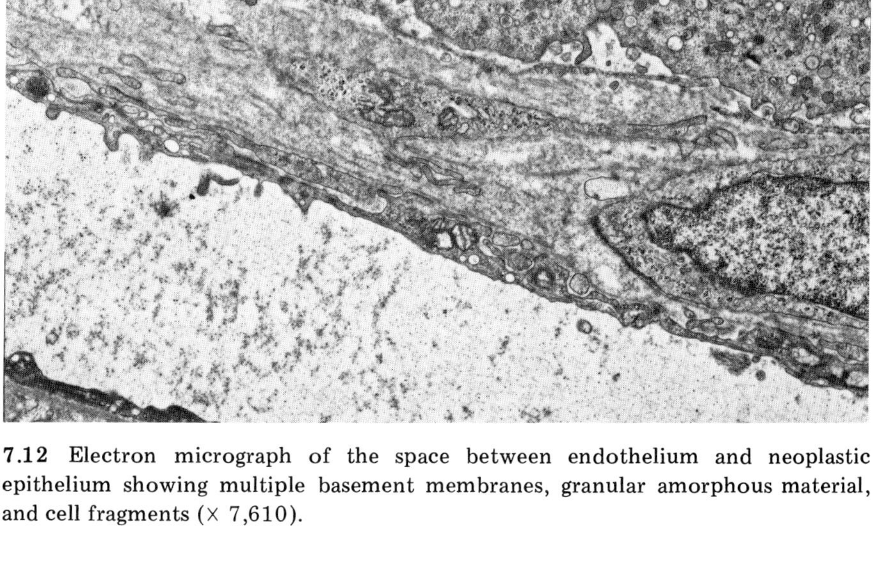

7.12 Electron micrograph of the space between endothelium and neoplastic epithelium showing multiple basement membranes, granular amorphous material, and cell fragments (× 7,610).

mon feature (fig. 7.13). Adjacent cells may have a sparse network of rough endoplasmic reticulum cisternae and widely dispersed free ribosomes (fig. 7.14). In the cells having the sparse rough endoplasmic reticulum, the cisternae often encircle mitochondria. The amount of smooth endoplasmic reticulum is highly variable; some cells contain large areas of it (fig. 7.15) and others have none. Tightly packed myelin structures made up of smooth membrane (fig. 7.16) and looser whorls of rough endoplasmic reticulum, (fig. 7.17) enclosing fat droplets and mitochondria, are often present. In addition, there is considerable variation in the number of Golgi complexes. In some cells they are much larger and more abundant than in normal hepatocytes (fig. 7.18). Electron-dense granules may be present within some Golgi cisternae. Annulate lamellae have been seen but are not common. Within the poorly differentiated carcinomas, cells contain few well-organized rough endoplasmic reticulum cisternae (fig. 7.19) or areas of smooth endoplasmic reticulum.

All trabecular carcinomas contain membrane-bound vacuoles containing disintegrating cell organelles. But without further characterization it is uncertain whether these are correctly designated as autophagic vacuoles. Bundles of microfilaments are present in most cells (fig. 7.20).

The mitochondria show considerable variation. In many cells this organelle is abundant and has a normal structure, except for an increase of matrical dense bodies (fig. 7.21). In other cells the mitochondria may be sparse and variable in size and form. Microbodies are rarely seen. Many cells within the well-differentiated carcinomas contain cytoplasmic glycogen; the amount varies greatly, however. Lipid droplets are found throughout all the tumors studied (fig. 7.22).

Between adjacent cells there are many structures that resemble bile canaliculi. They may be small, with multiple microvilli lying between two cells or at the junction of three or more cells (fig. 7.23). Desmosomes are present at the point of apposition of the plasma membranes. In other cases, the canalicular space is widely dilated, has sparse microvilli, and is lined by four or more parenchymal cells (fig. 7.24); again, a desmosome appears at each cell junction. In those cases in which large canaliculi are seen, cytoplasmic projections protrude into the canalicular space (fig. 7.25).

The nuclei of trabecular carcinomas are usually regular in shape with finely dispersed chromatin. The large nucleoli exhibit a prominent nucleonema (fig. 7.26). In this series, nuclear inclusions are uncommon.

7.3.2 Glandular Carcinomas

The well-developed glandular pattern is composed of irregular acini lined by columnar and cuboidal cells (fig. 7.27). The lumen is often empty but may contain cell debris and fibrin (fig. 7.28). Each acinus is bound by basement membrane, which may be multiple (fig. 7.29). In semithin sections, many dense, PAS-positive bodies can be seen scat-

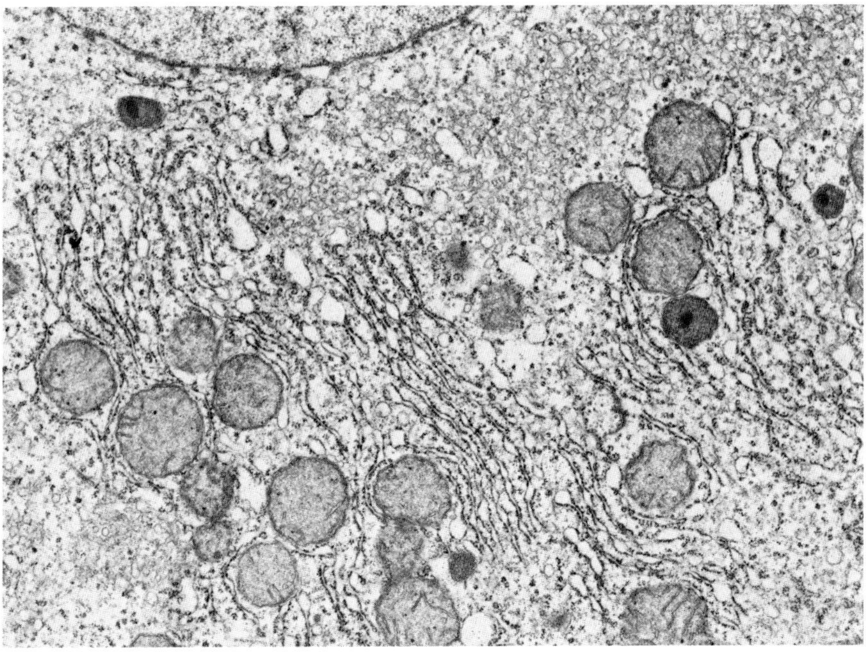

7.13 Cytoplasm of parenchymal cell from trabecular carcinoma showing well-organized stacks of rough endoplasmic reticulum (× 11,830).

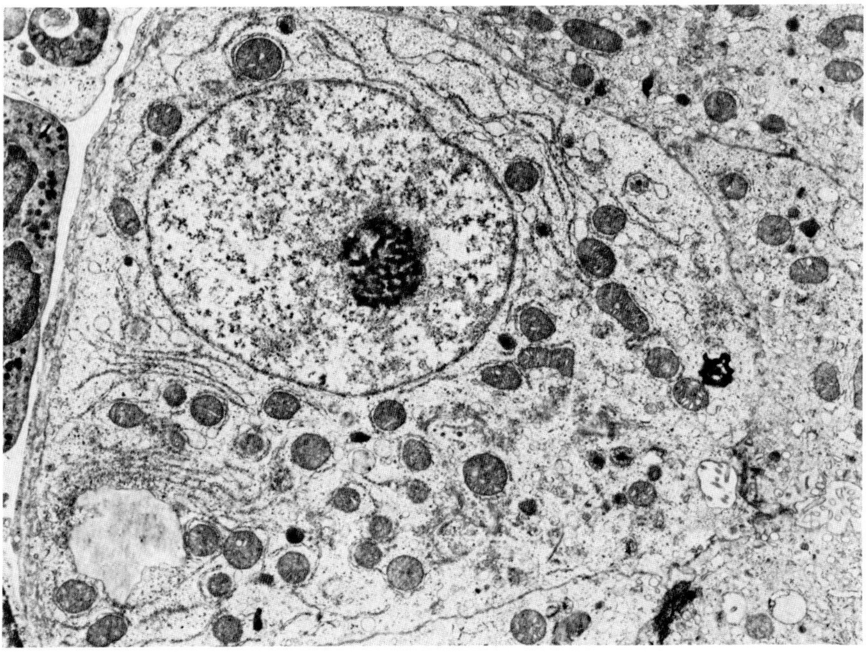

7.14 Electron micrograph of hepatic parenchymal cell of trabecular carcinoma showing sparse network of rough endoplasmic reticulum (× 5,920).

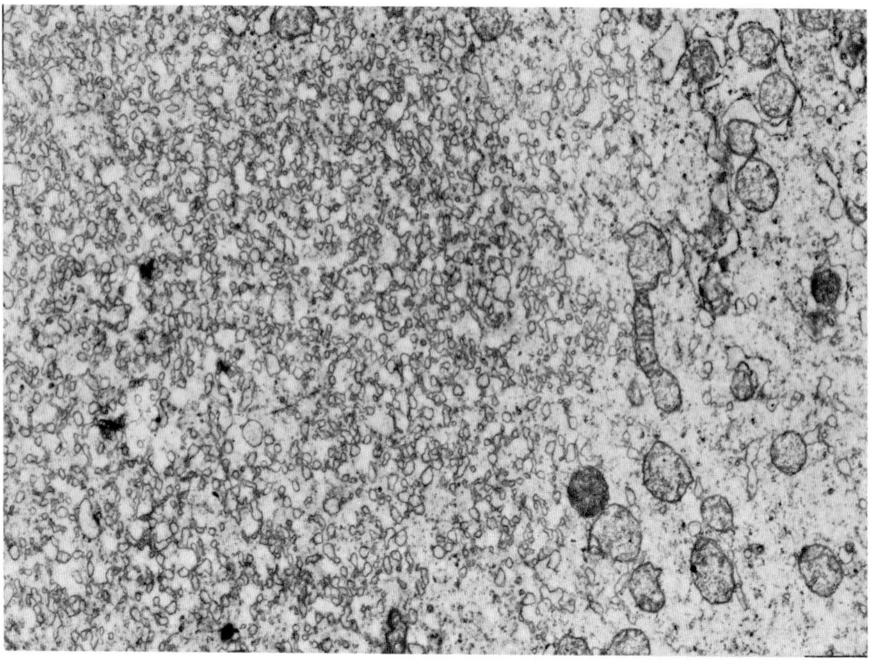

7.15 Cytoplasm of parenchymal cell from trabecular carcinoma showing abundant smooth endoplasmic reticulum (× 10,080).

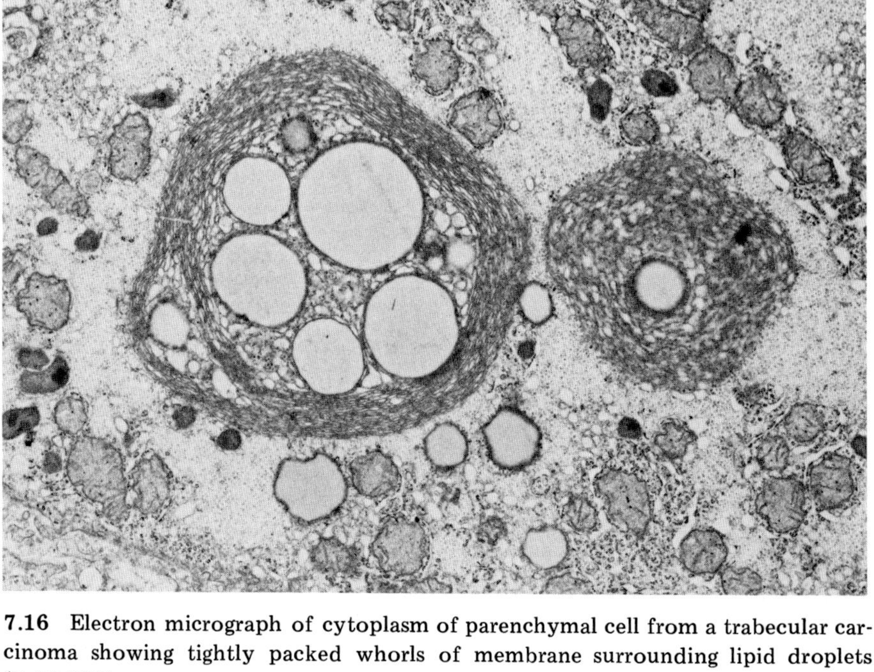

7.16 Electron micrograph of cytoplasm of parenchymal cell from a trabecular carcinoma showing tightly packed whorls of membrane surrounding lipid droplets (× 10,080).

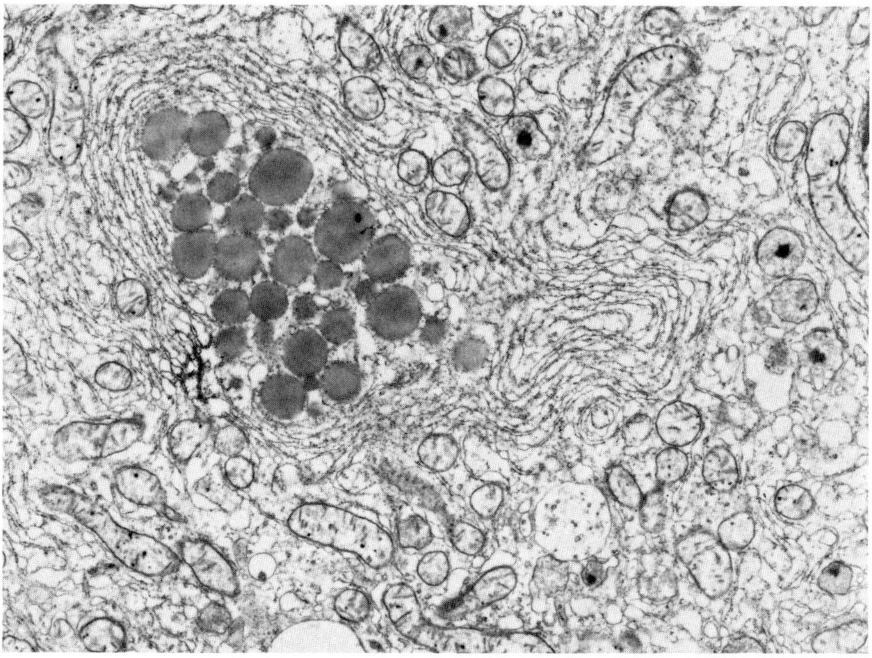

7.17 Electron micrograph of cytoplasm of trabecular carcinoma showing looser whorls of rough endoplasmic reticulum enclosing fat droplets (× 9,300).

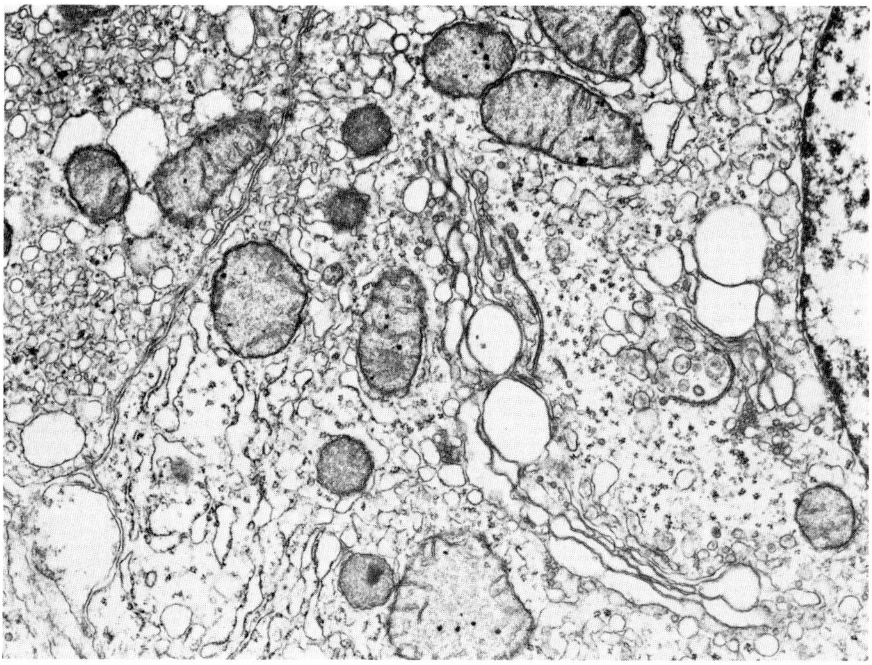

7.18 Electron micrograph of cytoplasm of parenchymal cell from a trabecular carcinoma showing prominent Golgi complexes (× 13,520).

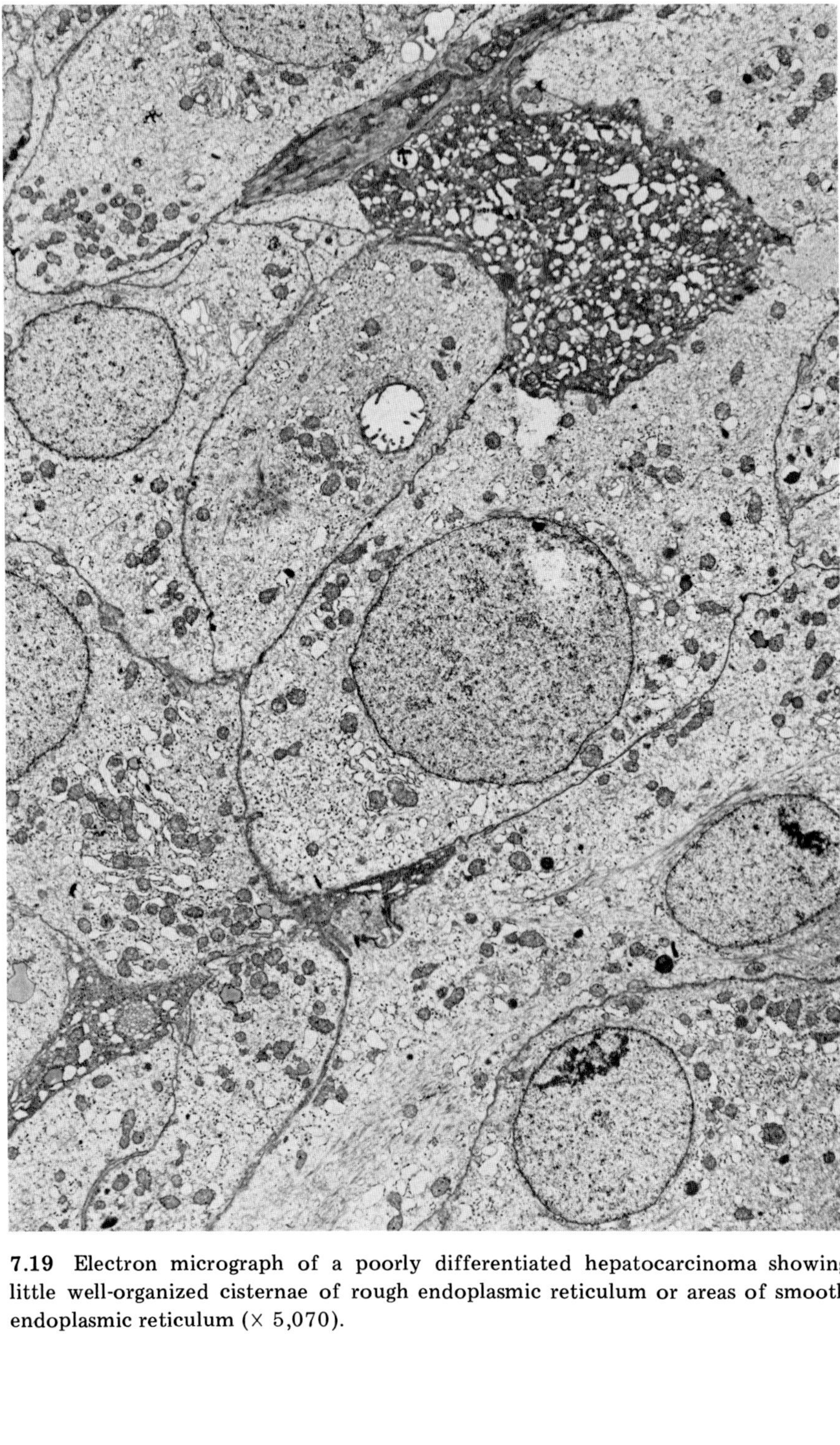

7.19 Electron micrograph of a poorly differentiated hepatocarcinoma showing little well-organized cisternae of rough endoplasmic reticulum or areas of smooth endoplasmic reticulum (× 5,070).

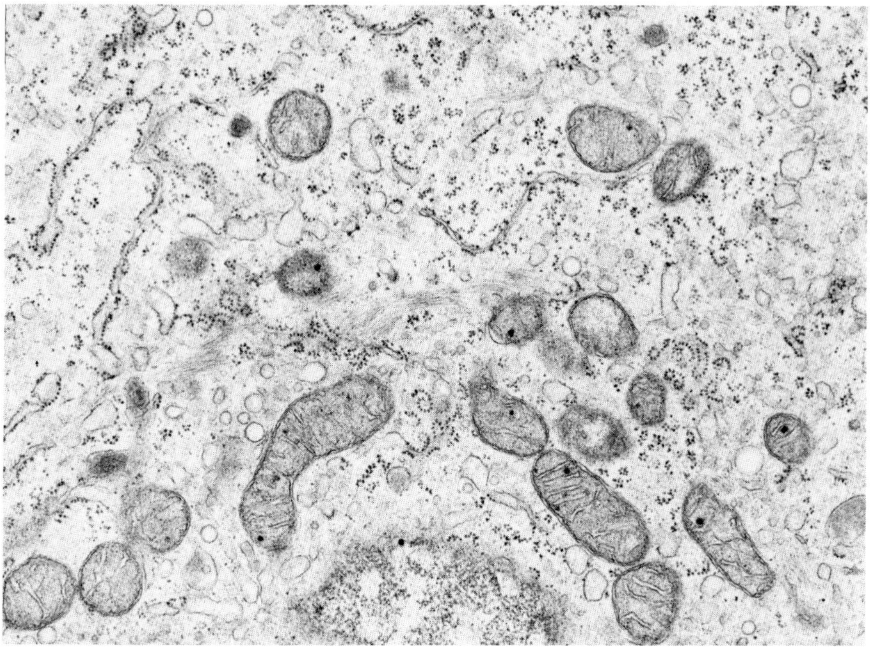

7.20 Electron micrograph of cytoplasm of parenchymal cell from a trabecular carcinoma showing bundles of microfilaments (× 21,130).

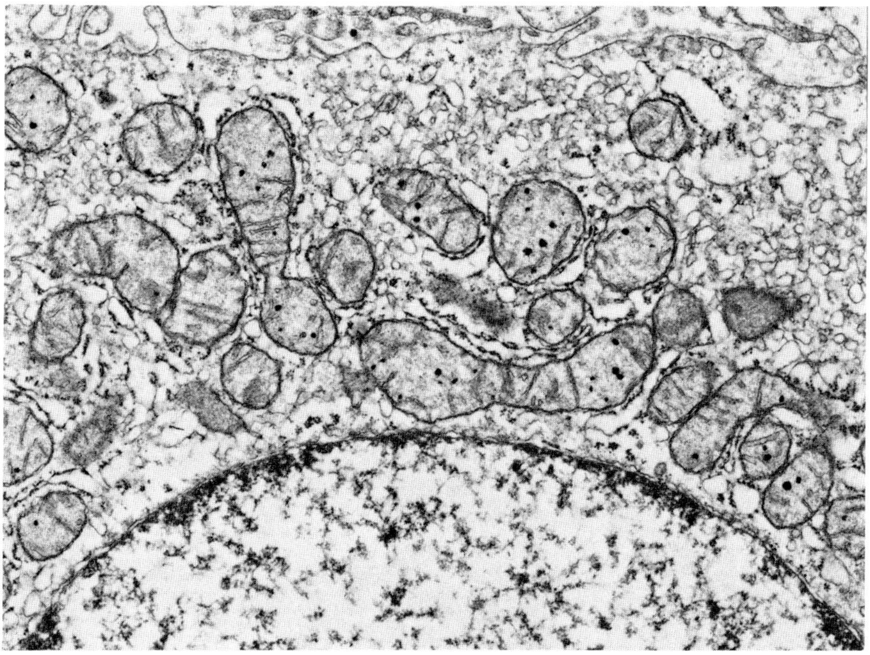

7.21 Electron micrograph of cytoplasm of parenchymal cell from a trabecular carcinoma showing mitochondria with increase of matrical dense bodies (× 17,640).

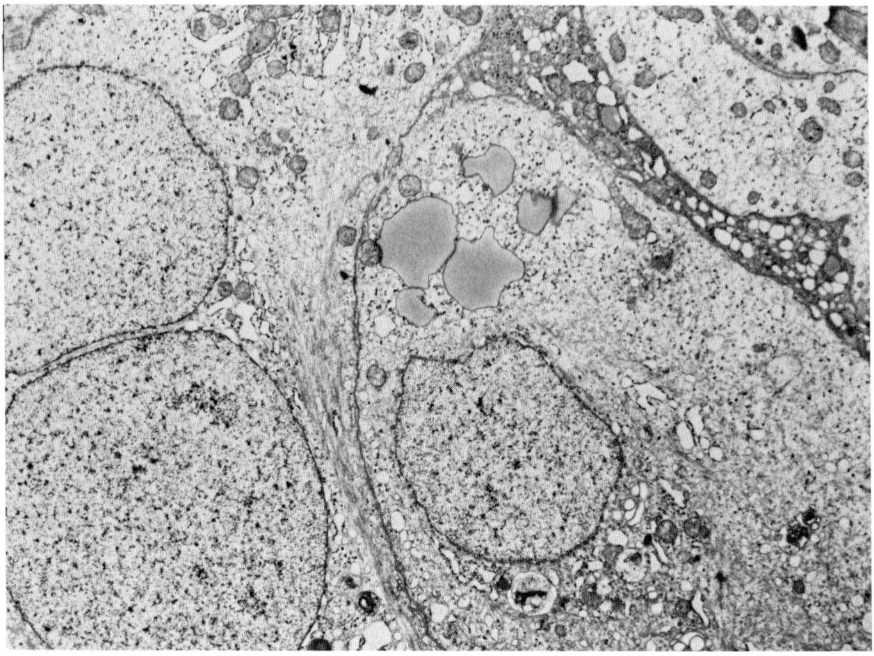

7.22 Electron micrograph of undifferentiated hepatocarcinoma showing lipid droplets (× 6,470).

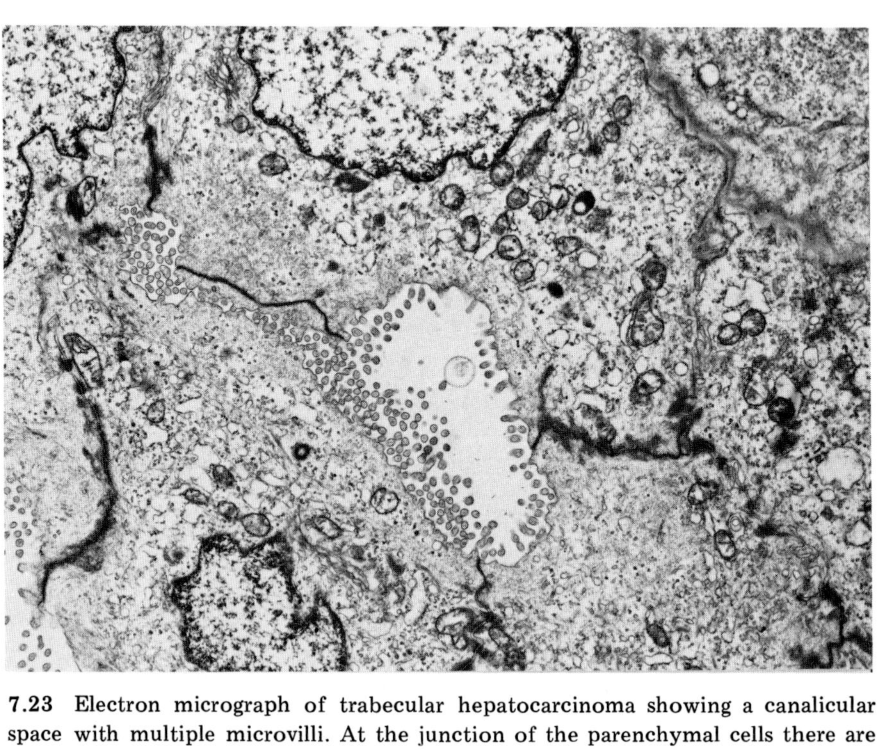

7.23 Electron micrograph of trabecular hepatocarcinoma showing a canalicular space with multiple microvilli. At the junction of the parenchymal cells there are prominent desmosomes (× 7,560).

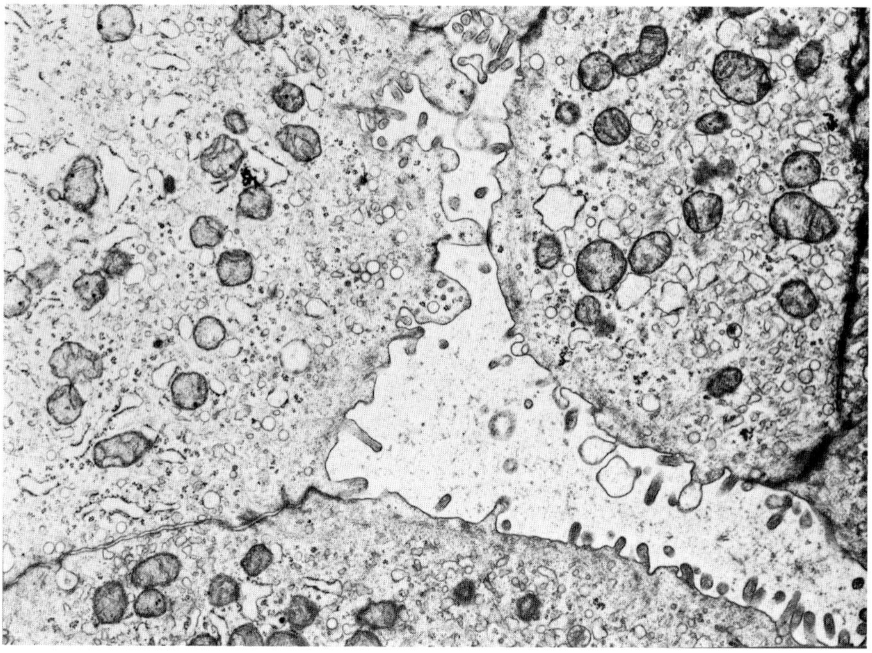

7.24 Electron micrograph of trabecular hepatocarcinoma showing dilated canalicular space bound by at least five cells showing sparse microvilli (× 10,500).

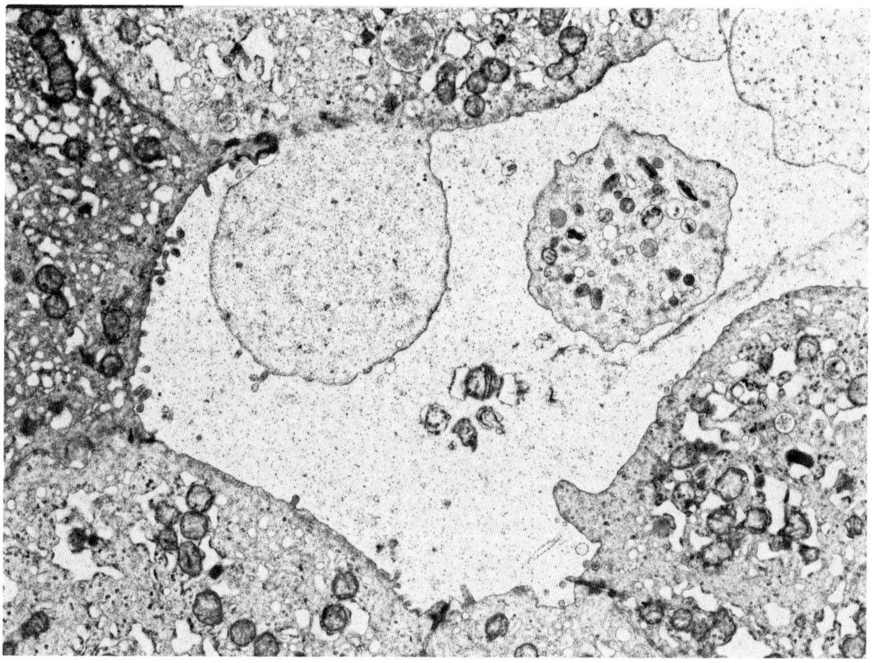

7.25 Electron micrograph of trabecular carcinoma showing large canalicular structure with cytoplasmic protrusions into the lumen (× 5,920).

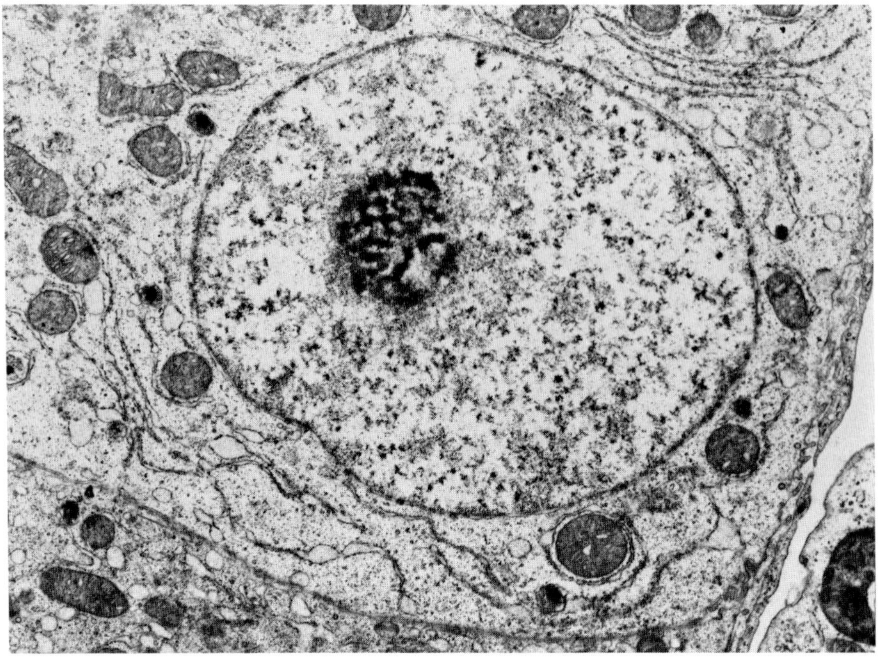

7.26 Electron micrograph of nucleus of parenchymal cell from trabecular carcinoma showing large nucleolus with prominent nucleolonema (× 8,450).

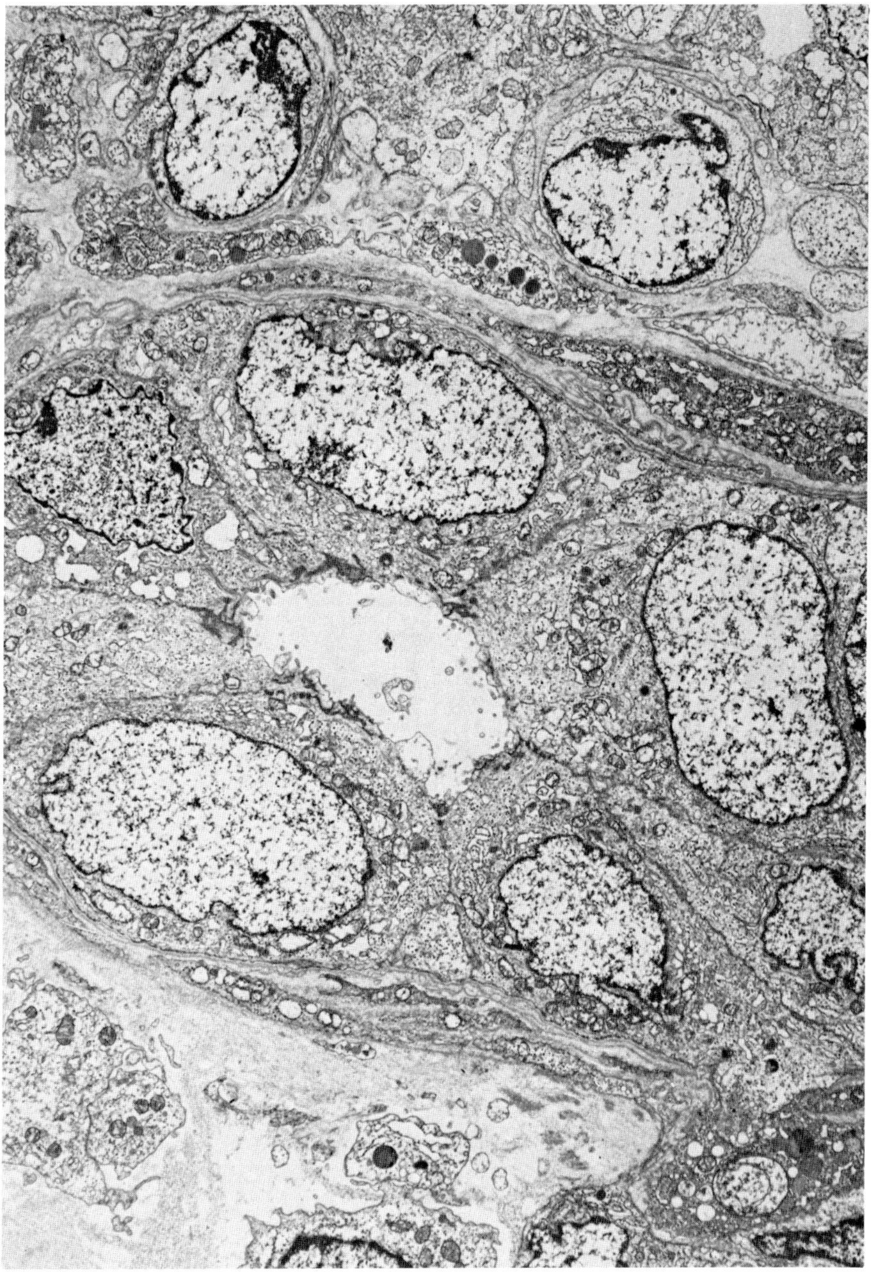

7.27 Low-power electron micrograph of hepatocarcinoma with a glandular pattern showing an irregular acinus lined by cuboidal cells (× 3,380).

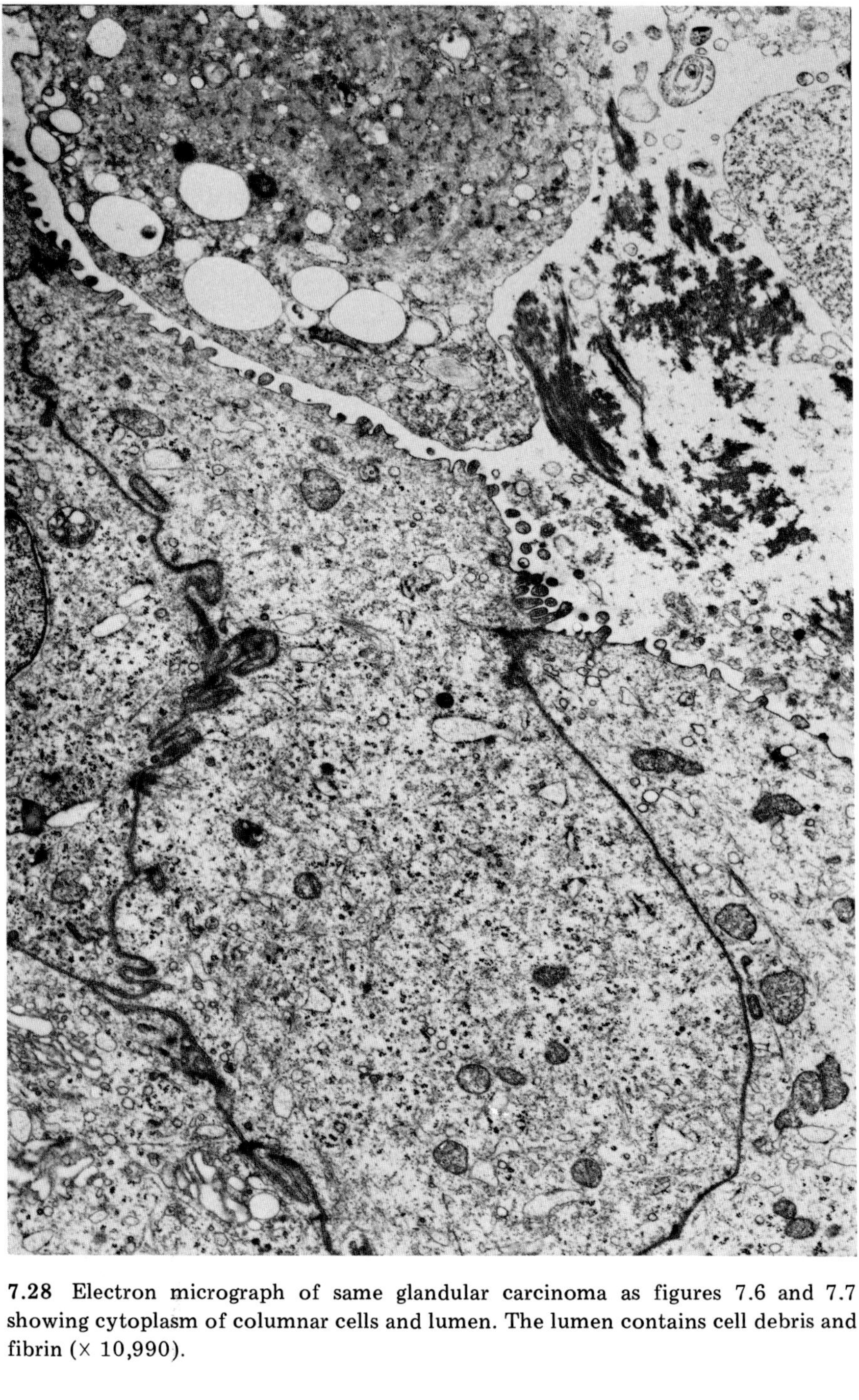

7.28 Electron micrograph of same glandular carcinoma as figures 7.6 and 7.7 showing cytoplasm of columnar cells and lumen. The lumen contains cell debris and fibrin (× 10,990).

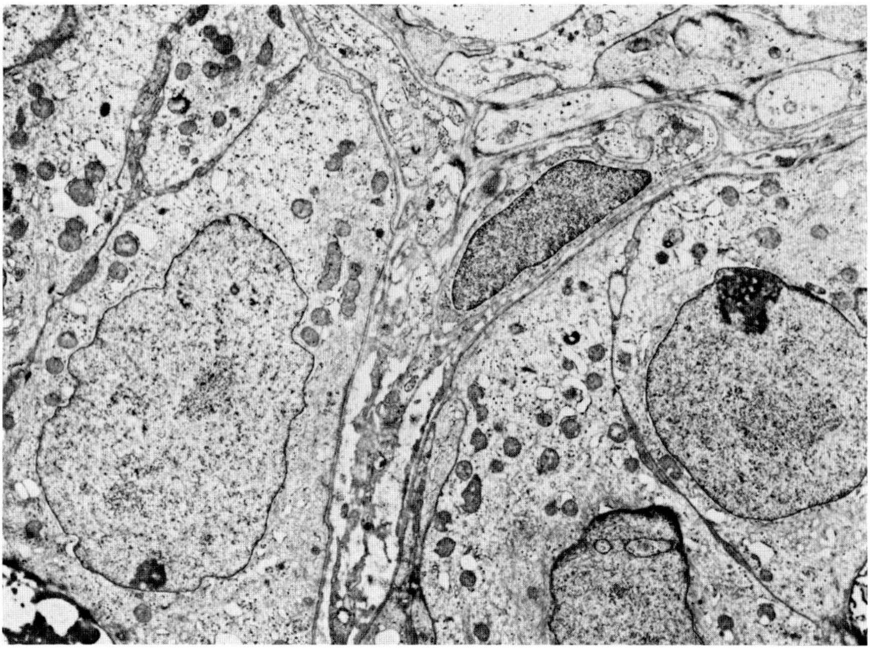

7.29 Electron micrograph of a glandular carcinoma showing vascular pole to be devoid of microvilli and bound by basement membrane (× 7,560).

tered between the columnar cells (figs. 7.6 and 7.7). At the ultrastructural level, these bodies are revealed as being composed of whole necrotic cells (fig. 7.30).

The plasma membrane abutting the basement membrane is devoid of microvilli (fig. 7.31), but where it is in contact with adjacent cells many interdigitations and a few desmosomes are present (fig. 7.28). No canalicular structures appear on this cell border. The luminal surface of the cell has a sparse population of microvilli (fig. 7.28). In both the columnar and cuboidal forms, cytoplasmic organelles are sparse. The rough endoplasmic reticulum is usually a loose network having many dispersed ribosomes (fig. 7.32). Smooth endoplasmic reticulum is seen only rarely, though large foci are present in a few cells (fig. 7.33). Golgi complexes are found in some cells (fig. 7.32). Mitochondria, although sparse, are usually round or oval, and microbodies are sometimes seen (fig. 7.32). In most cells bundles of fibrils (figs. 7.28 and 7.32), similar to those of the trabecular carcinomas are present. The nuclei are regular in size and shape within an individual tumor, and the nucleoli, when present, have prominent nucleonema like the ones seen in the trabecular carcinomas (figs. 7.27, 7.30, and 7.33). In the areas with flattened cuboidal epithelium, some nuclei have an irregular outline (fig. 7.34).

Surrounding the acini is a connective tissue stroma (fig. 7.6), within which are many spindle cells, some of which can be identified as fibroblasts containing variable amounts of collagen. It is also possible to recognize macrophages, occasional lymphocytes, vessels, and necrotic cells. However, many of the spindle cells are not readily identifiable. Basement membrane structures are present between many of the spindle cells (figs. 7.30 and 7.35).

It is not usual to find a neoplasm that is composed exclusively of either the trabecular or glandular form. Thus, cases of glandular carcinoma with an acinar structure are continuous with the trabecular form. The vascular pole in both instances is bound by basement membrane, while the canalicular margins limit a duct space of varying size and complexity (fig. 7.2). Necrotic cells are commonly found abutting directly onto the duct space, while debris lies within the duct.

7.3.3 Undifferentiated Carcinomas

Large, blood-filled spaces are present within the undifferentiated neoplasms. These spaces may be lined by normal endothelium or, as is more often the case, by neoplastic cells (fig. 7.36). The solid areas of these lesions may have the ultrastructural characteristics of the trabecular carcinomas described above, while others show considerable destruction and necrosis (fig. 7.37). Cells abutting directly onto the blood-filled space may be more elongated, but they retain many of the features of the trabecular lesions (fig. 7.38). Large spaces are seen to contain widely dispersed fibrillar and amorphous material. In some areas, this material is overlain with a single layer of cells lining a blood-

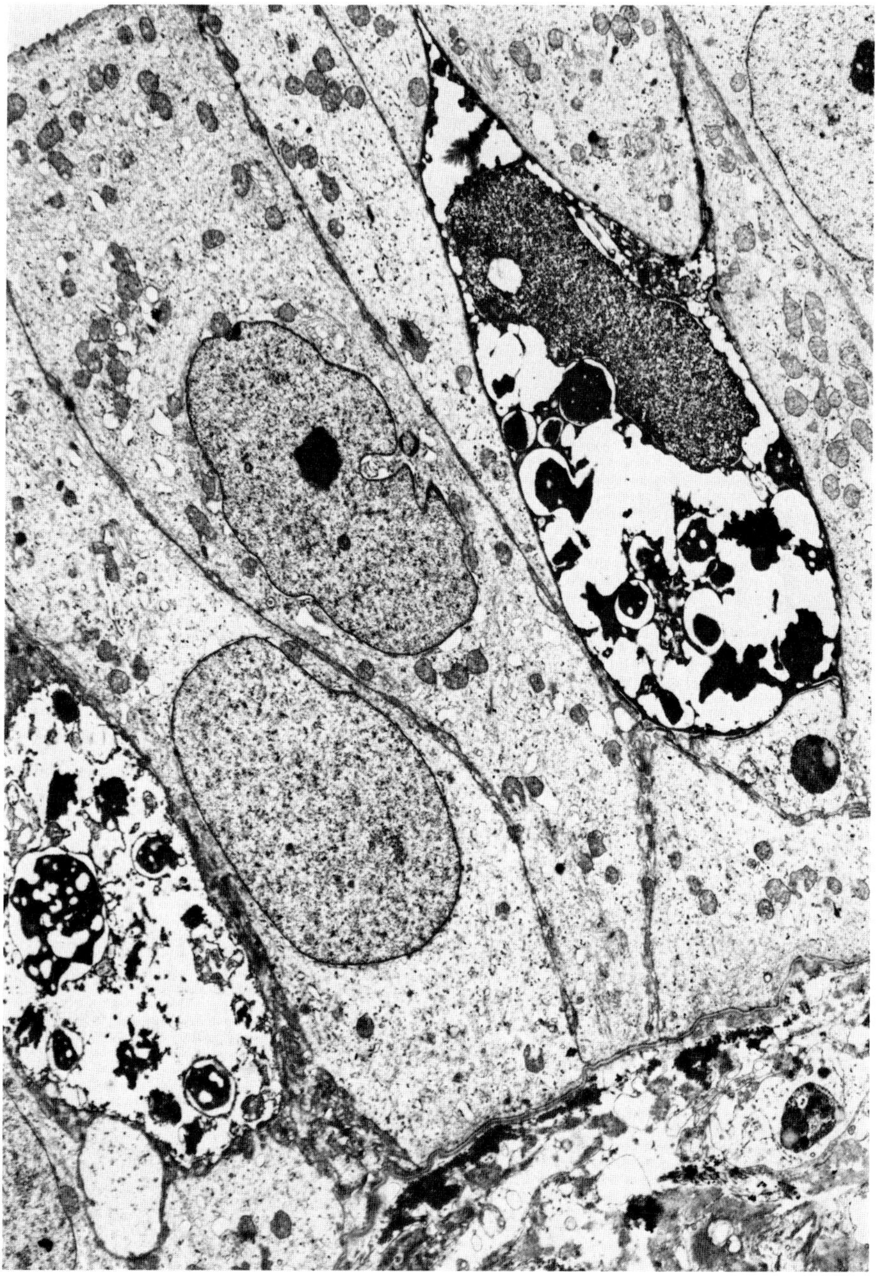

7.30 Low-power electron micrograph of the columnar cells lining the acinar space. The extremely electron-dense material, which corresponds to the PAS-positive material shown in figures 7.6 and 7.7, can be seen to be composed of necrotic whole cells (× 5,070).

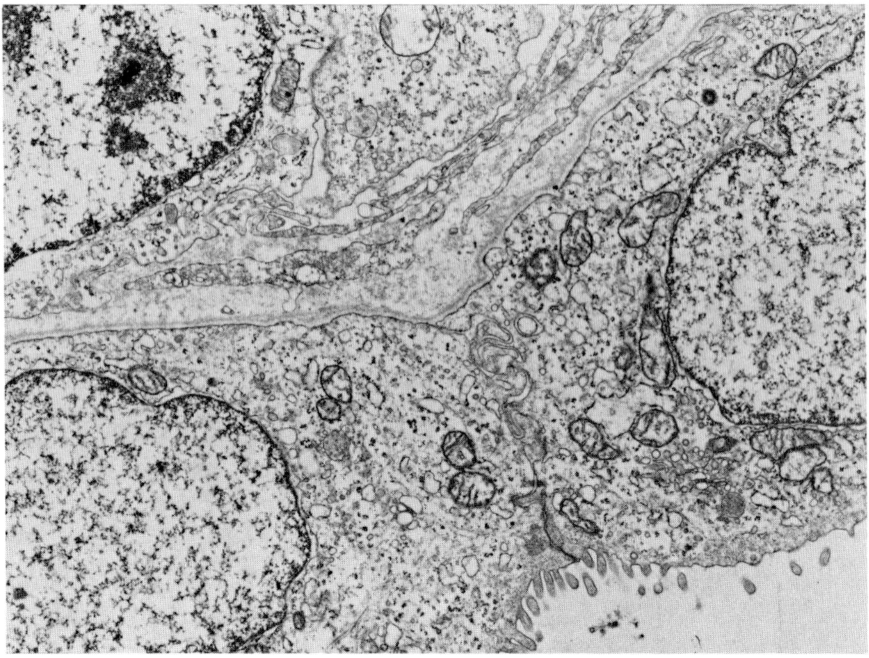

7.31 Electron micrograph of glandular carcinoma showing vascular pole to be devoid of microvilli and bound by basement membrane (× 5,040).

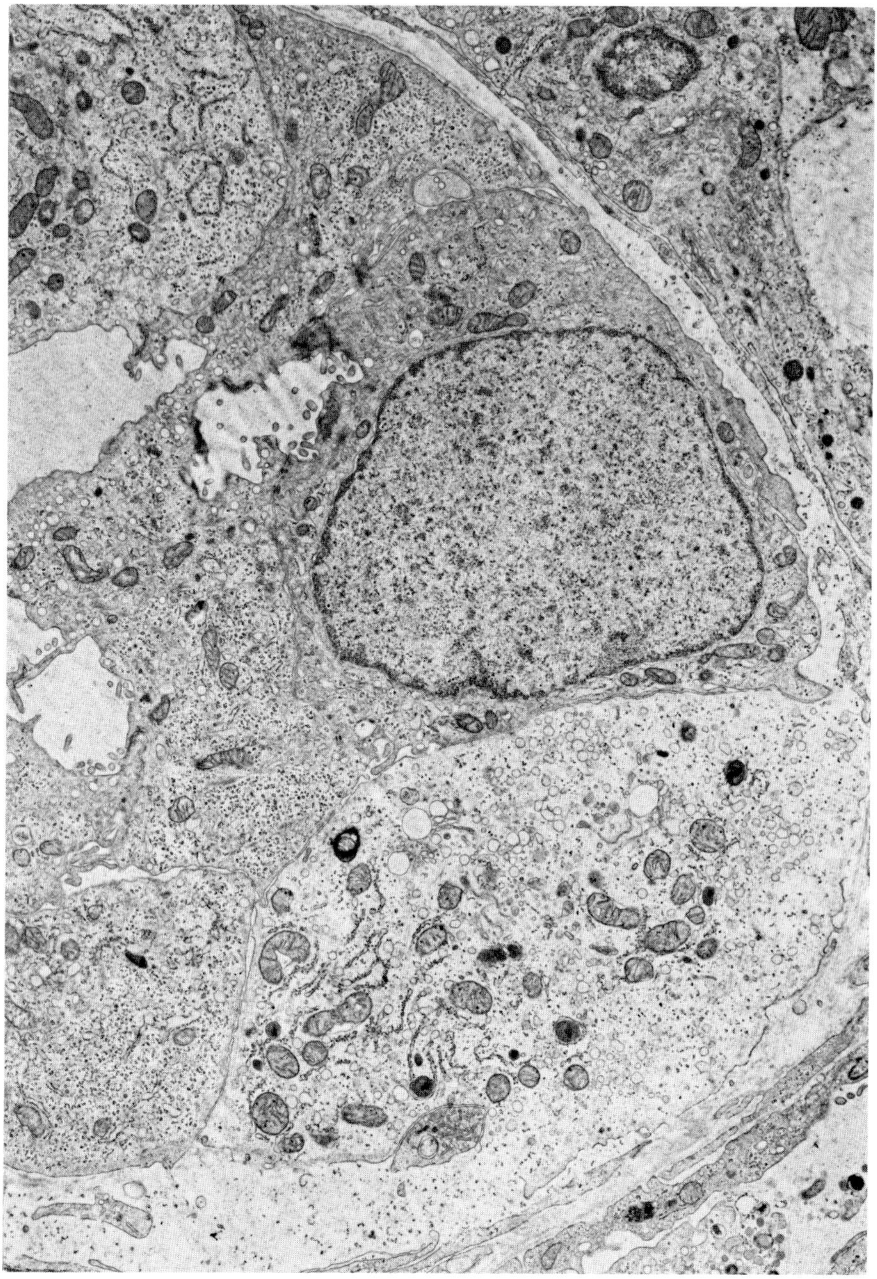

7.32 Low-power electron micrograph of an ascinus from glandular carcinoma showing only a loose network of rough endoplasmic reticulum and many dispersed ribosomes (× 5,460).

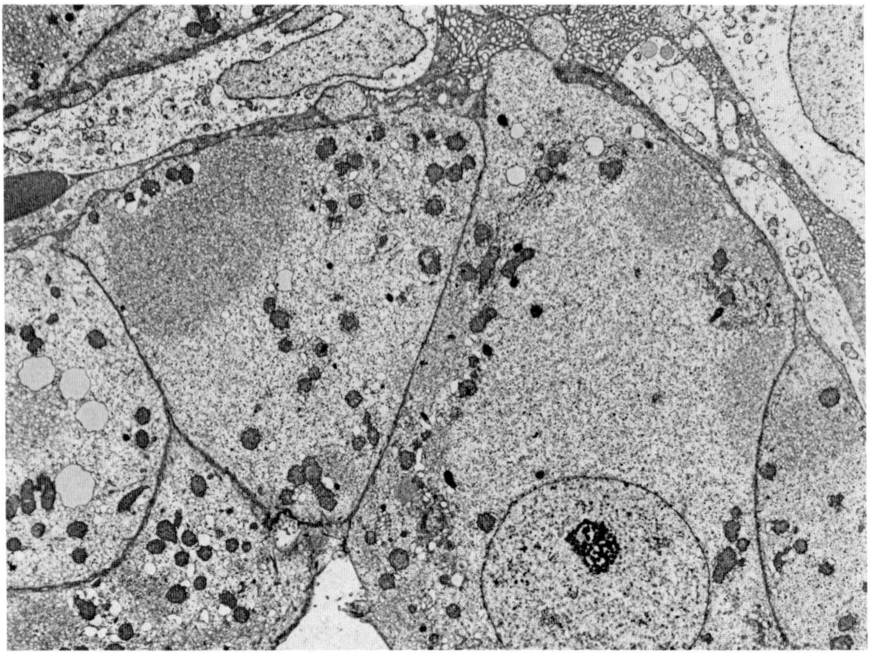

7.33 Low-power electron micrograph of glandular carcinoma showing epithelial cells with large foci of smooth endoplasmic reticulum (× 3,800).

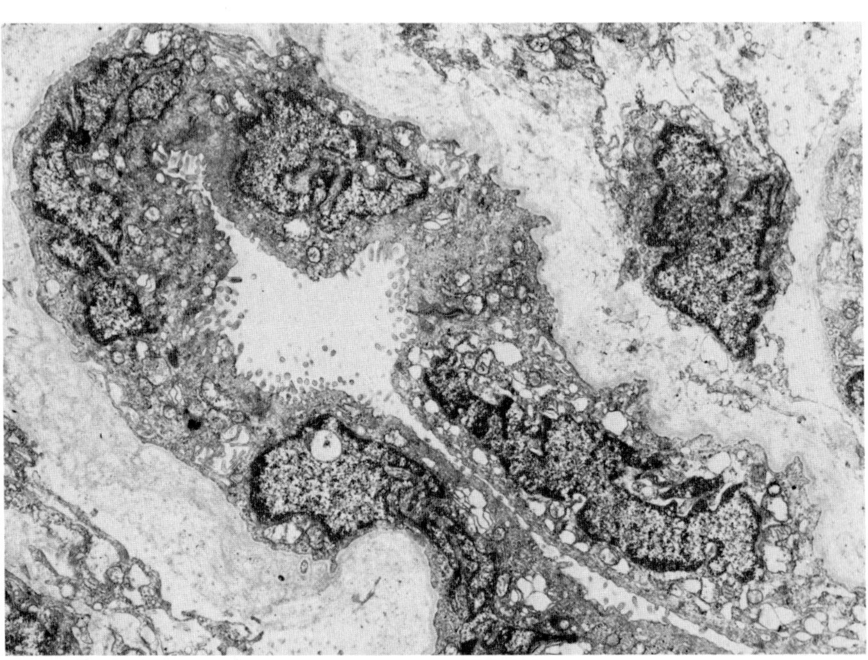

7.34 Low-power electron micrograph of an ascinus within glandular carcinoma showing the ascinus to be lined by cuboidal epithelium. The nuclei of these cells have very irregular outlines (× 3,360).

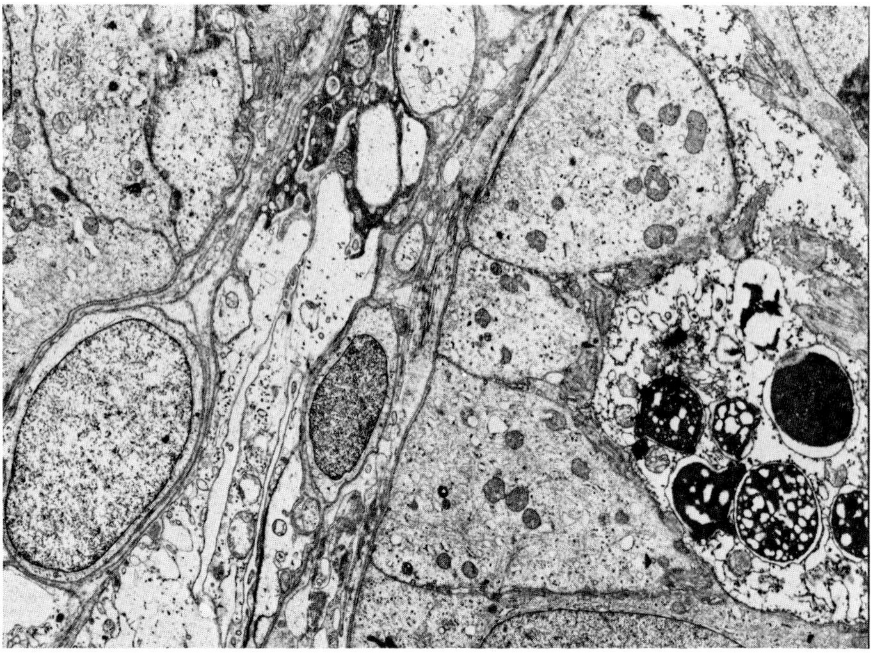

7.35 Electron micrograph of glandular carcinoma showing spindle cells in the stroma underlying the epithelium. Multiple basement-membrane structures are present between these spindle cells (× 3,890).

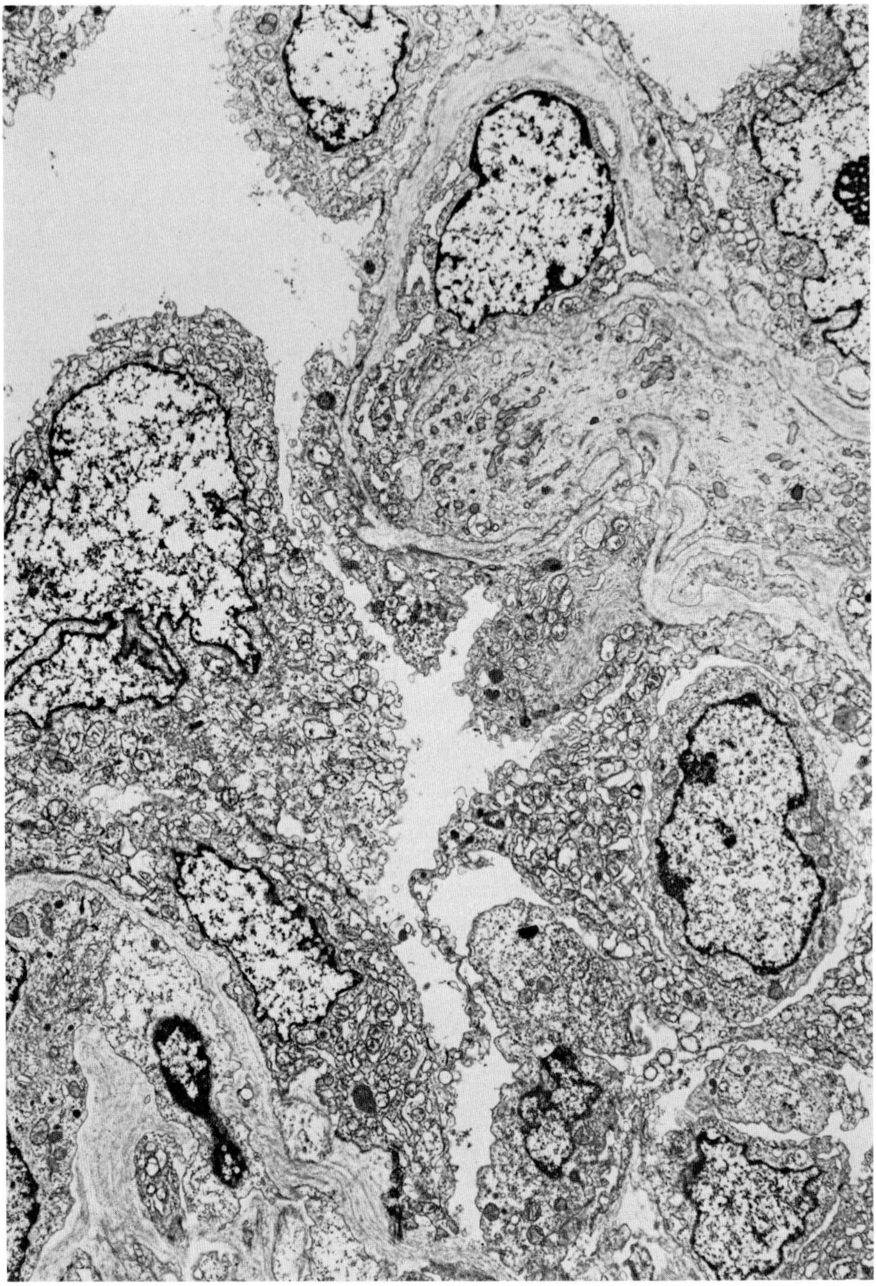

7.36 Low-power electron micrograph from undifferentiated carcinoma showing disorderly arrangement of cells lining the vascular space (× 4,230).

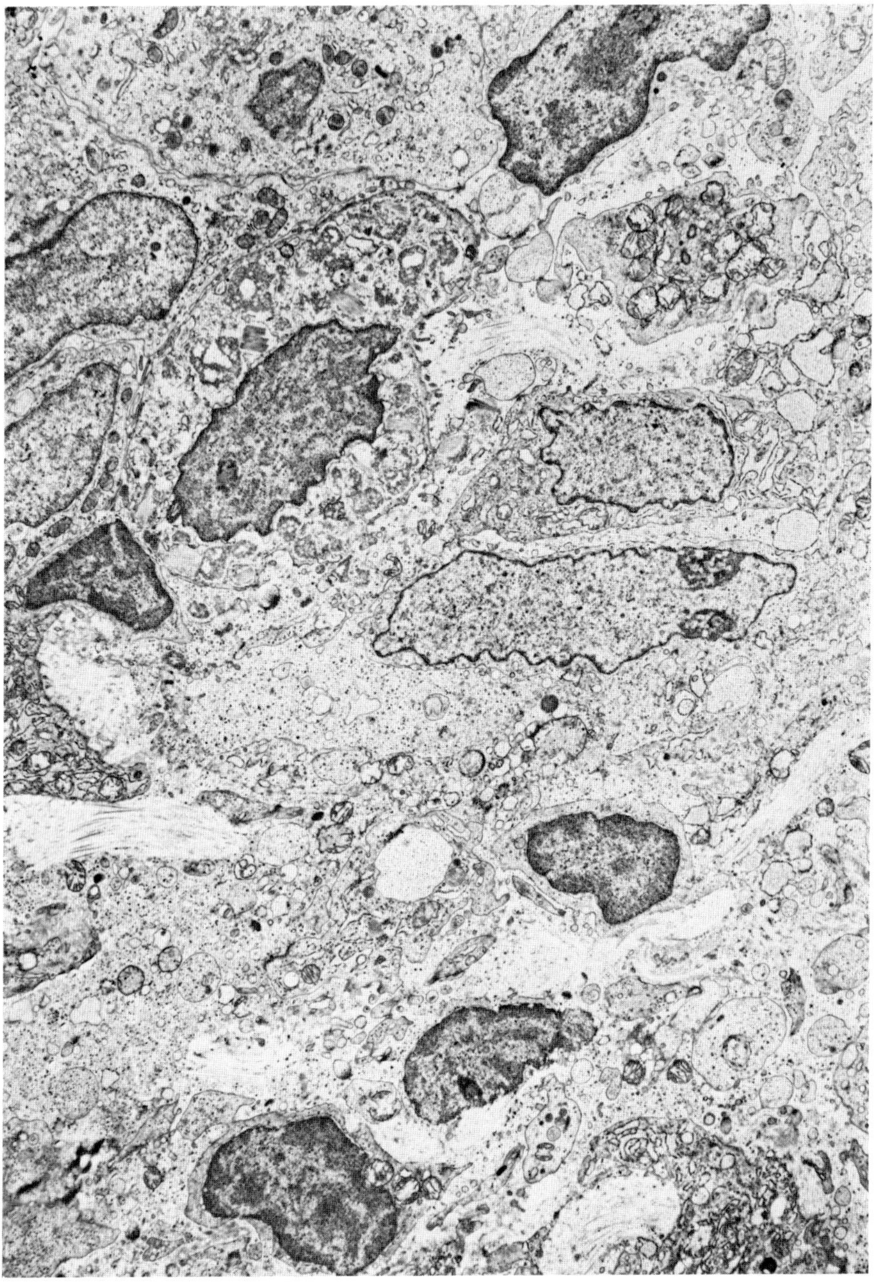

7.37 Low-power electron micrograph with undifferentiated hepatocarcinoma showing loss of organization and cell necrosis (× 3,800).

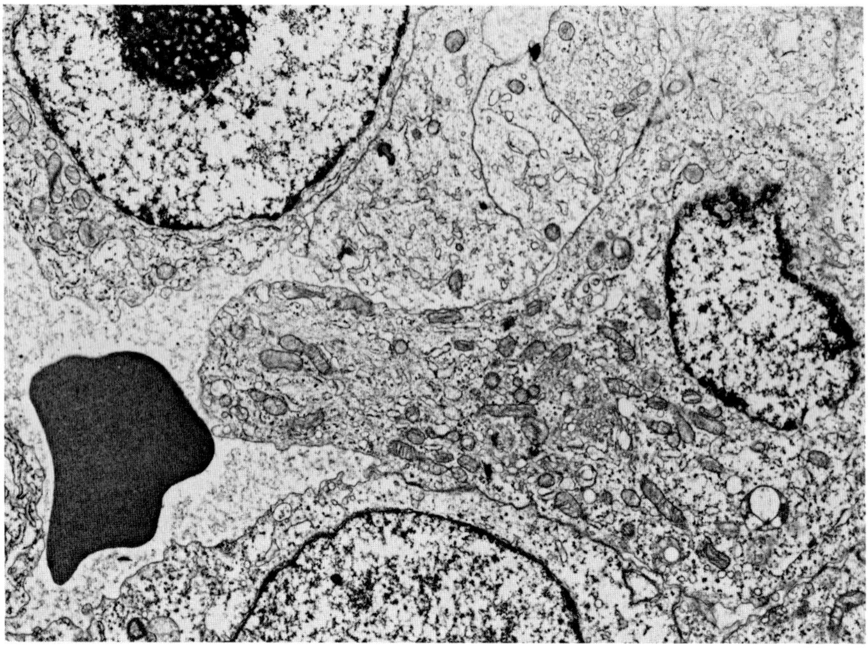

7.38 Electron micrograph of an undifferentiated hepatocarcinoma showing epithelial cells abutting directly onto a blood-filled space. These cells have the cytoplasmic characteristics of the trabecular carcinomas (× 3,380).

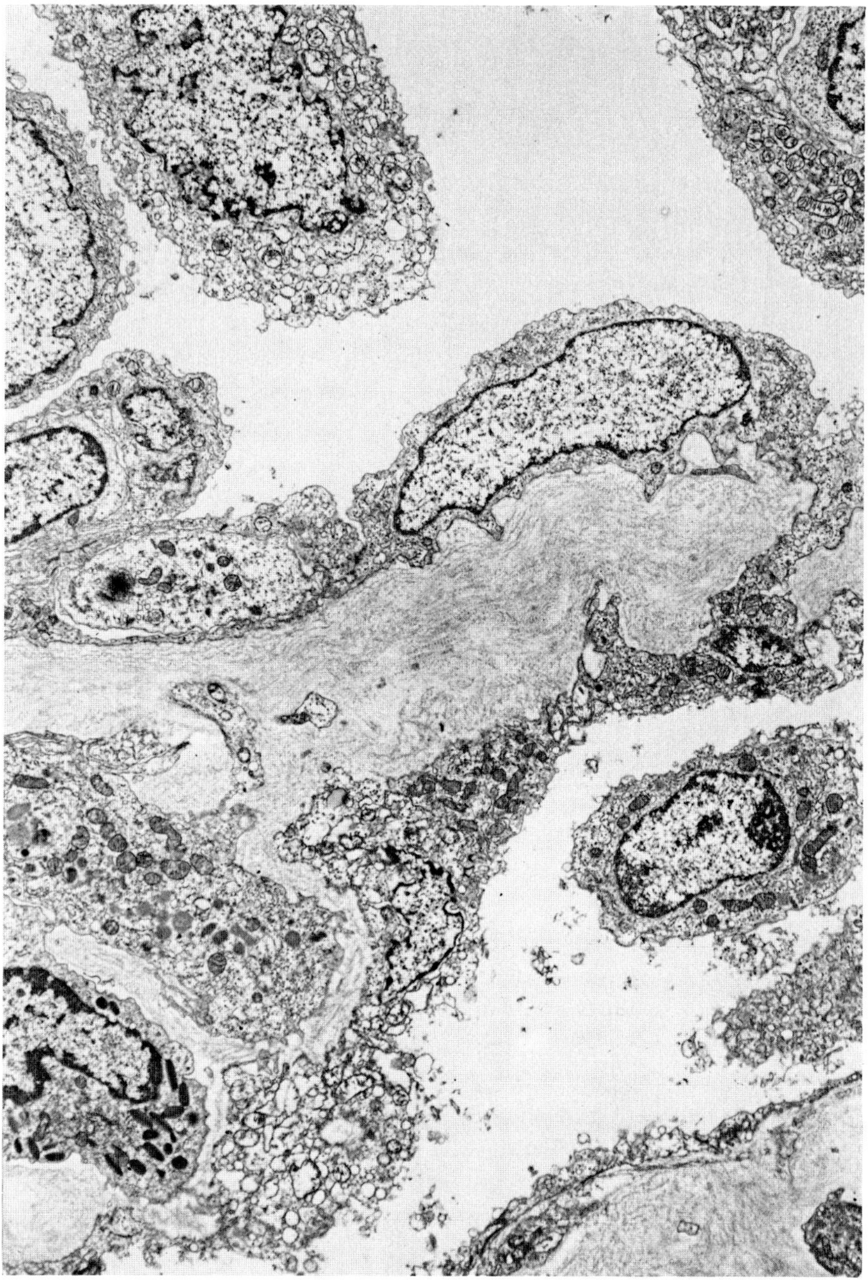

7.39 Low-power electron micrograph of a hemorrhagic undifferentiated hepatocarcinoma showing large blood spaces and abundant amorphous material overlaid by a single layer of cells. These cells are directly continuous and have the ultrastructural features of the remaining part of the neoplasm (× 3,380).

filled space (fig. 7.39). These cells do not have many of the characteristics of endothelium, such as fenestrations, but they may nevertheless represent an endothelial response to the fibrillar and amorphous material.

7.4 Conclusions

These observations suggest a lack of evidence, at the ultrastructural level, for the apparent pleomorphism of cell type within the tumors. The lesions that appear at a light microscope level to be trabecular hepatocellular carcinomas are revealed by electron microscope to be composed of cells retaining many of the features of normal hepatocytes. The overall structure of the liver cell plates is lost, and the vascular pole is always limited by basement membrane. This basement membrane appears to be associated with the endothelium or stromal cells, since multiple membranes may be seen between layers of the cells. Although relatively normal bile canaliculi are seen, some areas show greatly dilated canaliculi, which give rise to a tubular appearance (fig. 7.2 and 7.32). The mechanism of this dilation is uncertain but may be related either to the presence of individual cell necrosis and accumulation of debris or to the continued secretion of bile into blind-ending ducts. The degree of mesenchymal reaction may be related also to the individual cell necrosis. The continuity between the glandular and trabecular forms without any great difference in ultrastructure indicates that these cells are variants of the same neoplasm. We did not see scirrhous adenocarcinomas like those induced by DAB (Svoboda, 1964). In our studies of both aflatoxin- and nitrosamine-induced hepatic neoplasms, scirrhous adenocarcinomas have been uncommon. Whether they represent carcinoma of the intrahepatic biliary system is uncertain.

There do not appear to be many significant differences between the ultrastructures of hepatic carcinomas induced by various agents in the rat (Merkow et al., 1969 and 1972; Svoboda, 1964). Nuclear inclusions are somewhat less common in the aflatoxin-treated animals, but the large nucleoli with a very prominent nucleonema have been consistently found. The cytoplasmic changes of irregular strands of rough endoplasmic reticulum and dispersal of ribosomes are also consistently reported; however, in the present series, annulate lamellae were only rarely seen. The significance of this observation is not known. Foci of smooth endoplasmic reticulum and myelin fibers are another constant feature, as is the variation in mitochondrial morphology. The presence of the basement membrane lying between the endothelium and tumor cells as well as a simplification of the space of Disse are reported in trabecular carcinomas induced by DAB. However, this feature is not discussed in regard to neoplasms induced by ethionine or 2-FAA. Also, dialated canalicular structures, limited by desmosomes and bundles of cytoplasmic fibers, are present in the DAB trabecular carcinomas.

We have found evidence of one case in which similar tumors may arise from different cell populations. The hemorrhagic vascular neoplasm studied in this series is of obvious parenchymal cell origin. The neoplastic cells line the blood-filled spaces but retain many characteristics of hepatic parenchymal cells. The areas of fibrillar and amorphous material possibly represent precipitated serum proteins resulting from stasis, and the plump, endothelial cells may represent an attempt at recanalization. Although the rare, vascular tumors (hemangioendotheliosarcomas) described in the paper on light morphology have not been confirmed at an ultrastructural level as being derived from endothelium, Hadjilov and Markow (1973) have reported that the hemangioendotheliosarcomas induced by N-dimethylnitrosamine are composed of atypical endothelial cells, some of which have a structure similar to the angiosarcoma of the kidney, also induced by dimethylnitrosamine (Hard and Butler 1971). Gillman and Hallowes (1972a;b) have described the ultrastructure of another mesenchymal neoplasm, which is different from the hemangioendotheliosarcomas described by Hadjiolov and Markow (1973) and is believed to derive from Kupffer cells. It appears from these studies and also from studies of the sinusoid lining cells (Wisse, 1970 and 1974) that there are two distinct populations of cells that, on occasion, may give rise to hemangioendotheliosarcoma and Kupffer cell sarcomas as separate entities.

The ultrastructure of hepatocellular carcinoma in man (Creemers and Jadin, 1968; Ghadially and Parry, 1966; Ma and Blackburn, 1973; O'Conner et al., 1972; Theron, 1965; Toker and Trevino, 1966) does not differ from that in the rat. There is considerable variation between cells, but in both man and rat, most of the cytoplasmic organelles are reduced in number. The rough endoplasmic reticulum is irregularly arranged with dilated cisternae, but the dilation may be related to fixation. The smooth endoplasmic reticulum is sparse, and variable amounts of glycogen are present. Abnormal mitochondria are consistently reported. Bile canaliculi having a simplified microvillus border are present. The nuclei vary considerably in form, and many inclusions are reported. A striking difference from hepatic carcinomas of the rat is the presence of virus particles reported in some human cases (Ma and Blackburn, 1973; Creemers and Jadin, 1968).

The conclusions from ultrastructural studies are what would be expected. In the large majority of cases, induced hepatic carcinomas derive from hepatic parenchymal cells. The apparent glandular and anaplastic forms, like the trabecular form, are also of this origin and only rarely should be considered as derivative of the intrahepatic biliary or the vascular systems. This conclusion is supported by many studies of the sequential changes seen during development. Further, no great differences between carcinomas induced by aflatoxin, ethionine, 2-FAA, or DAB are seen. The ultrastructure of the rat hepatic carcinoma is very similar to that of the human carcinoma.

Although the ultrastructural organizations of neoplastic and normal hepatocytes differ greatly, it is not possible here or in studies of neoplasms of other sites to point to any single feature that identifies malignant neoplasia. At present, definitive diagnoses can still be based only on the biological behavior of the lesion.

7.5 References

Creemers, J., and Jadin, J. M. 1968. Ultrastructure of human hepatocellular carcinoma. *J. Microscopie* 7:257-264.

Ghadially, F. N., and Parry, E. W. 1966. Ultrastructure of a human hepatocellular carcinoma and surrounding non-neoplastic liver. *Cancer* 19:1989-2004.

Gillman, T., and Hallowes, R. C. 1972a. Ultrastructural and histochemical observations on a transplantable reticuloendothelial tumor in rats. *Cancer Res.* 32: 2383-2392.

Gillman, T., and Hallowes, R. C. 1972b. Ultrastructural changes in rat livers induced by repeated injections of trypan blue. *Cancer Res.* 32:2393-2399.

Hadjiolov, D., and Markow, D. 1973. Fine structure of hemangioendothelial sarcomas in the rat liver induced with n-nitrosodimenthylamine. *Arch. Geshwulstforsch* 42:120-126.

Hard, G. C., and Butler, W. H. 1971. Ultrastructural analysis of renal mesenchymal tumor induced in the rat by dimethylnitrosamine. *Cancer Res.* 31:348-365.

Jones, G., and Butler, W. H. 1974. A morphological study of the liver lesion induced by 2,3,7,8-tetrachorodibenzo-*p*-dioxin in rats. *J. Path.* 112:93-98.

Ma, M. H., and Blackburn, C. R. B. 1973. Fine structure of primary liver tumors and tumor bearing livers in man. *Cancer Res.* 33:1766-1774.

Merkow, L. P., Epstein, S. M., Farber, E., Pardo, M., and Bartus, B. 1969. Cellular analysis of liver carcinogenesis. III Comparison of the ultrastructure of hyperplastic liver nodules and hepatocellular carcinomas induced in rat liver by 2-fluorenylacetamide *J. Nat. Cancer Inst.* 43:33-63.

Merkow, L. P., Epstein, S. M., Slifkin, M., Farber, E., and Pardo, M. 1972. The cellular analysis of liver carcinogenesis. V. Ultrastructural alterations within hepatocellular carcinomas induced by ethionine. *Lab. Invest.* 26:300-305.

O'Conor, G. T., Tralka, T. S., Henson, E., and Vogel, C. L. 1972. Ultrastructural survey of primary liver cell carcinomas from Uganda. *J. Nat. Cancer Inst.* 48:587-603.

Svoboda, D. J. 1964. Fine structure of hepatomas induced in rats by *p*-dimethylaminoazobenzene. *J. Nat. Cancer Inst.* 33:315-339.

Theron, J. J. 1965. Fine structure of hepatic cancer. Hydrolytic enzymes and fine structure of Hurman hepatic cancer. In *Primary Hepatoma*, ed. W. J. Burdette, pp. 39-49. Salt Lake City: University of Utah Press.

Toker, C., and Trevino, N. 1966. Ultrastructure of human primary hepatic carcinoma. *Cancer* 19:1594-1606.

Wisse, E. 1970. An electron microscopic study of the fenestrated endothelial lining of rat liver sinusoids. *J. Ultrastruct. Res.* 31:125-150.

Wisse, E. 1974. Kupffer cell reactions in rat liver under various conditions as observed in the electron microscope. *J. Ultrastruct. Res.* 46:499-520.

7.6 Discussion

Bannasch What is your ultrastructural evidence for the statement that these glandular alterations, which are obviously not of bile duct origin, originate from swelling of hepatocytes?

Butler We have recently looked at these in semithin sections and at an ultrastructural level, and it is possible to see remnants of cytoplasm and parenchymal cell nuclei within the space. The space is lined by the plasma membrane, which may break, giving confluent areas which may communicate with the sinusoids. I therefore feel that these areas are the result of cell degeneration.

P. Newberne I think these may be lymphatic spaces. I am not sure that it makes any difference, but these may have been telangiectatic lesions. I do not consider this to be a compound response and see it in old, untreated rats.

Squires If you consider it a degenerative change, could it be a hemodynamic change, like cerebral edema, when it results in necrosis?

Butler I do not know the mechanism and do not know whether hemodynamic changes cause this type of swelling of liver cells.

Goyings I have seen this change as a compound response, and I think I have to agree with Dr. Butler that these are hepatic cells that have undergone some type of cell membrane change, and as a result, you get this tremendous change with the pushing of the cytoplasmic organelles and the nucleus off to the side. When they do rupture, you see red blood cells within the space.

Bannasch We have done some investigations on these lesions. We have injected India ink, which never gets into the holes. What we always can see is that these cysts are lined by mesenchymal cells which produce a very thick basement membrane material, which lines the whole cyst.

Squires Have electron microscope studies been done on angiosarcomas indicating their origin? Is there evidence from either endothelial or Kupffer cells?

Butler As far as I am aware, this is all done at the paraffin and light level. It is my impression that they were endothelial.

P. Newberne There is a difference in the morphological appearance of the vinyl chloride tumors and what we see in angiosarcomas induced with nitrosomorpholine and also in the kinds that you see with thorotrast, which I think is a really Kupffer cell tumor.

Dr. Merkow, what is your interpretation of the enormous increase in smooth endoplasmic reticulum in the cancer cell as opposed to other cells? Is there some functional difference?

Merkow I think, in general, the presence of increased smooth endoplasmic reticulum is a relatively nonspecific change. I think it is the result of degranulation of the rough endoplasmic reticulum, as well as a hyperplasia of the smooth endoplasmic reticulum. Studies show that the protein present in the early nodule was about the same as in normal liver. Possibly the presence of clustered smooth endoplasmic reticulum may have some relation to the presence of increased amounts of glycogen.

Laqueur What does Dr. Merkow think the significance is of the cytoplasmic filaments? Do you see similar structures in other malignant tumor cells of other organs?

Merkow I think that these are probably a product manufactured by malignant liver cells. I did not examine them for any periodicity to see what kind of fiber it is, but I would presume that it is some kind of fiber similar to collagen, possibly immature collagen. The liver cell has the potential to make this, but it does not do so until it becomes malignant. I have seen them in other malignant cells, but they are usually cells that have this capability of manufacturing collagen. We attempted to culture the hyperplastic liver nodule in vitro, and the electron microscopy of the in vitro nodular cells then showed these same fibers to an even greater degree.

Butler Fibrils are very common in these carcinomas, but they are also present in all the other liver cells. In normal liver they do not occur in bundles but are randomly distributed in the cytoplasm. They do not appear to have banding.

Dr. Becker talked about reversible and irreversible nodules. Dr. Merkow, are there any ultrastructural differences at completion of feeding compared with later lesions?

Merkow Dr. Becker admitted he could not distinguish histologically between the two types of nodules; they do disappear also in our experimental model using FAA and ethionine, but you know this afterward. If I may compare the hyperplastic liver nodule for a moment to the ovary during the first half of the menstrual cycle, many graaflian follicles mature, and at the time of ovulation only one and possibly two ovulate and all the others become atretic follicles. Similarly, this may be related to the one nodule that is on the way to cancer. I am sure that there are some rats that regress all of their hyperplastic liver nodules and do not get cancer. Our criterion here was more whether a nodule may become a malignancy, and if it does, how do you define the malignancy? We defined it by metastasis.

Grice If you saw a series of changes similar to the early nodules in an invasive but nonmetastasizing tumor, would you consider the lesion to be malignant?

Merkow Yes, I would. But I do not know how convincing that would be to the other people.

8 Hepatocellular Growth Control

H. L. Leffert
Cell Biology, Laboratory
The Salk Institute for Biological Studies
Post Office Box 1809
San Diego, California 92112

8.1 Introduction

My colleagues and I have been using a primary monolayer tissue culture system of rat hepatocytes to analyze hepatocellular growth control (Leffert and Paul, 1972). Three conclusions from these studies may prove to be of biological importance:

1. It appears that control of DNA synthesis and mitosis are independent processes that involve both stimulatory and inhibitory serum factors, including specific hormones, lipids, and lipoproteins.
2. Quiescent cells may consist of heterogeneous subpopulations in defined "states," resulting from differing equilibrium interactions between cell receptors and one or more components of *sets* of growth-controlling factors.
3. Some differentiated functions may depend on the state of growth.

This report will summarize current evidence supporting these conclusions, and will consider their possible bearing upon the control of hepatocellular growth, especially in regard to studies in vivo (Short et al., 1972 and 1973).

8.2 The In Vitro System

8.2.1 Plating Methods

The cultures are derived from term fetal rat liver that is dispersed enzymatically with a crude collagenase preparation. The resulting single-cell suspension is plated into plastic dishes, together with arginine-free medium supplemented with dialyzed fetal bovine serum. After about ten days, it forms a monolayer cell population with a heterogeneous morphology (fig. 8.1). These plating conditions appear to *select* against nonparenchymal cells and to enrich (Leffert and Paul, 1972 and 1973) the remaining population with arginine-synthesizing cells. Thus, the morphological heterogeneity may reflect different states of growth.

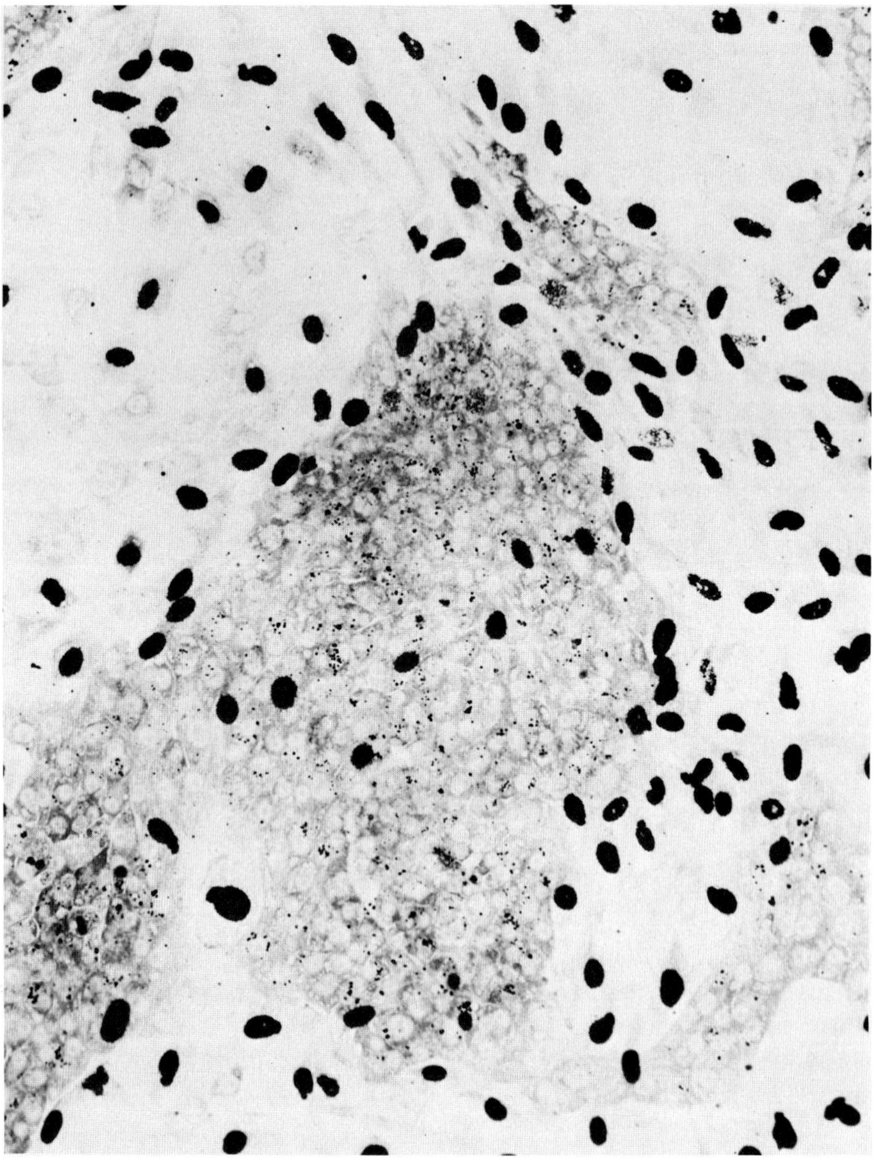

8.1 Photomicrograph (radioautograph) of cultured fetal rat liver cells 10 days after plating (× 62). (For details see Leffert and Paul, 1972, *J. Cell Biol.* 52:559–568.)

As shown in the photomicrograph of cultures labeled with ^{3}H-thymidine and then subjected to radioautography, DNA is synthesized by a greater fraction of the cells situated between and at the periphery of monolayer aggregates (Leffert and Paul, 1972). Direct evidence provided by time-lapse microcinematography also indicates that the majority of cells begin to divide only after they move away from a cellular aggregate. Evidence from other cell culture systems suggests that this phenomenon is not the result of breaking cell contacts (Lipton et al., 1971) but is perhaps the result of increased accessibility to growth factors in the medium.

8.2.2 Functional Characterization

Several differentiated functions have been examined. An intact urea cycle is present, as indicated by the cellular conversion of 3-^{3}H-L-ornithine to arginine (fig. 8.2). The cells were tested by incubation in the presence of ^{3}H-ornithine in arginine-deficient medium of twenty-four hours and by release into the culture medium of ^{14}C-urea synthesized from ^{14}C-guanido-L-arginine (unpublished results). At least four proteins specific to the liver are synthesized and secreted (Sell et al., 1975); they are albumin, haptoglobin, hemopexin, and alpha$_1$-fetoprotein (α_1F) (Leffert and Sell, 1974). Albumin and haptoglobin appear with first-order kinetics; the curves for α_1F and hemopexin are complex. Different curves of secreted protein also may reflect differences in the growth state of the population rather than in cell types. Although we have not demonstrated lipoprotein secretion, lipoprotein synthesis cannot be ruled out. Glycogenolytic and gluconeogenic capacities also may be present, as indicated by the observation that slowly growing cultures synthesize DNA in glucose-deficient medium (Leffert and Paul, 1972).

8.2.3 Growth Cycles and Conditioning

Growth cycles of hepatocytes plated without and with L-ornithine are shown in figures 8.3 and 8.4, respectively. The cells grow sooner, faster, and to higher densities in the presence of L-ornithine and increasing concentrations of serum (Leffert and Paul, 1973; Leffert, 1974a). But if the culture media, under both conditions, are removed within forty-eight hours of plating and are replaced with fresh media, the cells fail to grow. When conditioned medium is reintroduced, growth is restored (fig. 8.5). Evidently, freshly plated hepatocytes rapidly condition the medium by synthesizing and secreting low levels of arginine, and this process is stimulated by both the presence of exogenous ornithine (but not putrescine or uridine) and by higher initial inoculum densities (Koch and Leffert, 1974).

Arginine, therefore, at levels of 2–8 μM, is an absolute growth requirement, and the stimulatory effects of L-ornithine cannot be obtained in arginine-free medium (Koch and Leffert, 1974).

8.2 Assay for the transformation of ^{3}H-ornithine into ^{3}H-arginine in liver cells. The cells to be tested were incubated in the presence of ^{3}H-ornithine in arginine-deficient medium for 24 hours. Amino acid analysis was performed on hydrolysate and the radioactivity was determined by amino acid analysis. (For details see Leffert and Paul, 1972, *J. Cell Biol.* 52:559–568.)

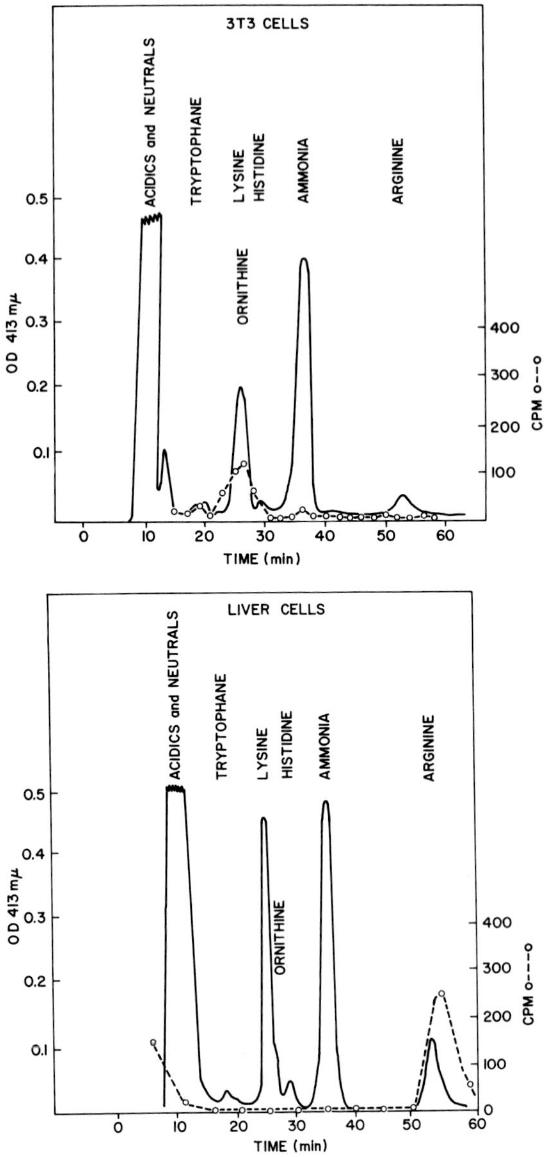

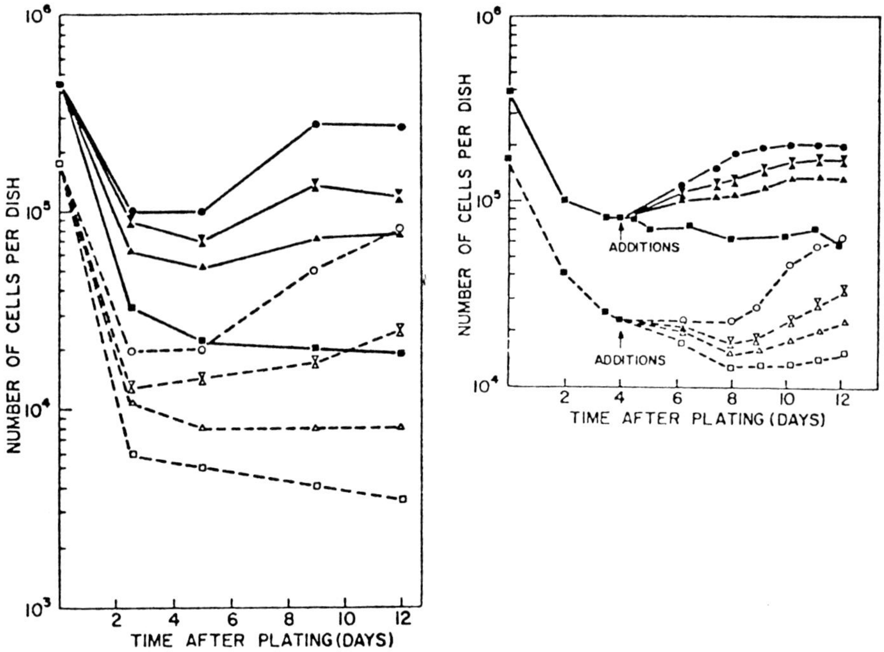

8.3 Serum-dependent multiplication and attachment or survival of fetal rat liver cells in arginine-deficient medium. (top) Cells plated at different concentrations of dialyzed serum. The cells were seeded at either 4×10^5 (solid lines) or 1.8×10^5 (dashed lines) cells per dish. For both high- and low-cell plating densities, the initial serum concentrations (v/v) were 0.1 percent. ▪ high-density plating; ▫ low-density plating. 1 percent (▴, ▵); 5 percent (⧗, ⧖); and 10 percent (•, ◦). (bottom) Cell multiplication assays in fetal rat liver cell cultures plated in 1 percent serum. The cells were seeded at either 4×10^5 (solid lines) or 1.8×10^5 (dashed lines) cells per dish. Initial serum concentration: 1 percent. At 4 days (arrows) dialyzed fetal bovine serum was added. Final serum concentrations: 2.5 percent (▴ high-density plating, ▵ low-density plating); 5 percent (⧗, ⧖); and 15 percent (•, ◦). No addition (▪, ▫).

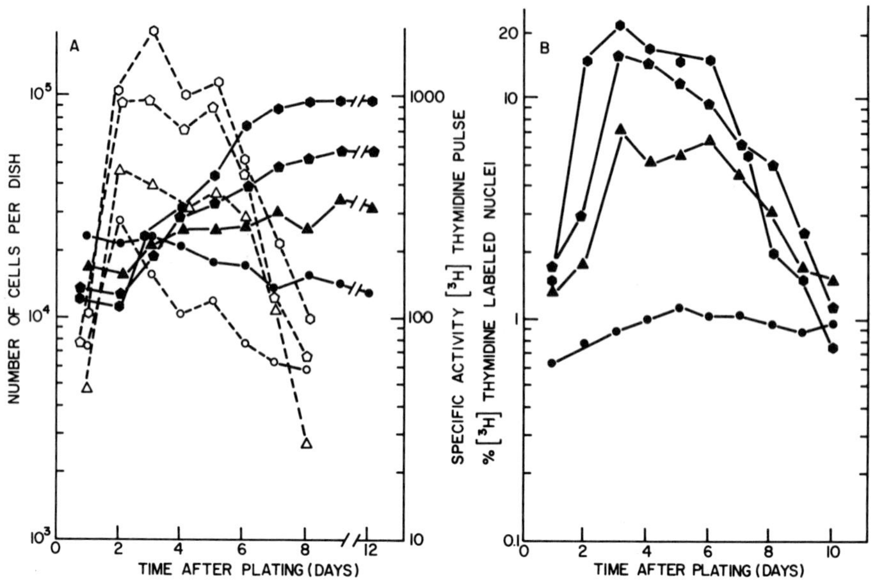

8.4 Growth cycle studies with fetal rat hepatocytes cultured in L-ornithine-supplemented (0.2 mM) medium. (a) Growth curves and DNA synthesis rate as a function of initial serum concentrations. Fetal rat hepatocytes were plated (2×10^5 cell/dish) with arginine-free L-ornithine-supplemented medium (2 ml/dish) containing the following initial quantities (percent vol/vol) of dialyzed fetal bovine serum: 0.0 (circles), 0.5 (triangles), 2.0 (pentagons), and 10.0 (hexagons). At various times after plating, cell multiplication (solid lines) was determined by counting the number of attached cells recovered by trypsinization. DNA synthesis rates (dashed lines) were determined by pulse labeling the cultures for 2 hours with ^{3}H-thymidine (^{3}H-dT) (1.25 μCi/ml, 3×10^{-6} M dT) and measuring the quantity of radioactivity incorporated into TCA-(5 percent vol/vol) insoluble material. Abscissa: time after plating (days). Left ordinate: number of cells per dish. Right ordinate: TCA-insoluble cpm per 10^5 cells per 2-hour pulse. (b) Proportion of DNA-synthesizing cells as a function of initial serum concentrations. Parallel cultures were established under conditions identical to those indicated above (fig. 8.4a). At various times after plating, the proportion of DNA-synthesizing cells per culture was determined by pulse labeling the culture for 2 hours with ^{3}H-dT (1.25 μCi/ml), after which autoradiography was performed using stripping film. One thousand cells per culture were scored. Abscissa: time after plating (days). Ordinate: percent labeled nuclei per culture. (For details, see Leffert, 1974, *J. Cell Biol.* 62:767–779.)

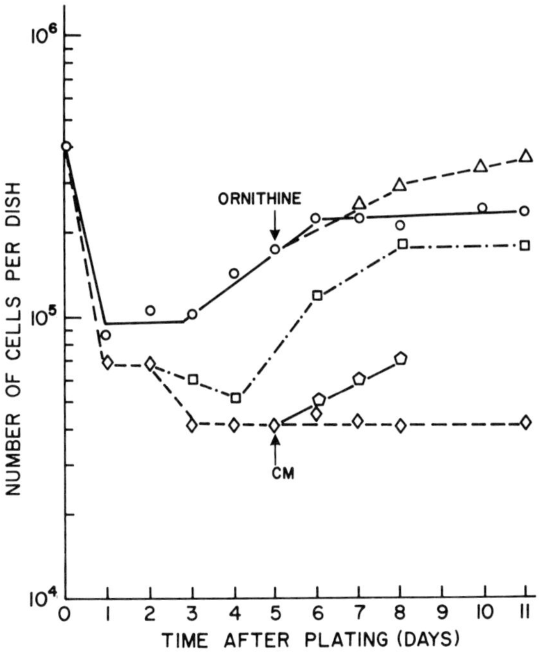

8.5 Obligatory growth requirement of conditioned medium by cultured fetal hepatocytes of rats. Hepatocytes were plated (4×10^5 cell/dish) into 2 ml argine-free medium containing 10 percent vol/vol dialyzed fetal bovine serum (dFBS). No further changes were made on one group of cultures (○ — ○); a similar group received 0.4 mM DL-ornithine 5 days postplating (△ --- △). Two other groups of cultures underwent a change of medium hours after plating and received 2 ml identical plating medium. At 48 hours, some of these cultures (□ -·- □) received 10 percent vol/vol dFBS-containing conditioned medium and the rest (◇ --- ◇) received fresh plating medium. Five days postplating, some cultures from this group were again subjected to a medium change to similar 10 percent vol/vol dFBS-conditioned medium (pentagons). Cell multiplication was determined by counting attached cells recovered by trypsinization. Abscissa: time after plating (days). Ordinate: number of cells per dish. (For details, see Koch and Leffert, 1974, *J. Cell Biol.* 62:780-791.)

8.2.4 Growth Assays

Three kinds of assays are used to study growth-promoting effects of serum factors. The simplest assay consists of plating cells together with serum fractions to be tested and then using routine procedures (Leffert, 1974a) to measure the numbers of attached cells, DNA synthesis rates, and percentages of DNA-synthesizing cells. Experiments of this type are shown in figures 8.4 and 8.6. For the second kind of assay, slowly growing cultures are set up, and serum fractions are added about four days after plating, as shown in figure 8.3 (bottom). Finally, one can study growth-promoting effects by using quiescent cultures generated either by plating cells into ornithine-free medium and waiting twelve to fourteen days (Leffert and Paul, 1973) or by a series of "step-down" treatments using serum-deficient, conditioned medium (Koch and Leffert, 1974). This type of assay is shown in figure 8.7.

8.2.5 Multiple Serum Factor Requirements in the Control of Hepatocellular Proliferation

Initial studies carried out with ammonium sulfate fractions of dialyzed fetal bovine serum suggested that fetal hepatocyte proliferation required not only the presence of a fraction that is unstable at pH 4–10 and is precipitable at 50 percent salt saturation but also the presence of an acid-stable, "survival" fraction soluble at 50 percent salt saturation (Leffert and Paul, 1973). These assays, however, were carried out under conditions where serum fractions were added to slowly growing cultures (fig. 8.3) in which a large proportion of the cells (30 to 40 percent) were already engaged in DNA synthesis. Subsequent results (Leffert, 1974a) indicate that the pH 4–10 stable material is required for stimulating the initiation of DNA synthesis [$G_{0,1} \rightarrow S$ transitions] (G_0 = zero growth; G_1 = first stage of growth; S = synthesis), whereas the pH-labile material is required for stimulating net increases in cell numbers [$S \rightarrow M$ transitions] (M – mitosis); the labile material may also stimulate DNA synthesis rates per se. The appropriate control experiments have been discussed in detail elsewhere (Leffert, 1974a).

These results have two important implications. First, they suggest that DNA synthesis (or events associated with it) is coupled with cell survival. Additional supporting evidence comes from observations in our laboratory that continuous exposure to 50 μg/ml cytosine-arabinoside, which inhibits DNA synthesis $\geqslant 95$ percent and protein synthesis $\leqslant 5$ percent, is cytotoxic toward these cells. Second, these results suggest that cellular events that initiate DNA synthesis depend on a different factor or set of factors than do events that permit continued progression through mitosis. A single dissociable complex has not been excluded (Leffert, 1974a).

The bulk of the growth-stimulating material present in the precipitated ammonium sulfate fraction was eluted from a Bio-Gel P200

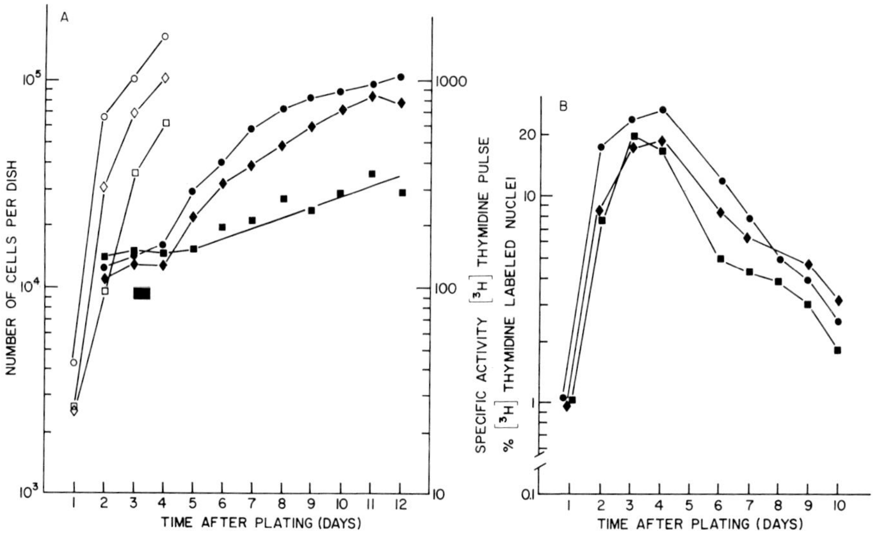

8.6 Growth cycle studies with fetal rat hepatocytes cultured in L-ornithine-supplemented (0.2 mM) medium together with ammonium sulfate fractions of dialyzed fetal bovine serum (dFBS). (a) Growth curves and DNA synthesis rates. Fetal rat hepatocytes were plated (2×10^5 cell/dish) together with argine-free, ornithine-supplemented medium (2 ml/dish) containing the following initial quantities of either dFBS (10 percent vol/vol [circles] or material precipitated by ammonium sulfate at 50 percent saturation equivalent to 10 percent vol/vol unfractionated serum (diamonds)) or material soluble in ammonium sulfate at 50 percent saturation equivalent to 10 percent vol/vol unfractionated serum (squares). At various times thereafter, cell multiplication (solid points) was determined by counting the numbers of attached cells recovered by trypsinization. DNA synthesis rates (open points) were determined by labeling the cultures for 2 hours with ^{3}H-thymidine (^{3}H-dT) (1.25 μCi/ml, 3×10^{-6} M dT) and measuring the quantity of radioactivity incorporated into TCA-(5 percent vol/vol) insoluble material. Abscissa: time after plating (days). Left ordinate: number of cells per dish. Right ordinate: TCA-insoluble cpm/10^5 cells per 2-hour pulse. (b) Proportion of DNA synthesizing cells. Parallel cultures were established under conditions identical to those indicated above (fig. 8.6). At various times after plating, the proportion of DNA-synthesizing cells per culture was determined by labeling the culture for 2 hours with ^{3}H-dT (1.25 μCi/ml) after which autoradiography was performed using stripping film. The percentage of labeled cells was based on a count of 1000 cells per culture. Abscissa: time after plating (days). Ordinate: percent labeled nuclei per culture (For details, see Leffert, 1974, *J. Cell Biol.* 62:767–779.)

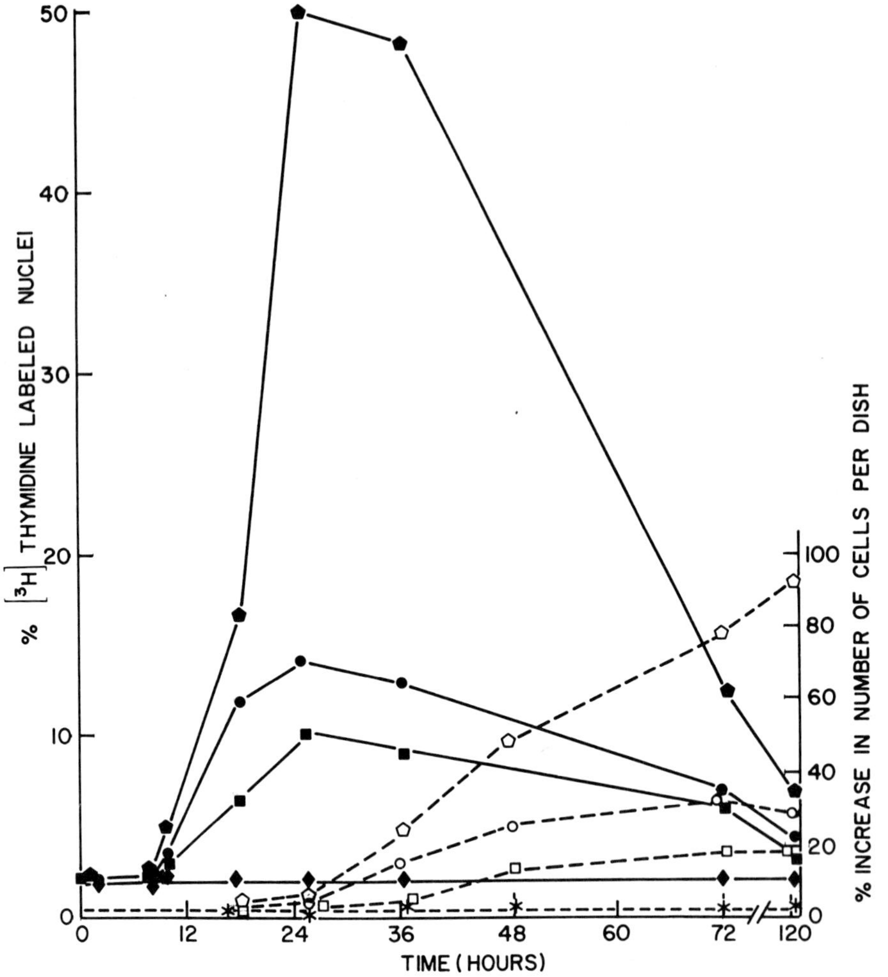

8.7 Stimulation of the initiation of DNA synthesis and cell division in quiescent fetal rat hepatocyte cultures by serum-free (0.00005 percent vol/vol) conditioned medium. At 216 hours postplating (time zero), some cultures (0.29×10^5 cell/dish) received an addition of 0.2 ml isotonic saline solution (*); others were left undisturbed (♦). Another group of cultures received a medium change to one of the following: 2 ml fresh basal medium ±10 percent vol/vol dialyzed fetal bovine serum (dFBS); 2 ml CM derived from 3T3 cultures (also ♦ or *); 2 ml 30 percent CM (▪, ▫); 2 ml 50 percent CM (•, ○); 2 ml 50 percent CM supplemented with 10 percent vol/vol dFBS (closed pentagons, open pentagons). At various times the percentage of DNA-synthesizing cells (solid lines) was determined by autoradiography; cell multiplication (dashed lines) was determined by counting attached cells recovered by trypsinization. Abscissa: time after additions (hours). Left ordinate: percent labeled nuclei. Right ordinate: percent increase in number of cells per dish. (For details, see Koch and Leffert, 1974, *J. Cell Biol.* 62:780–791, 1974.)

column with 0.05 M phosphate, pH 7.4. It had an apparent molecular weight of ⩾120,000 daltons (fig. 8.8) and is about fifteen-fold purified. This column also resolved growth-inhibitory material, about twelve-fold purified, that had a molecular weight between 40,000 and 80,000 daltons. Mixing experiments indicate that stimulatory (SF I) and inhibitory (SF II) fractions interact to determine the quantity of DNA synthesis and mitosis, hence the concept that a ratio (or balance) between positive and negative regulatory substances may control growth (Leffert, 1974a,b).

These types of interactions are not species-specific, for they also occur in studies with homologous rat serum. For example, in an attempt to extend our observations that nonpolar lipids also are involved with initiating DNA synthesis (Koch and Leffert, 1974), we prepared various lipoprotein fractions and tested them for ability to initiate DNA synthesis in quiescent cultures. The results were surprising, for we found instead that material present in the very low density lipoprotein (VLDL) fraction markedly inhibits the initiation of DNA synthesis (under noncytotoxic conditions). DNA synthesis was stimulated more markedly by delipidated rat serum (obtained from normal adult rats) than by native unfractionated serum but was suppressed upon readdition of the VLDL (fig. 8.9). Subsequent studies indicate that inhibitory activity resides mainly within the lipid portion of the VLDL fraction (Leffert and Weinstein, 1976).

Serum obtained from partially hepatectomized rats enhanced DNA synthesis (Paul et al., 1972). This serum also enhanced initiation of DNA synthesis in quiescent cultures; delipidated serum, however, was comparatively ineffective. Thus the low levels of VLDL found in serum of partially hepatectomized rats may be responsible for the serum's stimulatory effects (Leffert and Weinstein, 1976). Whether or not the inhibitory material present in SF I is in any way related to the inhibitory lipid(s) of VLDL remains to be determined. Preliminary studies with ammonium sulfate (fig. 8.10) and gel chromatography fractions of rat serum (Leffert and Weinstein, 1976) suggest that the inhibitory material differs from that found in fetal bovine serum in regard to molecular weight (25,000 to 100,000 daltons), although it too is heat and acid stable (boiling 1 minute; pH 4). Possible similarity between this material and other macromolecular peptide "mitogen"-containing fractions has not been excluded (Taylor et al., 1974; Van Wyk et al., 1971).

The acid stability of material initiating DNA synthesis in both rat and fetal bovine sera also suggested that hormones implicated in growth processes, such as insulin (Temin, 1967) and highly purified insulin-free somatomedin C (Van Wyk et al., 1971), might be active in this system. We found this to be true; in comparison to unfractionated serum, both hormones stimulated a small but significant percentage of cells to enter synthesis. A variety of pituitary hormones (thyroid-stimulating hor-

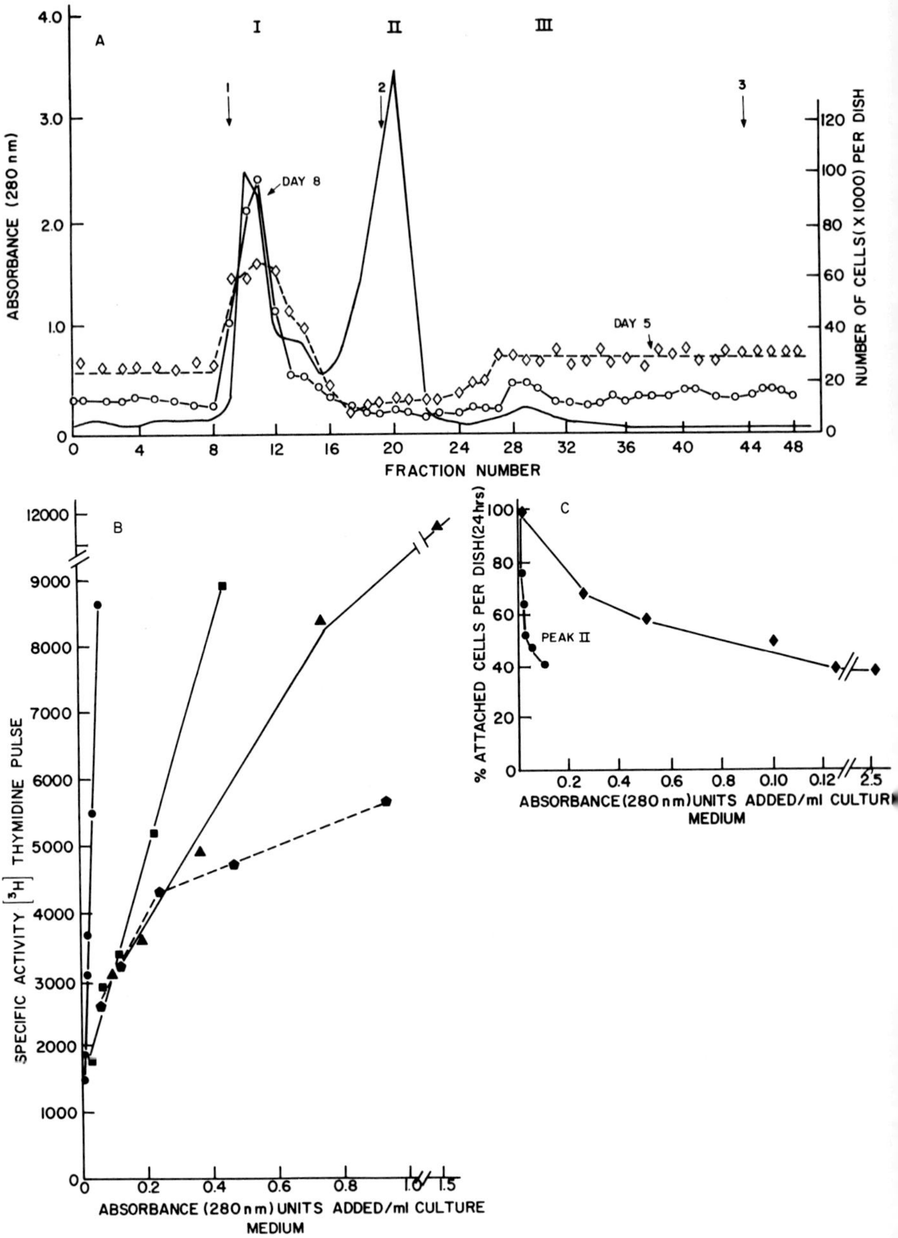
A
I
II
III
DAY 8
DAY 5
ABSORBANCE (280 nm)
NUMBER OF CELLS (X 1000) PER DISH
FRACTION NUMBER
B
SPECIFIC ACTIVITY [3H] THYMIDINE PULSE
ABSORBANCE (280 nm) UNITS ADDED/ml CULTURE MEDIUM
C
% ATTACHED CELLS PER DISH (24 hrs)
PEAK II
ABSORBANCE (280 nm) UNITS ADDED/ml CULTURE MEDIUM

8.8 Simultaneous occurrence of growth-stimulatory and growth-inhibitory material in the precipitate-derived ammonium sulfate fraction of dialyzed fetal bovine serum (dFBS). (a) Gel filtration and cell multiplication assays. Material precipitated by ammonium sulfate at 50 percent saturation (8 ml 127 A_{280} units) was layered onto a Bio-Gel P200 column (2 × 120 cm, 100–200 mesh) and eluted with 0.05 M phosphate buffer, pH 7.4. Absorbance at 280 nm of each fraction (6.2 ml) (left ordinate, continuous solid line) was monitored; arabic numbers 1, 2, and 3 represent, respectively, the following markers; blue dextran (mol wt ⩾ 2 × 10^6 daltons), bovine albumin (mol wt = 67,000 daltons), and phenol red (mol wt 354 = daltons), 1 ml of each fraction is equivalent to 2.5 ml serum. Abscissa: fraction number. Fetal rat hepatocytes were plated (2 × 10^5 cell/dish) into 2 ml arginine-free, L-ornithine-supplemented (0.2 mM) medium together with the fractions (0.13 ml/dish equivalent to 15 percent vol/vol serum). Cell multiplication was determined by counting attached cells recovered by trypsinization. Right ordinate: number of cells (× 1000) per dish (b) DNA synthesis rate: dose response curves. Fetal rat hepatocytes were plated (2 × 10^5 cell/dish) into 2 ml arginine-free, L-ornithine-supplemented (0.2 mM) medium together with the fractions (0.1 ml/dish) to be tested (● — ●, [peak I] pooled fractions 10–12 (see a); ■ — ■, ammonium sulfate fraction derived from material precipitable at 50 percent salt saturation; ▲ — ▲, dFBS; shaded pentagons, material soluble in ammonium sulfate at 50 percent saturation). Beforehand the fractions were adjusted by dilution with standard buffer so that their addition to each culture represented the following serum equivalents (percent, vol/vol): 0.32, 0.64, 1.25, 2.5, and 5.0. Control addition was standard buffer (0.1 ml/culture, "zero" absorbance). DNA synthesis rates were determined by labeling the cultures for 8 hours (64–72 hours after plating) with ^{3}H-thymidine (1.25 μCi/ml, 3 × 10^{-6} M) and measuring the radioactivity incorporated into TCA-insoluble material. The number of attached cells was also determined at this time. Abscissa: absorbance (280nm) units added permilliter culture medium. Ordinate: cpm radioactivity incorporated per 10^5 cells per 8-hour pulse. (c) Cytotoxicity of peak II: dose response. Fetal rat hepatocytes were plated into 2 ml arginine-free L-ornithine-supplemented (0.2 mM) medium together with the fractions (0.1 ml—0.2 ml/dish) to be tested (● — ●), fraction 21, (see a) and ♦ — ♦, dFBS. Beforehand the fractions were adjusted by dilution with buffer so that their addition to each culture represented the following serum equivalents (percent vol/vol): 0.63, 1.25, 2.50, 5.0, and 10.0. Control addition was standard buffer (0.1 ml/culture). Twenty-four hours after plating, the attached cells recovered by trypsinization were counted (100 percent = control = 22,000 cells per culture). Abscissa: absorbance (280 nm) units per milliliter culture medium. Ordinate: percent attached cells per dish, 24 hours postplating. (For details, see Leffert, 1974, *J. Cell Biol.* 62:767–779.)

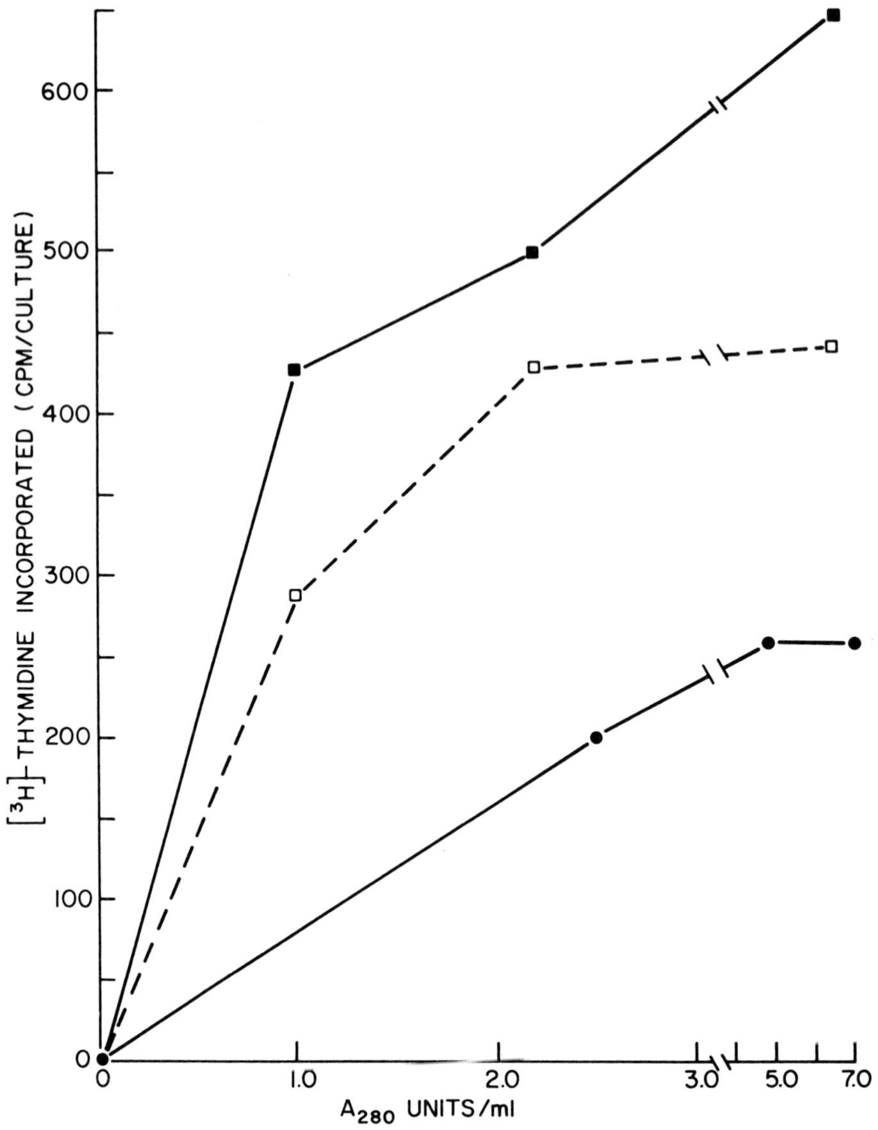

8.9 Stimulation and inhibition by delipidated rat serum and VLDL-reconstituted rat serum, respectively, of the initiation of DNA synthesis in quiescent fetal hepatocyte cultures. (Native, ● — ●; delipidated, ■ — ■; and VLDL-reconstituted rat serum, □ — □). (For details, see Leffert and Weinstein, 1976, *J. Cell Biol.* 70: 20–32.)

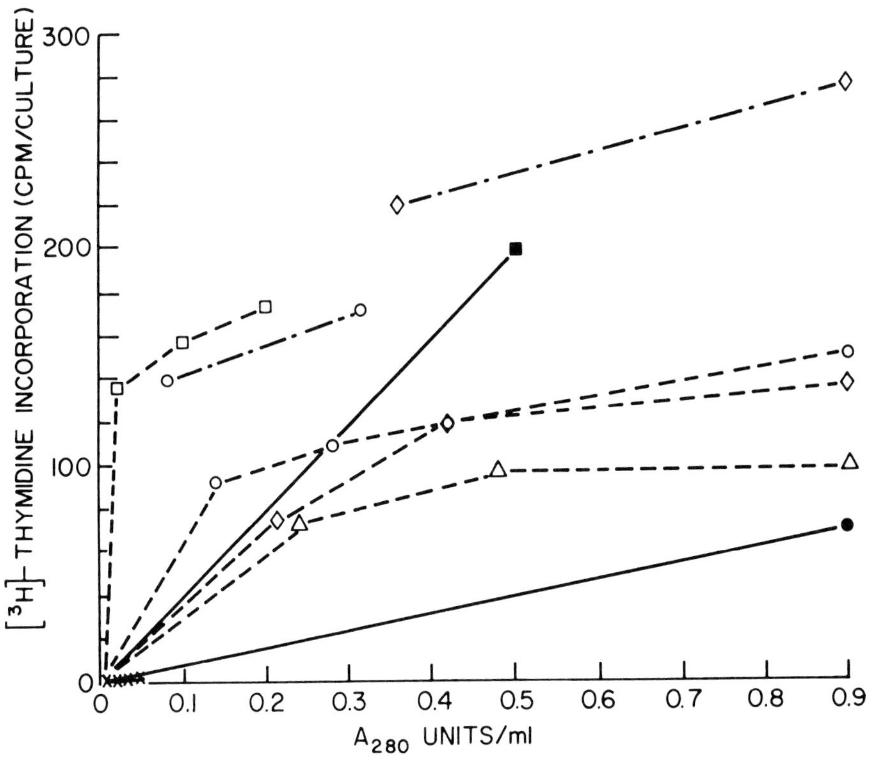

8.10 Stimulation of the initiation of DNA synthesis in quiescent fetal hepatocyte cultures by ammonium sulfate fractions of delipidated rat serum. (Native, ● —— ●; delipidated, ■ —— ■; 0–30 percent, X —— X; 30–50 percent, △ --- △; 50–70 percent, ◇ --- ◇; (unheated) and ◇-●-●-◇ (heated); 70–80 percent, ○ --- ○ (unheated) and ○-●-●-○ (heated); and ⩾ 80 percent, □ --- □.)

mone, luteinizing hormone, follicle-stimulating hormone, adrenocorticotrophic hormone, prolactin, growth hormone, and fibroblast growth factor) were not active when tested alone. This stimulation showed a "commitment" phenomenon (inset, fig. 8.11) although the kinetics differed from those of whole serum and suggested the involvement of additional factors. The commitment experiments with whole serum indicate that twelve hours of exposure is required for near-maximal DNA synthesis in the first wave of DNA synthesis; however, about 30 percent and 60 percent of the maximal level can be obtained with four and eight hours of serum exposure, respectively.

Of special interest, then, were observations that glucagon (fig. 8-12) and VLDL (fig. 8.13) inhibited the initiation of DNA synthesis. Furthermore, inhibition of DNA synthesis by material present in the VLDL fraction seems to occur only during an interval *prior* to DNA synthesis; cells that have completed these prereplicative events appear to be unaffected (fig. 8.13). These types of inhibition experiments are consistent with the commitment experiments in that the quantity of serum-depleted DNA synthesis obtained by a given exposure time equals the amount of DNA synthesis inhibited by an equivalent VLDL exposure time (Leffert and Weinstein, 1976). These experiments are difficult to do with insulin and glucagon because the quantity of stimulation is only 10 to 20 percent that of whole serum.

Commitment kinetics and the experiments shown in figure 8.14 suggest that many serum factors are required for initiation of DNA synthesis. This need can be demonstrated with hormones known to affect liver protein synthetic metabolism in vivo (John and Miller, 1969). Whereas somatotrophin preparations and hydroxycortisone failed to stimulate DNA synthesis (either alone or combined), both together potentiated the effects of insulin, permitting lower levels of insulin to stimulate (fig. 8.15). Similarly, thyroid hormones (T_3, T_4) failed to stimulate DNA synthesis, except when added together with fetal bovine serum depleted of thyroid hormone (Leffert, 1974b).

Because of evidence that diminishes the possibility that these cultures are a mixture of qualitatively heterogeneous cell types (Leffert and Paul, 1973), the observations of multiple-factor requirements and "quantized" responses (fig. 8.15) for initiation of DNA synthesis in quiescent cultures are puzzling; one explanation is that G_0 (and/or early G_1) heterogeneity exists. Under these culture conditions, insulin and/or insulinlike factors would be most limiting because they alone stimulate some DNA synthesis. But their responding cells, therefore, must not have been limited by the additional factors and must be different from the nonresponding cells. The additional factors would be postulated to affect metabolism in such a way as to "prepare" the cells for responding to insulinlike factors. Thus, it appears that there exists a *set* of factors ($\{f_1, f_2, \ldots, f_n\} G_{0,1} \rightarrow S$) that when present in excess, stimulate maximally the rates of entrance into S by recruiting the entire

8.11 Initiation of DNA synthesis by insulin in chemically defined medium. (a) Time course. Fresh medium containing insulin (♦ — ♦, 0.01 μg/ml; • — •, 10.0 μg/ml), 10 percent dialyzed fetal bovine serum (dFBS) (□ — □), or 0.2 ml isotonic NaCl (hexagons) was added to quiescent cultures (0.5×10^5 cell/dish). At various times, ^{3}H-thymidine (^{3}H-dT) (1.25 μCi/ml, 3×10^{-6} M dT) was added, and after a 2-hour pulse (points shown), the incorporation of TCA-precipitable radioactivity determined. Abscissa: time (hours) after fresh medium change. Ordinate: counts per minute ^{3}H-dT incorporated per culture per 2-hour pulse. (b, inset) Insulin or serum exposure periods required for stimulation of DNA synthesis. Fresh medium containing insulin (10 μg/ml) or 10 percent vol/vol dFBS was added to quiescent cultures (at time zero, cpm/culture/2 hour pulse = 80). At 4, 8, or 12 hours thereafter, the dishes were washed twice with insulin-free and serum-free medium, and media were replaced. Control dishes received continuous exposure. All cultures were pulse-labeled and assayed as in (a). Absolute incorporation rates (100 percent maximal response) for dishes receiving continuous exposure were: insulin, 240 cpm; 10 percent serum, 1640 cpm. Abscissa: replacement time (hours), that is, time at which dishes were washed and replaced with fresh insulin-free and serum-free medium. Ordinate: percent of maximal response. (For details, see Leffert, 1974, *J. Cell Biol.* 62:792–801.)

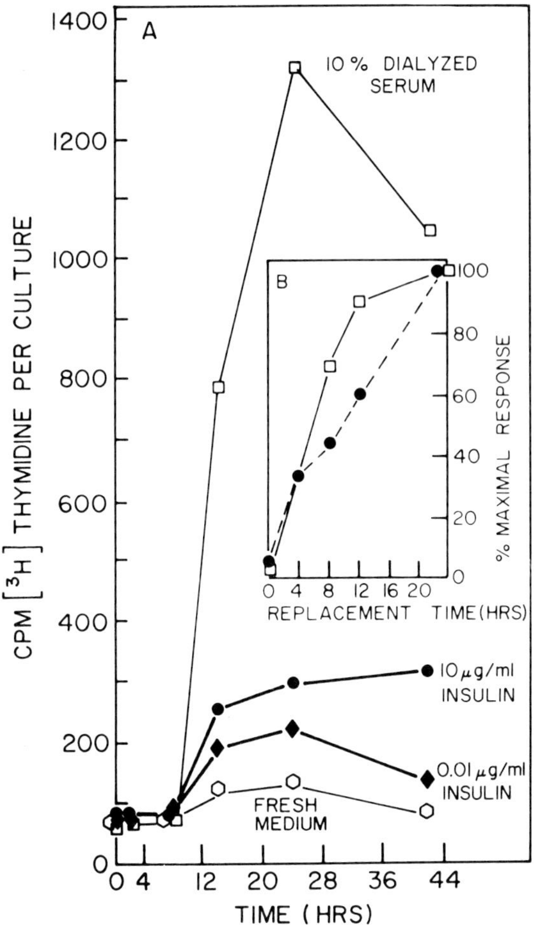

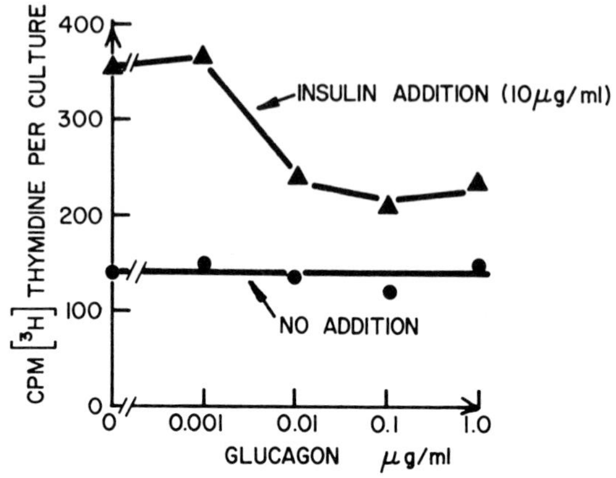

8.12 Inhibition of insulin-initiated DNA synthesis by glucagon. Fresh medium containing increasing concentrations of glucagon (0–1.0 μg/ml) with (▲) or without (●) insulin (10 μg/ml) was added to quiescent cultures (0.57 × 10^5 cell/dish). All cultures were pulse-labeled with ^{3}H-thymidine (^{3}H-dT) (1.25 μCi/ml, 3 × 10^{-6} M dT) for 22–24 hours and the radioactivity incorporated into TCA-precipitable material per culture was determined. Abscissa: glucagon added to culture medium (μg/ml). Ordinate: cpm ^{3}H-dT incorporated per culture per 2-hour pulse. (For details, see Leffert, 1974, *J. Cell Biol.* 62:792–801.)

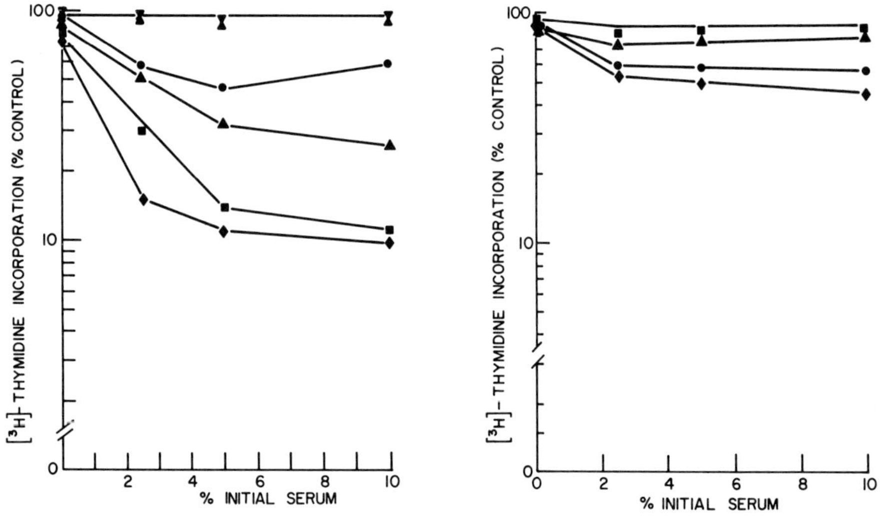

8.13 VLDL exposure times required to inhibit the initiation of DNA synthesis. (left) Exposure (time zero) followed by washout at 22–24 hours, ⧗ — ⧗; 0–4 hours, ● — ●; 0–8 hours, ▲ — ▲; 0–12 hours, ■ — ■; and 0–24 hours, ♦ — ♦. (right) Additions made at 12 hours, ■ — ■; 8 hours, ▲ — ▲; 4 hours, ● — ●; and 0–24 hours, ♦ — ♦. (For details, see Leffert and Weinstein, 1976, *J. Cell Biol.* 70:20–32.

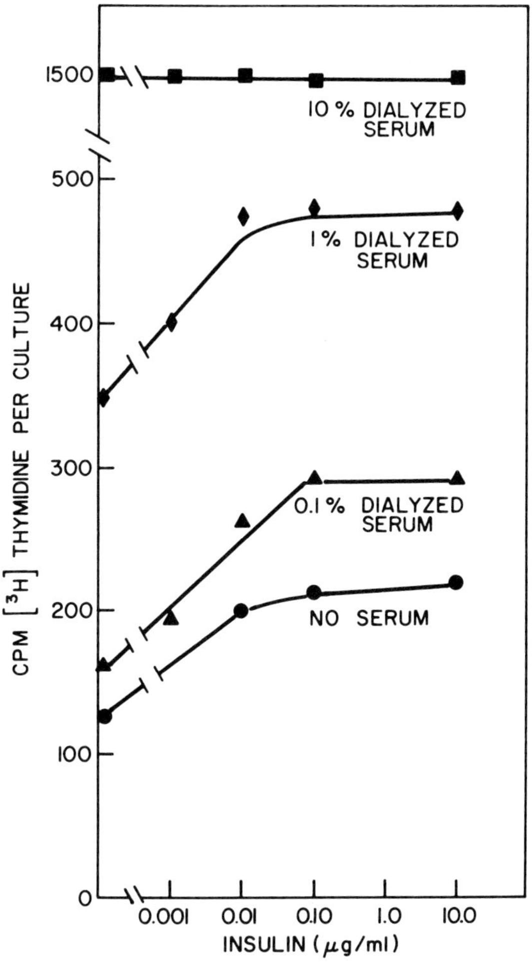

8.14 Serum enhancement of insulin-initiated DNA synthesis. Fresh medium containing varying concentrations of insulin (0–10 μg/ml) and supplemented with either isotonic saline (● — ●, 0.2 ml/dish) or dialyzed fetal bovine serum (▲ — ▲, 0.1 percent, vol/vol containing 1–2 pg immunoreactive insulin; ♦ — ♦, 1.0 percent, vol/vol containing 10–20 pg immunoreactive insulin; ■ — ■, 10 percent vol/vol containing 300 pg immunoreactive insulin) was added to quiescent cultures (0.53×10^5 cell/dish). All cultures were pulse-labeled with ^{3}H-thymidine (^{3}H-dT) (1.25 μCi/ml, 3×10^{-6} M dT) between 22 and 24 hours and the radioactivity incorporated into TCA-precipitable material per culture was determined. Abscissa: insulin added to culture medium (μg/ml). Ordinate: cpm ^{3}H-dT culture. For details, see Leffert, 1974, *J. Cell Biol.* 62:792–801.)

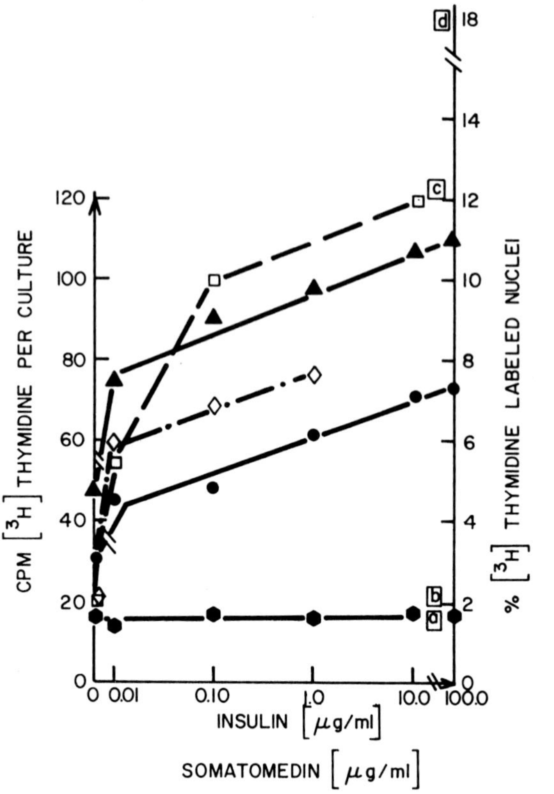

8.15 Quantitative initiation of DNA synthesis by insulin and somatomedin; enhancement by growth hormone preparations and hydroxycortisone and by L-arginine. Fresh medium containing 0.4 mM L-arginine was added to quiescent cultures (0.29×10^5 cell/dish) together with insulin (□ — □) or somatomedin (◇-•-•-◇) at the concentrations (μg/ml) indicated by the abscissa. Eighteen to 24 hours later, these cultures were incubated for 6 hours with ^{3}H-thymidine (^{3}H-dT) (1.25 μCi/ml), and the proportion of DNA-synthesizing cells was determined by autoradiography (1000 cell/dish scored as indicated by the right ordinate). Similar experiments were conducted with the following additions: growth hormone preparation from the NIH Endocrine Study Section (0.025 μg/ml) and hydroxycortisone-succinate (0.025 μg/ml), [a]; growth hormone (0.025 μg/ml) or hydrocortisone-succinate (0.025 μg/ml), [b]; growth hormone (0.025 μg/ml, hydroxycortisone-succinate (0.025 μg/ml, and insulin (0.05 μg/ml), [c]; growth hormone (0.025 μg/ml), hydroxycortisone - succinate (0.025 μg/ml), and insulin (1.0 μg/ml), [d]; and 10 percent vol/vol dialyzed fetal bovine serum (not shown, = 60 percent. In separate experiments (solid lines), insulin of various concentrations (abscissa) was added to quiescent cultures along with fresh medium containing no L-arginine (solid hexagons) or L-arginine at 0.004 mM (• — •) or 0.4 mM (▲ — ▲). These cultures were pulse-labeled with ^{3}H-dT (1.25 μCi/ml, 3×10^{-6} dT) for 22–24 hours, and the radioactivity incorporated in TCA-precipitable material per culture was determined (left ordinate). (For details, see Leffert, 1974, *J. Cell Biol.* 62:792–801.)

compliment of responding subpopulations. Quantitative and qualitative similarities between such a hypothetical set with a second *set* ($F_1, F_2, \ldots, F_n$ $S \rightarrow M$) required to stimulate progression through M remains to be determined, although qualitative differences are expected.

8.2.6 Mechanisms of Growth Factors

Our studies indicate that factors that initiate DNA synthesis do so by stimulating utilization rates of growth-limiting nutrients such as the amino acid L-arginine. This relationship is shown in figure 8.16, where it may be seen that both SF I and insulin stimulate the incorporation of 3-^{3}H-L-arginine into the trichloroacetic acid (TCA)-soluble and hot acid-precipitable cell fractions. Evidence presented elsewhere shows that exogenous serum and arginine interact to determine the fraction of cells entering S or the quiescent state (Koch and Leffert, 1974).

Studies with cultured fibroblasts (Griffiths, 1972; Holley, 1972; Everhart and Prescott, 1972) have led to the suggestion that growth processes are controlled at the level of transport systems controlling intracellular levels of limiting nutrients (Holley, 1972). Thus far, the evidence in vitro argues against this hypothesis; glucagon and VLDL, both of which inhibit initiation of DNA synthesis, do not inhibit incorporation of 3-^{3}H-L-arginine into the acid soluble fraction but do inhibit its incorporation into protein. These observations suggest that processes that affect the *disposition*, as opposed to the transportation, of exogenous arginine also are critical. Such processes may be those affecting the integrity (or activity) of the protein-synthesizing apparatus. However, counterarguments that transport of still other obligatory substances might have been inhibited under these conditions have not been ruled out (for a detailed discussion, see Leffert and Koch, 1977).

We have observed stimulation of DNA synthesis by one or more nonpolar lipids isolated from conditioned medium (Koch and Leffert, 1974), which may be identical with material synthesized by the cells from ^{14}C-acetate (K. S. Koch, personal communication). In addition, preliminary experiments indicate that prostaglandin-E_1, which has not been detected in the culture fluids (L. Levine, personal communication), also may have stimulatory properties (fig. 8.17). We do not know yet whether these lipids alter intracellular metabolism or effect membrane phase transitions (Ramwell and Rabinowitz, 1971) or both; in any case, they may alter membrane-related processes that affect arginine utilization (Koch and Leffert, 1974).

8.3 Implications of In Vitro Studies for In Vivo Hepatocellular Growth Control

Based upon direct observations with the in vitro system, one might

8.16 Stimulation of arginine utilization by insulin and serum fraction I (SF I). Two ml fresh, arginine-free medium was supplemented with 3-^{3}H-L-arginine (0.2 μCi/ml) and 0.1 ml isotonic NaCl (● — ●) or insulin (△ — △, 1.0 μg/ml) or SF I (4) (□ — □, equivalent to 5 percent vol/vol whole dialyzed fetal bovine serum and then added to quiescent cultures (0.41 × 10^5 cell/dish). At different times, cumulative TCA-soluble (ordinate in part a) and cold TCA-precipitable (ordinate in part b) radioactivity per culture were determined. (For details, see Koch and Leffert, 1974. *J. Cell Biol.* 62:780–791.)

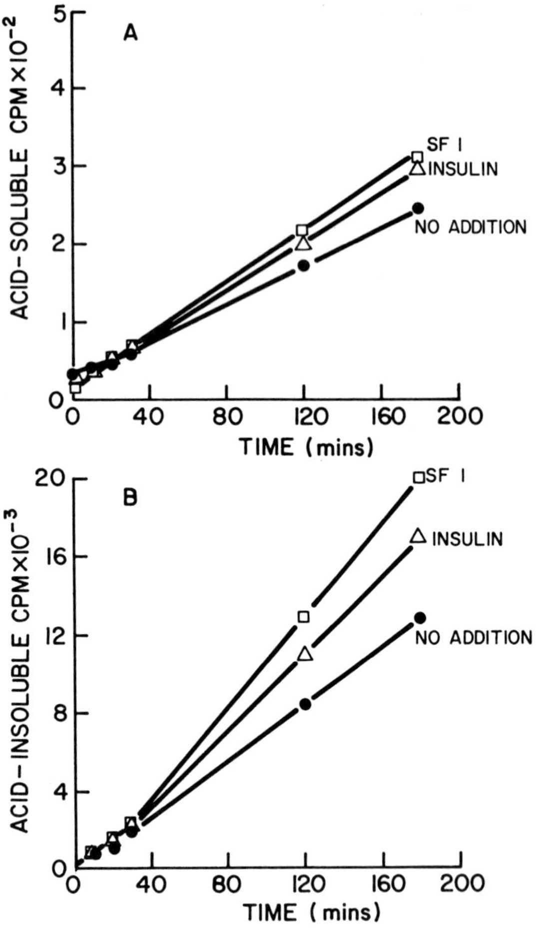

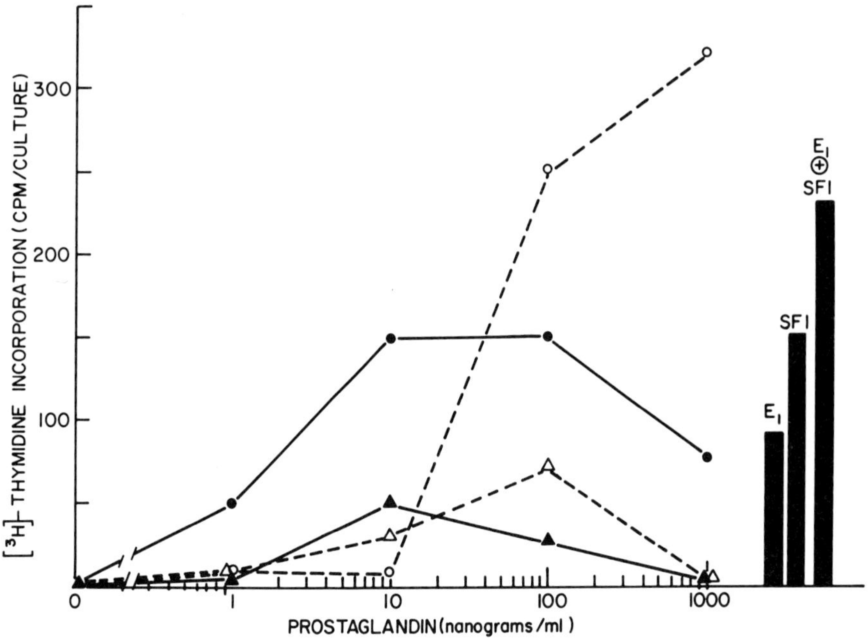

8.17 Prostaglandin-E_1-stimulated DNA synthesis (triangles, prostaglandins A_1 and A_2; circles, prostaglandin E_1; solid lines, ⊖ serum; dashed lines, ⊕ serum). (For details, see Leffert et al., 1976, *Cancer Res.* 36:4250.)

predict that a stimulus such as partial hepatectomy would cause blood levels of stimulatory and inhibitory factors to increase and decrease, respectively, in the proximate environment (for instance, in the sinusoid) of the hepatic parenchymal cell.

We have begun in vivo experiments by measuring peripheral (left ventricular blood levels of insulin, glucagon, and thyroxine in 70 percent hepatectomized rats. The results, shown in figures 8.18 and 8.19 and elsewhere (Leffert et al., 1975), indicate that hyperglucagonemia, hypoinsulinemia, and hypothyroxinema occur specifically as the result of partial hepatectomy and in proportion to the amount of liver tissue excised. Hypoglycemia cannot alone explain the changes of pancreatic hormone levels under fasted conditions, because exogenously administered glucose does not entirely abolish hyperglucagonemia. Similar results occur in fed animals. Additional studies (Alexander and Leffert, 1976) have suggested that hypothyroxinemia arises at least partially from increased hepatic utilization of thyroid hormones. The nonglucoregulatory mechanisms underlying hyperglucagonemia and hypoinsulinemia are yet to be determined; obviously, many explanations are possible.

Studies in vitro (fig. 8.12) suggest that a *reversal* of the intrahepatic ratio of insulin to glucagon (that is, from a low I:G to a high I:G) would play a growth-regulatory role (Leffert, 1974b). That hypothesis receives support from in vivo studies (Starzl et al., 1973). Although direct proof will require measurements of intraheptic hormone levels, we approached the problem in a different way by testing whether or not the regenerating liver alters its capacity to respond to these hormones, as indicated by the capacity of plasma membrane preparations from caudate and accessory lobe to bind ^{125}I-insulin and ^{125}I-glucagon. Earlier studies (Leffert et al., 1975) indicated that, whereas insulin-binding capacity (at high levels of ^{125}I-labeled insulin, about 10^{-9} *M*) remained relatively unaltered, glucagon-binding capacity was diminished 40 to 60 percent in membrane preparations derived twenty-four (but not six) hours postoperatively. Increased binding capacity for insulin was observed, however, at concentrations of 10^{-10} *M* ligand (Leffert and Koch, in press; Leffert and Rubalcava, unpublished observations). No similar changes have been seen in laparotomized or fasted controls.

The unproved implication is that, like hyperinsulinemia (Gavin et al., 1974), hyperglucagonemia and hypoinsulinemia *may* decrease and increase the concentration of hepatic glucagon and insulin receptors, respectively. Thus, "reversal of intrahepatic I:G" (Leffert, 1974b) may be accomplished in proliferating cells at the receptor level. It remains to be determined whether this is a direct effect and whether such putative receptor changes are causally linked with proliferation.

Other interesting predications can be made from the results of experiments shown in figures 8.18 and 8.19. Clearly, one must think in terms of two compartments: *peripheral*, that is, the nonhepatic organ

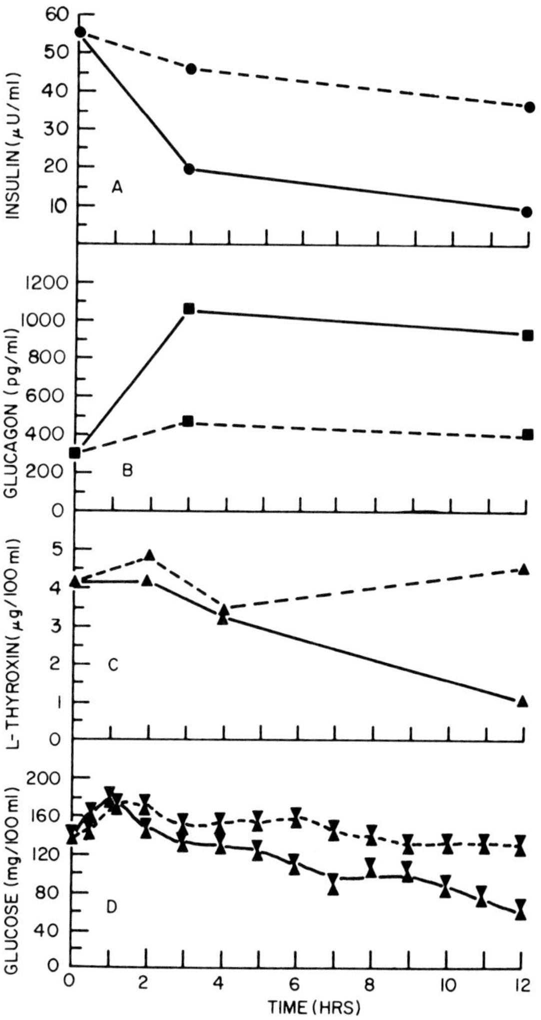

8.18 Specific alterations of plasma insulin, glucagon, thyroxine, and glucose levels following partial hepatectomy in adult rats (------ controls; —— partial hepatectomy). (For details, see Leffert et al., *Proc. Nat. Acad. Sci. USA* 72: 4033–4036, 1975.)

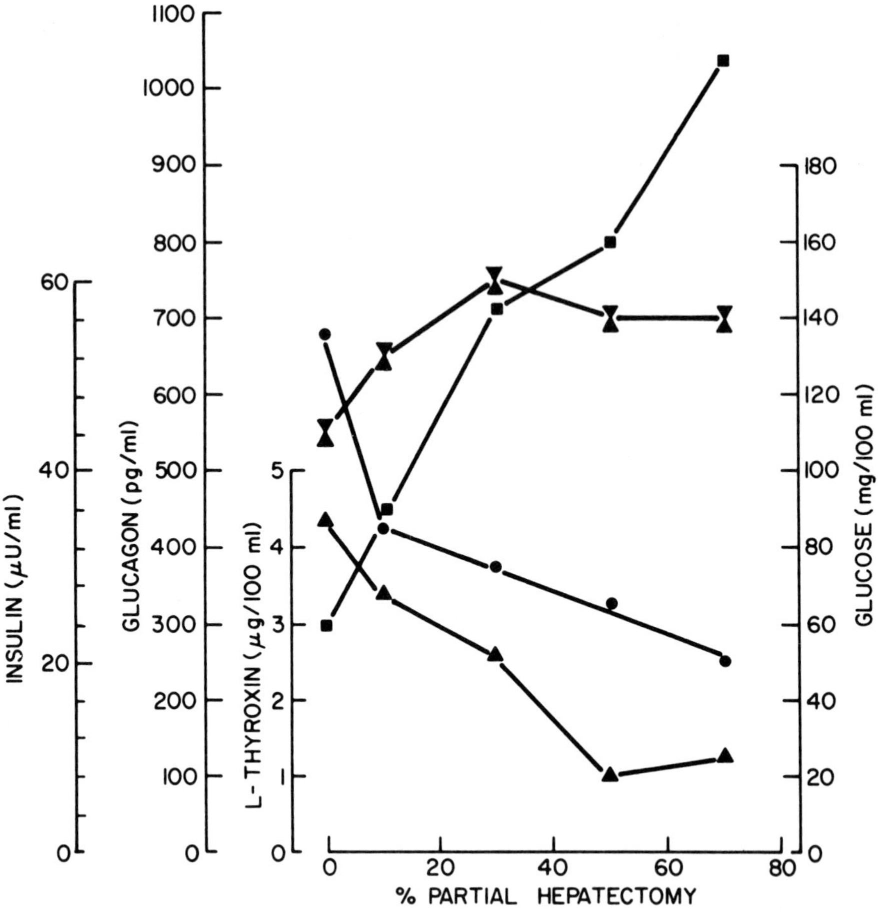

8.19 Alterations of plasma insulin, glucagon, thyroxine, and glucose levels as a function of liver ablation (● — ● insulin; ■ — ■ glucagon; ▲ — ▲ L-thyroxine; ⧗ — ⧗ glucose. (For details, see Leffert et al., 1975, *Proc. Nat. Acad. Sci. USA* 72:4033–4036.)

systems; and *central*, that is, the liver. It also seems reasonable to divide the prereplicative period in vivo into two phases, phase I being defined by retention of glucagon and elevation of insulin binding capacity and phase II by loss of glucagon-binding capacity.

From the standpoint of the peripheral compartment, its function during both phases could be to *supply* the remnant with fuels and putative membrane regulatory substances (perhaps fatty acids) and nutrients (amino acids, perhaps derived from glucocorticoid-induced muscle catabolism), as well as additional hepatoregulatory hormonal substances. Peripheral hypoinsulinemia (Simek et al., 1967) and hyperglucagonemia are ideally suited to permit adipose tissue lipolysis to occur. Perhaps some of these lipids are prostaglandins (Ho and Sutherland, 1971) derived from splanchnic adipose reserves. In support of this possibility we have measured portal blood levels of prostaglandin-E metabolites eleven to twelve hours postoperatively in 70 percent partially hepatectomized rats and found them to be elevated two- to three-fold. Whether these compounds participate in stimulation of DNA synthesis as suggested by studies in vitro (fig. 8.17), remains to be determined. We do not know yet whether inhibition of hyperglucagonemia, perhaps with α-adrenergic antagonists such as phentolamine, will affect the portal levels of prostaglandin-E metabolites, but there is evidence that phentolamine blocks the initiation of DNA synthesis in regenerating liver when administered late in the prereplicative period (MacManus et al., 1973).

As for the central compartment, other studies (Sigel et al., 1968) lead us to expect that parenchymal cells that proliferate earliest respond in phase I to the peripheral hyperglucagonemia, as indicated by increased glycogenolysis and lysosomal activation (Becker and Bitensky, 1969). It remains to be determined whether, at that point, putative lysosomal proteases affect the mucopolysaccharide moieties of the cell surface (Miyamoto et al., 1973) and they in turn play a growth-controlling role (presumably by altering surface receptors). Increased hepatic amino acid utilization and ureogenesis also occur at this time (Ferris and Clarke, 1972), but to our knowledge there is no tissue localization of these processes to the region of the portal triad. Indirect evidence from in vitro experiments with fetal hepatocytes (Leffert and Paul, 1972) suggests that urea cycle functions may be growth-state dependent because many, but not all, cells that enter synthesis phase lose the capacity to incorporate exogenous 3-^{3}H-L-ornithine into acid-precipitable grains, whereas monolayer aggregates are densely labeled (as determined by radioautography; see fig. 8.1). Assuming no inhibition of ornithine *transport* in *S*-phase cells, such observations would be consistent with either decreased arginine biosynthetic capacity (and subsequent incorporation into cell proetin) and/or increased ornithine decarboxylase activity (Russell and Snyder, 1969).

The low plasma I:G during phase I may indicate that circulating VLDL levels are lowered, possibly because of decreased liver output (Kohout et al., 1971). This relationship is suggested by the data in table 8.1 and from additional studies (Leffert et al., 1976). However, hepatic Kuppfer cells may be induced to secrete heparin by partial hepatectomy, which also could activate lipoprotein lipase (de Pury and Collins, 1972), which in turn may contribute to a lowering of VLDL levels. Recent observations are consistent with this prediction (Leffert et al., 1976). Obviously, it is crucial to determine whether (as suggested by the data in fig. 8.13) exogenously administered VLDL will retard liver regeneration; the reported effects of lipoproteins on alterations of membrane permeability are also interesting (Bruckdorfer et al., 1968). In this regard, membrane permeability may be controlled and linked to growth by alterations in hepatic cholesterol and fatty acid metabolism suspected of occurring during phase I. For this reason, we are currently examining hydroxymethyglutaryl-CoA reductase activity in phase-I-regenerating liver because there is evidence implicating both glucagon (Nepokroeff et al., 1974) and thyroid (Ness et al., 1973) hormones in regulatory control of this enzyme. Thyroid hormones may also selectively alter mitochondrial fatty acid metabolism (Wolff and Wolff, 1964).

Phase II can be expected to coincide with increased protein synthetic rates, about four to six hours postoperatively (Short et al., 1972). Our studies do not indicate the threshold for detecting diminished hormone binding, however; more detailed kinetic studies are in progress. In any event, it seems reasonable for insulin also to stimulate increased hepatic utilization of amino and fatty acids. In all likelihood, additional insulinlike factors (perhaps one or more of the somatomedins—fig. 8.15) are involved. Finally, endogenous hepatic somatomedin synthesis is another possibility. Detailed discussions of these points are presented elsewhere (Leffert ct al., 1976; Leffert and Koch, 1977 and in press).

Table 8.1 Plasma lipoprotein levels in partially hepatectomized rats (11–12 hours postoperatively).[a]

Manipulation	Plasma glucose (mg/100ml)	VLDL (μg/ml)	HDL (μg/ml)	LDL (μg/ml)
70% Partial hepatectomy, fasted	138 ± 7.6	140	125	1320
70% Partial hepatectomy, fed	155 ± 3.1	125	120	1440
Laparotomy, fasted	167 ± 3.6	235	NM[b]	1600
Laparotomy, fed	157 ± 9.8	200	NM	NM
Nothing, fasted	163 ± 5.1	250	75	1560
Nothing, fed	187 ± 7.5	235	60	1660

[a]For details, see Leffert and Weinstein, *J. Cell Biol.* 70:20–32, 1976.
[b]Not measured.

A heterogeneous G_0 cell population would have survival value. Were the liver to divide synchronously, it is conceivable that differentiated functions coupled to growth state could be compromised for periods of time sufficient to be life threatening. Alternatively, asynchronous proliferation would permit maintenance of function while the liver regenerates. As previously reported (Sigel et al., 1968), proliferation in regenerating liver is, in fact, spatially determined and the cells further from the incoming blood supply proliferate later. This phenomenon may be due to the existence of hormonal gradients across the hepatic lobule. The corollary of such a hypothesis is that hepatocytes proximal to the portal triad would comprise, in line with the above arguments, a subpopulation of G_0 cells having fewer and/or reduced serum factor requirements of the set $\{ f_1, f_2, \ldots, f_n \}$ $G_{0,1} \rightarrow S$; whereas distal hepatocytes (central vein) may be postulated to exist in a different G_0 state and therefore have greater and/or increased serum factor requirements from the *same* set of growth factors. Pursuing this model, restoration of the steady-state liver mass in partially hepatectomized rats would result from restoration of the steady-state intrahepatic hormone gradients, as well as the extrahepatic peripheral hormone levels. Consequently, membrane permeability would have to be reversed, as would the delivery to the liver of peripheral nutrients.

8.4 Alpha$_1$-Fetoprotein (α_1 F): Functional Retrogression in Regenerating Liver and Hepatic Carcinoma

Evidence presented elsewhere (Leffert and Sell, 1974; Sell et al., 1974) suggests that α_1 F may be a "growth cycle" protein in the sense that it is synthesized by normally proliferating hepatocytes, that is, in neonatal rats (fig. 8.20) and in adult rats following partial hepatectomy (fig. 8.21). Quiescent fetal hepatocytes in culture also resume detectable α_1 F synthesis following growth induction by serum factors, as shown in figure 8.22, and with a time course remarkably similar to that observed in vivo. These results suggest that the protein is made prior to the second stage of growth (G_2) and released prior to mitosis. In short-term studies, in vitro inhibitors of DNA (but not of protein) synthesis and secretion of α_1 F, so perhaps ongoing DNA synthesis per se does not require this function. Moreover, high levels (1 mg/ml) of α_1 F neither stimulate nor inhibit growth in vitro.

What, then, is the function of α_1 F? Unless proliferating cells generally must synthesize similar proteins—and there is as yet no evidence that this is the case—why is α_1 F uniquely a hepatocyte cell cycle protein? The answer may come from further studies on the relationship of this protein with hepatic lipid metabolism. Perhaps α_1 F clears estradiol (or some other inhibitory lipid) from the hepatocyte; such a function could affect the control of VLDL and/or glyceride synthesis. This effect is implied by reports that α_1 F may specifically bind estradiol (Savu et al.,

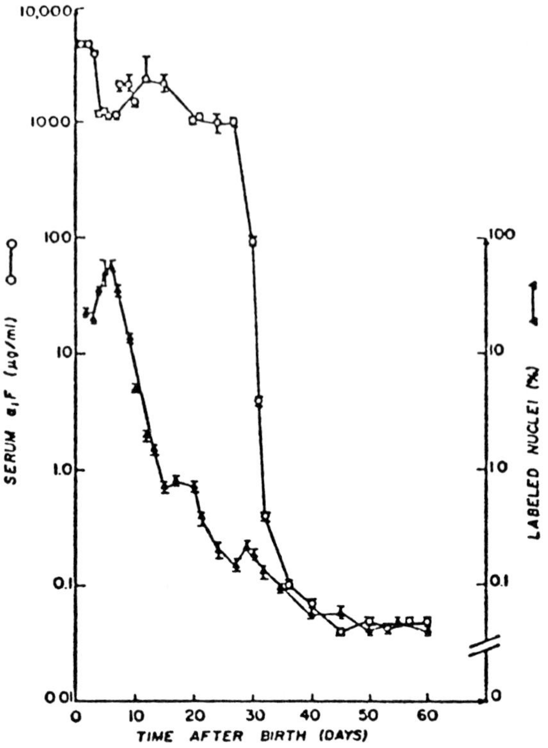

8.20 Serum α_1F concentration and hepatic cellular proliferation in newborn Fischer rats. Each *point* represents the average of 2 animals, 2 samples per animal, and 2000 nuclei counted per rat; the *bars* signify the range of values for the 2 animals. These values rarely differed by more than ±10 percent. An oscillating pattern is observed in the serum α_1F concentrations. Waves of cell proliferation at 3–7, 15–20, and 27–30 days precede elevations or leveling off of serum α_1F. (For details, see Sell et al., 1974, *Cancer Res.* 34:865–871.)

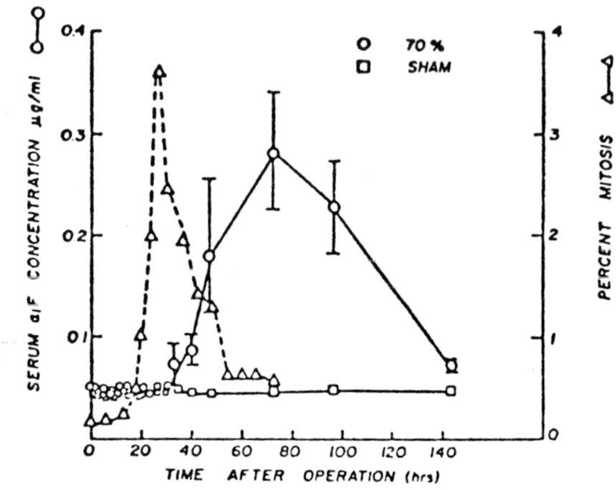

8.21 The relationship of serum α_1F to hepatic mitosis in adult male Fischer rats after 70 percent hepatectomy (○). Bars illustrate the range of points from 2 hepatectomized rats; □ value of sham-operated animals; △ percentage of mitotic cells in the regenerating liver as redrawn from the data of Grisham et al. (*Cancer Res.* 22:842–849, 1962). (For details, see Sell et al., 1974, *Cancer Res.* 34:865–871.)

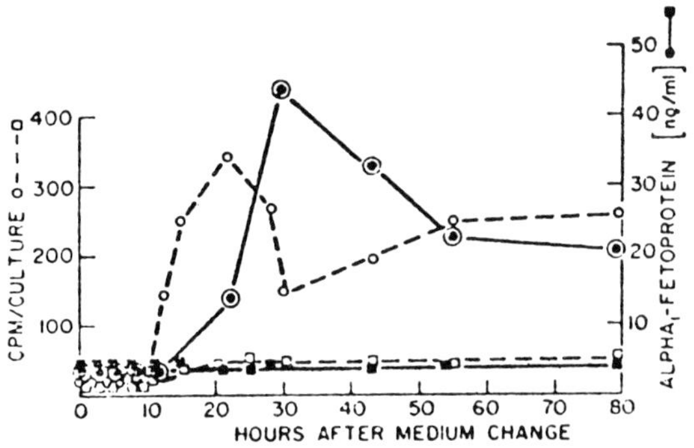

8.22 α_1F in the culture medium of quiescent fetal rat hepatocytes after the induction of DNA synthesis. Quiescent cultures were prepared by step-down conditions. At time zero (9 days postplating), the spent culture medium was aspirated and the cultures were washed twice with 2 ml arginine-free medium. These cultures (0.29×10^5 cell/dish) then received either 2 ml fresh medium containing dialyzed fetal bovine serum (10 percent vol/vol) and L-arginine (0.4 mM) (circles); or similar fresh medium without serum (squares). At intervals thereafter, DNA synthesis rates were determined in one set of cultures labeled with ^{3}H-thymidine (^{3}H-dT) (1.25 μCi/ml, 3×10^{-6} M dT) for 2 hours and by measuring the radioactivity incorporated into 5 percent vol/vol TCA-insoluble material (dashed lines). In the parallel set of cultures, media were sampled from unlabeled cultures at identical times (solid lines). During the 80-hour interval of this experiment, there was no detectable increase in radioimmunoassayable α_1F in culture media derived from control dishes (that is, quiescent cultures remaining in day-9 spent medium). Abscissa: time (hours) after the medium change. Left ordinate: counts per minute CH_3-^{3}H-dT incorporated per culture per 2-hour pulse. Right ordinate: α_1F (nanograms per milliliter culture medium) by radioimmunoassay (For details, see Leffert and Sell, 1974, *J. Cell Biol.* 61:823–829.)

1972) and that hepatic estradiol receptors possibly are involved in the control of protein and glyceride synthesis (Blythe et al., 1971; Young, 1971). There is further evidence that estradiol stimulates hepatic VLDL output in ovariectomized rats (Watkins et al., 1972). Perhaps the synthesis of VLDL is curtailed by the secretion of α_1 F; the putative failure to synthesize sufficient apoprotein could then account for the accumulation of neutral lipid seen after partial hepatectomy (Delahunty and Rubinstein, 1970). The implied *causal* relationship—that lowering VLDL removes inhibitory influences over metabolic events required to initiate hepatic DNA synthesis (fig. 8.13)—will be more difficult to prove (Koch and Leffert, 1976).

If the reorientation of hormone patterns induced by partial hepatectomy is a major determinant of hepatocellular growth control, it follows that functional retrogression during liver regeneration and tumor growth may result from a hormonally controlled growth-cycle function and/or a constitutive mutation in one or more of the genes controlling a (set of) hormone receptor system(s) involved in a set of proliferative events.

8.5 The TAGH Solution: Speculation on Its Mode of Action

A mixture of pharmacological amounts of tri-iodothyronine, amino acids, glucagon, and heparin (TAGH solution) has been shown to stimulate hepatocellular proliferation when injected into the tail vein of rats (Short et al., 1972). Possible conceptual ways of organizing some of the controversial results (Leffert, 1974) have been suggested (table 8.2) (Leffert et al., 1976). It is important to bear in mind that this perfusion phenomenon may not be entirely tissue-specific (Malamud and Perrin, 1974).

Table 8.2 TAGH solution: Speculated mode of action.

	Compartment	
Component(s)	Central	Peripheral
L-T_3	Mitochondrial fatty acid oxidation Regulation of HMG-CoA reductase	Insulin release
Amino acids	Protein synthesis (somatomedin; histones, polymerases, etc.) Gluconeogenesis Polyamine synthesis	Insulin release Glucagon release Somatotrophin release
Glucagon	Inhibition of HMG-CoA reductase Inhibition of VLDL production Increased lysozomal activity (perhaps in membrane proteases)	Insulin release Lipolysis
Heparin	Chromatin perturbations	Plasma lipoprotein lipase activation

8.6 Acknowledgments

Portal blood levels of prostaglandin-E metabolites were determined by L. Levine, Graduate Department of Biochemistry, Brandeis University, Waltham, Mass. The technical assistance of Anne Claysmith is gratefully acknowledged. This work was supported by grants from the NIH (CA14312 and CA 14195). This paper is no. VIII in a series; paper no. VII appeared in *J. Cell Biol.* 62:792–801, 1974.

Many of these studies could not have been accomplished without the collaboration of K. S. Koch (Molecular Biology Laboratory, Salk Institute), D. Weinstein (Department of Medicine, UCSD School of Medicine, La Jolla), and B. Rubalcava (Department of Medicine, Southwestern Medical School, University of Texas, Dallas).

8.7 References

Alexander, N., and Leffert, H. 1976. Thyroid hormone metabolism during liver regeneration in rats. *Endocrinology* 98:1241.

Becker, F. F., and Bitensky, M. W. 1969. Glucagon and epinephrine responsive adenyl cyclase activity of regenerating rat liver. *Proc. Soc. Exp. Biol. Med.* 130:983–986.

Blyth, C. A., Freedman, R. B., and Rabin, B. R. 1971. The effects of aflatoxin B_1 on the sex-specific binding of steroid hormones to microsomal membranes of rat liver. *Eur. J. Biochem.* 20:580–586.

Bruckdorfer, K. R., Graham, J. M., and Green, C. 1968. The incorporation of steroid molecules into lecithin sols, beta-lipoproteins and cellular membranes. *Eur. J. Biochem.* 4:512–518.

Delahunty, T. J., and Rubinstein, D. 1970. Accumulation and release of triglycerides by rat liver following partial hepatectomy. *J. Lipid Res.* 11:536–543.

de Pury, G. G., and Bollins, F. D. 1972. Very low density lipoproteins and lipoprotein lipase in serum of rats deficient in essential fatty acids. *J. Lipid Res.* 13:268–275.

Everhart, L. P., and Prescott, D. M. 1972. Reversible arrest of Chinese hamster cells in G_1 by partial deprivation of leucine. *Exp. Cell Res.* 75:170–174.

Ferris, G. M., and Clark, J. B. 1972. Early changes in plasma and hepatic free amino acids in partially hepatectomized rats. *Biochim. Biophys. Acta* 273:73–79.

Gavin, J. R., Roth, J., Neville, D. M., Jr., De Meyts, P., and Buell, D. N. 1974. Insulin-dependent regulation of insulin receptor concentrations: a diet demonstration in cell culture. *Proc. Nat. Acad. Sci. USA* 71:84–88.

Griffiths, J. B. 1972. Role of serum insulin and amino acid concentration in contact inhibition of growth of human cells in culture. *Exp. Cell Res.* 75:47-56.

Ho, R.-J., and Sutherland, E. W. 1971. Formation and release of a hormone antagonist by rat adipocytes. *J. Biol. Chem.* 246:6822–6827.

Holley, R. W. 1972. A unifying hypothesis concerning the nature of malignant growth. *Proc. Nat. Acad. Sci. USA* 69:2840–2841.

John, D. W., and Miller, L. L. 1969. Regulation of net biosynthesis of serum albumin and acute phase plasma proteins. Induction of enhanced net synthesis of fibrinogen, $alpha_1$-acid glycoprotein, $alpha_2$ (acute phase)-globulin, and haptoglobin by amino acids and hormones during perfusion of the isolated normal rat liver. *J. Biol. Chem.* 244:6134–6142.

Koch, K., and Leffert, H. L. 1974. Growth control of differentiated fetal rat hepatocytes in primary monolayer culture. VI. Studies with conditioned medium and its functional interactions with serum factors. *J. Cell Biol.* 62:780–791.

Koch, K. S., and Leffert, H. L. 1976. Control of hepatic proliferation. A working hypothesis involving hormones, lipoproteins and novel nucleotides. *Metabolism* 25:1419–1422.

Kohout, M., Kohoutova, B., and Heimberg, M. 1971. Regulation of hepatic triglyceride metabolism by free fatty acid. *J. Biol. Chem.* 16:5067–5074.

Leffert, H. L. 1974a. Growth control of differentiated fetal rat hepatocytes in primary monolayer culture. V. Occurrence in dialized fetal bovine serum of macromolecules having both positive and negative growth regulatory functions. *J. Cell Biol.* 62:767–779.

Leffert, H. L. 1974b. Growth control of differentiated fetal rat hepatocytes in primary monolayer culture. VII. Hormonal control of DNA synthesis and its possible significance to the problem of liver regeneration. *J. Cell Biol.* 62:792-801.

Leffert, H. L., Alexander, N., Faloona, G., Rubalcava, B., and Unger, R. 1975. Specific endocrine and hormonal receptor changes associated with liver regeneration in adult rats. *Proc. Nat. Acad. Sci. USA* 72:4033-4036.

Leffert, H. L., and Koch, K. S. 1977. Control of animal cell proliferation. In *Growth, Nutrition and Metabolism of Cells in Culture,* vol. 3, eds. V. J. Cristofalo and G. H. Rothblatt, pp. 225-294. New York: Academic Press.

Leffert, H. L., and Koch, K. S. 1977. Proliferation of hepatocytes. In *CIBA Foundation Symposium on Hepatotrophic Factors,* May 1977 (in press).

Leffert, H. L., Koch, K. S., and Rubalcava, B. 1976. Present paradoxes in the environmental control of hepatic proliferation. *Cancer Res.* 36:4250-4255.

Leffert, H. L., and Paul, D. 1972. Studies on primary cultures of differentiated fetal liver cells. *J. Cell Biol.* 52:559-568.

Leffert, H. L., and Paul, D. 1973. Serum dependent growth of primary cultured differentiated fetal rat hepatocytes in arginine-deficient medium. *J. Cell Physiol.* 81:113-124.

Leffert, H. L. and Sell, S. 1974. $Alpha_1$-fetoprotein biosynthesis during the growth cycle of differentiated fetal rat hepatocytes in primary monolayer culture. *J. Cell Biol.* 61:823-829.

Leffert, H. L., and Weinstein, D. B. 1976. Growth control of differentiated fetal rat hepatocytes in primary monolayer culture. IX. Specific inhibition of DNA synthesis initiation by very low density lipoprotein and possible significance to the problem of liver regeneration. *J. Cell Biol.* 70:20-32.

Lipton, A., Klinger, I., Paul, D., and Holley, R. W. 1971. Migration of mouse 3T3 fibroblasts in response to a serum factor. *Proc. Nat. Acad. Sci USA* 68:2799-2801.

MacManus, J. P., Braceland, B. M., Youdale, T., and Whitfield, J. F. 1973. Adrenergic antagonists, and a possible link between the increase in cyclic adenosine $3',5'$-monophosphate and DNA synthesis during liver regeneration. *J. Cell. Physiol.* 82:157-164.

Malamud, D., and Perrin, L. 1974. Stimulation of DNA synthesis in mouse pancreas by triiodothyronine and glucagon. *Endocrinology* 94:1157-1160.

Miyamoto, M., Terayama, H., and Ohnishi, T. 1973. Effects of protease inhibitors on liver regeneration. *Biochem. Biophys. Res. Commun.* 55:84-90.

Nepokroeff, C. M., Lakshmanan, M. R., Ness, G. C., Dugan, R. E., and Porter J. W. 1974. Regulation of the diurnal rhythm of rat liver beta-hydroxy-beta-methylglutaryl coenzyme A reductase activity by insulin, glucagon, cyclic AMP and hydrocortisone. *Arch. Biochem. Biophys.* 160:387-396.

Ness, G. C., Dugan, R. E., Lakshmanan, M. R., Nepokroeff, C. M., and Porter, J. W. 1973. Stimulation of hepatic beta-hydroxy-beta-methylglutaryl coenzyme A reductase activity in hypophysectomized rats by L-triiodothyronine. *Proc. Nat. Acad. Sci. USA* 70:3839-3842.

Paul, D., Leffert, H., Sato, G., and Holley, R. W. 1972. Stimulation of DNA and protein synthesis in fetal-rat liver cells by serum from partially hepatectomized rats. *Proc. Nat. Acad. Sci. USA* 69:374-377.

Ramwell, P. W., and Rabinowitz, U. 1971. Interaction of prostaglandins in cyclic AMP. In *Effects of Drugs on Cellular Control Mechanisms,* eds. B. R. Rabin and R. B. Freedman. pp. 207 -236. London: MacMillan.

Russell, D. H., and Snyder, S. H. 1969. Amine synthesis in regenerating rat liver: effect of hypophysectomy and growth hormone on ornithine decarboxylase *Endocrinology* 84:223–228.

Savu, L., Crepy, M. A., Guerin, M. A., Nunez, E., Engelmann, F., Benassayag, C., and Jayle, M. F. 1972. Etude des constantes de liason entre les oestrogenes et l'α,-foetoproteine de rat. *FEBS Lett.* 22:113–116.

Sell, S., Leffert, H. L., Meuller-Eberhard, U., Kida, S., and Skelly, H. 1975. Relationship of the biosynthesis of alpha$_1$-fetoprotein, albumin, hemopexin, and haptoglobulin to the growth state of fetal rat hepatocyte cultures. *Ann. N.Y. Acad. Sci.* 259:45–58.

Sell, S., Nichols, M., Becker, F. F., and Leffert, H. L. 1974. Hepatocyte proliferation and alpha$_1$-fetoprotein in pregnant, neonatal, and partially hepatectomized rats. *Cancer Res.* 34:865–871.

Short, J., Armstrong, N. B., Zemel, R., and Lieberman, I. 1973. A role for amino acids in the induction of deoxyribonucleic acid synthesis in liver. *Biochem. Biophys. Res. Commun.* 50:430–437.

Short, J., Brown, R. F., Husakova, A., Gilbertson, J. R., Zemel, R., and Lieberman, I. 1972. Induction of deoxyribonucleic acid synthesis in the liver of the intact animal. *J. Biol. Chem.* 247:1757–1766.

Sigel, B., Baldia, L. B., Brightman, S. A., Dunn, M. R., and Price, R. I. M. 1968. Effect of blood flow reversal in liver autotransplants upon the site of hepatocyte regeneration. *J. Clin. Invest.* 47:1231–1237.

Šimek, J., Chmelař, Vl., Mělka, J., Pazderka, J., and Charvat, Z. 1967. Influence of protracted infusion of glucose and insulin on the composition and regeneration activity of liver after partial hepatectomy in rats. *Nature* 213:901–911.

Starzl, T. E., Francavilla, A., Halgrimson, C. G., Francavilla, F. R., Porter, K. A., Brown, T., and Putnam, C. W. 1973. The origin, hormonal nature, and action of hepatotrophic substances in portal venous blood. *Surg. Gynecol. Obstet.* 137:179–199.

Taylor, J. M., Mitchell, W. M., and Cohen, S. S. 1974. Characterization of the binding protein for epidermal growth factor. *J. Biol. Chem.* 249:2188–2194.

Temin, H. M. 1967. Studies on carcinogenesis by avian sarcoma viruses. VI. Differential multiplication of uninfected and of converted cells in response to insulin. *J. Cell Physiol.* 69:377–384.

Van Wyk, J. J., Hall, K., Van den Brande, J. L., and Weaver, R. P. 1971. Further purification and characterization of sulfation factor and thymidine factor from acromegalic plasma. *J. Clin. Endocr. Metab.* 32:389–403.

Watkins, M. L., Fizette, N., and Heimberg, M. 1972. Sexual influences on hepatic secretion of triglyceride. *Biochim. Biophys. Acta* 280:82–85.

Wolff, E. C., and Wolff, J. 1964. In *The Thyroid Gland,* eds. R. P. H. Rivers and W. R. Trotter, p. 237. London: Butterworth.

Young, D. L. 1971. Estradiol- and testosterone-induced alterations in phosphatidylcholine and triglyceride synthesis in hepatic endoplasmic reticulum. *J. Lipid Res.* 12:590–595.

8.8 Discussion

Squires Have you correlated the α_1 F production with the cell cycle?

Leffert So far, in the in vitro system, any set of factors that stimulate cell division stimulates some amount of production of protein. It seems that once cells are stimulated protein can be produced, but the complications to interpreting these data are that we do not have perfect synchrony and that there can be a small fraction of the cells that are in G_2, for example, that could be the producers. There is some evidence in fetal and neonatal liver that the transient appearance of α_1 F in animals treated with low doses of carcinogens could be due to stimulation of the G_2 population. Production of the protein is a concomitant to proliferation, but it does not seem to be a necessary concomitant. One may ask the question: Do all cells that enter the cell cycle need to make this kind of a protein? There is no available evidence yet that other cells make α_1 F. Maybe they make another protein like it, but I do not know of any evidence for that. It seems that under some conditions you can produce protein and not get proliferation, so it seems from the available data so far that it is not an obligatory event for cell division; but if it is a concomitant function, then it seems to me, for reasons stated in my presentation, that it might be interesting to ask: Is it related to lipid metabolism?

Squires Several years ago I recall a paper on the effect of a thymus factor in regeneration. Has any more work come out or has that been disproved?

Leffert All I know is that thymectomy does not seem to interfere with liver regeneration.

Baldwin I may have missed it, but you were talking about the serum levels of α_1 F in the young adults. Do you have information as to why these levels decrease?

Leffert The correlation is with decreased percentages of cells dividing. The levels fall, apparently exponentially (which suggests a random process) and quite abruptly. In vitro, we see this kind of effect when the cells get so confluent that protein synthesis rates go down markedly. I would put it more along the lines Dr. Bannasch brought up—that there must be some kind of change in the membrane of the hepatocytes so that perhaps messenger RNA molecules function differently. That is to say, in line with the current ideas about messenger stability, it would appear that the efficiency of transcribing and/or translating the α_1 F messenger changes as the liver assumes its pathological changes. Perhaps the carcinogen is in some way having a direct effect on the translation of the messenger. Maybe it stimulated attachment to the membrane, or maybe it stimulates its stability or something like that. I do not know if anybody has yet shown that α_1 F messengers exist in the adult liver after carcinogen treatment.

Baldwin Can you reverse the proliferative process in vitro by adding α_1 F?

Leffert I have added large amounts of it and it does not seem to have any effect in the cultures. If you calculate the rates at which these cells produce α_1 F, they seem to be more active and efficient than tumors. I have tried things like adding antibody, but that, too, has been without effect.

Rogers I would like to ask a question about the regeneration studies. Which of the hormones or VLDL changes were early enough to determine the initial alterations? A lot of your data was at two hours or twelve hours. Is anything changing in the first hour when, presumably, things are being started up?

Leffert The glucogen changes were already elevated threefold after twenty minutes. The insulin also falls during the first hour; thyroid levels do not fall for about three hours. The VLDL levels start within three hours. I would expect that the fall would be inversely correlated with the rise of the glucagon.

Rogers We were discussing the role of VLDL, which is really a very interesting and exciting one if it is an inhibitor. In many of the studies of parabiotics using infusion of serum, stimulation of cell division in normal animals has not been affected; but almost invariably there is some degree of inhibition of regeneration in the hepatectomized animal, as if some inhibitor was derived from the normal animals via serum.

Butler In carcinogenesis, it depends upon which chemical you use whether or not hypophysectomy is effective.

Leffert That is why I think that one of the interesting questions, and I may be going out of my field here, is the question of whether different cells are selected by carcinogens. Does one carcinogen select a single cell type or do cells have to be in different physical states? I do know little about carcinogens and am waiting and hoping to learn and get answers to those types of questions. The capacity to respond to a given carcinogen may be determined by where that cell is in its metabolic scheme, which in turn could be determined by the hormones that are in the environment.

Weisburger We can take different liver carcinogens, dimethyl- or diethylnitrosamines, and acetylaminofluorene or azo dyes, or aflatoxin. Removing the pituitary inhibits liver cancer from aflatoxin; with acetylaminofluorene one can even go to the stage of nodules and then remove the pituitary and inhibit tumor development. But with the nitrosamines, hypophysectomy does not seem to have any effect. The tumors from those different carcinogens are, in many respects, very similar.

Leffert Can the response be restored by giving hormones?

Weisburger I have no information on that.

9 Biochemical Studies on Cultured Epitheliumlike Cells

Elizabeth K. Weisburger
Jane Idoine
Jerry M. Elliot
Carcinogen Metabolism and Toxicology Branch
National Cancer Institute
Bethesda, Maryland 20014

9.1 Introduction

Because 90 percent of human tumors are carcinomas derived from epithelial cells, the development of well-characterized and replicable epithelial cell cultures is a useful goal in studies of carcinogenesis. Cells derived from liver are of special interest due to their special capacity to metabolize and/or activate carcinogens. Williams et al (1971 and 1973) have reported malignant conversion in vitro of epitheliumlike cells from rat liver following exposure to chemical carcinogens. They also reported histochemical evidence of a range of enzyme activity in some of these cell populations, or "lines."

Using the same method (Williams et al., 1971 and 1973), we have cultured epithelial cells derived from livers of seven- to ten-day-old Fischer (F344) rats; the technique of physically removing fibroblasts was supplemented by transferring colonies ofepithelial cells, detached with a rubber scraper and suspended in the culture media, to new culture flasks. When biochemical characterizations were attempted in such lines, carried in continuous culture, several problems arose. Some lines transformed "spontaneously," one as early as three months after initiation of the primary cultures. Transformation was detected, in retrospect, by the capacity of cells to produce tumors when inoculated into syngeneic animals, though no significant changes were observed in the morphology of the cells in culture or in the diploid number of chromosomes characteristic of these lines. Experiments were difficult to replicate in lines in continuous culture because several subcultures arose between experiments, and initiation of lines was very tedious and time-consuming.

9.2 Recent Experiments

In an effort to eliminate these problems, we have been studying several cell lines of epithelial morphology and monitoring them in respect to spontaneous transformation and the activity of an enzyme especially associated with the activation of one group of carcinogens, namely, aryl hydrocarbon hydroxylase (AHH) (Gelboin, 1967).

Studies conducted for over 12 months were initiated with eight separate cell lines; seven of them were sublines, designated TRL-12, from one initial pool of liver from ten-day old rats. The eighth, TRL-13, was from liver tissue of seven-day-old rats. As soon as these lines were replicating adequately to provide enough cells, all the lines were tested for transformation in soft agar (MacPherson and Montagnier, 1964) and in animals, and their basal AHH activity was determined (Nebert and Gelboin, 1968). The TRL-12 lines had been in culture for four and a half months and the TRL-13 for two and a half months, and all were at passage levels from 4 to 8. On initial assay, the lines demonstrated 8.3 to 13.4 units of specific AHH activity (per mg protein per minute of incubation at 37°C). Enzyme assays were performed on homogenates of whole cells, and assays of duplicate cell pools from parallel but separate cultures gave results that agreed within 10 percent. These levels are comparable to those observed in liver homogenates prepared from eight- and nine-day-old rats: 7.5 to 15 units AHH activity.

Stocks of all lines were frozen at passage levels 8 to 11, and in addition four of the lines were kept in continuous culture for further study. These four lines were used in various biochemical investigations and were tested at intervals for AHH activity and evidence of transformation. Shown in figure 9.1 are data for the base levels of AHH activity observed in three of these continuously cultured lines over a six-month period. Activity in the fourth line, TRL-12 (21), was similar to that of TRL (13). As seen at the end of six months, TRL-12 (15) demonstrated a high level of activity in culture. Trl-12 (13) had only trace levels. The TRL-12 (15) line was lost after this time, but activity of the two other lines has remained at the levels shown here through passage 43. Animals inoculated with cells from these lines at the last passage shown have yielded no tumors in nine months, in eight animals each for TRL-12 (15) and TRL-13 (1). However, one animal of eight inoculated with cells from TRL-12 (13) developed a probable osteosarcoma on the flank that may have originated from the injected cells. In addition, one animal of eight inoculated with TRL-13 (1), P-17, developed a mammary tumor described as possibly an anaplastic tumor. All soft agar tests have been negative. Open figures in the chart represent AHH activity of cultures of the three lines grown from stocks frozen at passage levels 8–11, stored for twelve months, and then subcultured two or three times after thawing. Photomicrographs of lines TRL-12 (15) at P-8 and P-29 and TRL-12 (13) at P-5 and P-27 are shown in figure 9.2.

AHH activity was induced to higher levels in cells from continuous cultures and in thoses grown from frozen stocks when cultures were treated with benz(a)anthracene, 7,12-dimethylbenz(a)anthracene, and

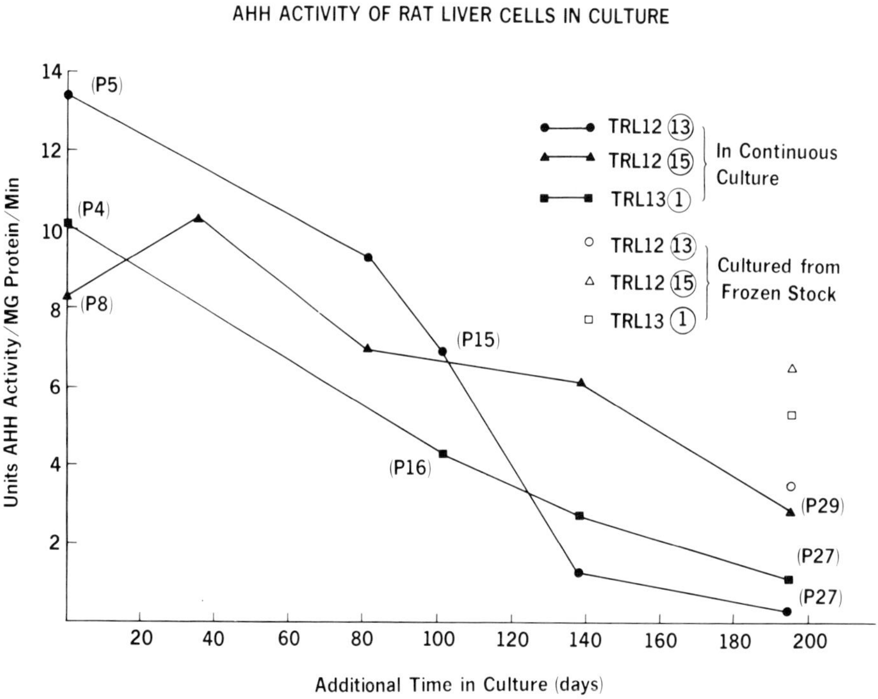

9.1 Aryl hydrocarbon hydroxylase (AHH) activity in three cell lines, either in continuous culture or reconstituted from the frozen state.

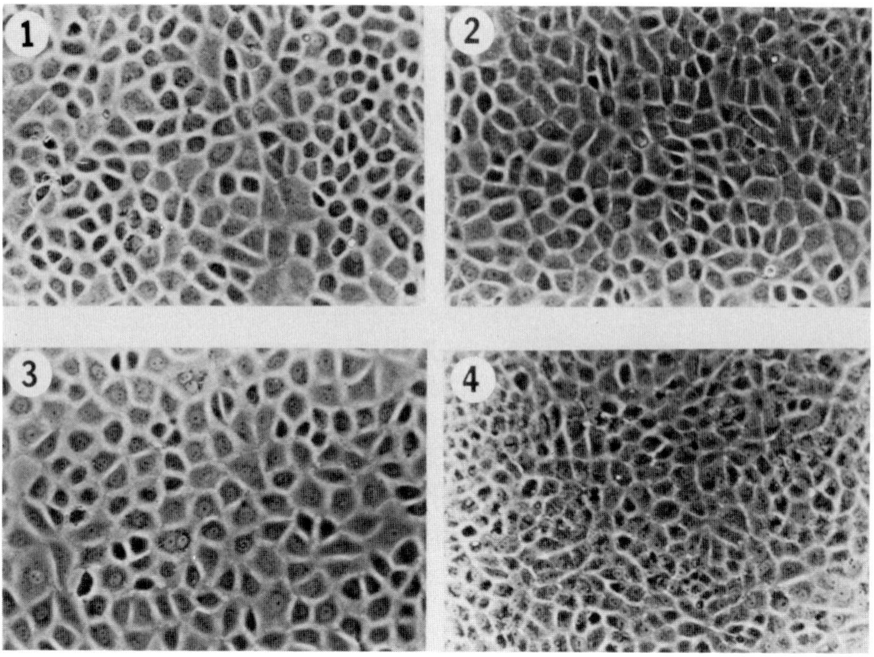

9.2 (1) TRL-12 (15) line in culture at eighth passage. (2) TRL-12 (15) line after passage 39. (3) TRL-12 (13) line after passage number 5. (4) TRL-12 (13) line after passage 27.

3-methylcolanthrene (Nebert and Gelboin, 1969). Slight but replicable increases in activity were observed after phenobarbital treatment.

Figure 9.3 shows data from studies af AHH activity in TRL-8, a line that had transformed spontaneously by passage 35. This transformation was not apparent in the cell cultures for several months, but animals inoculated with cells from P-35 showed palpable nodules within seven weeks of injection; three out of three animals inoculated developed carcinomas.

A comparison of AHH activity in normal, spontaneously transformed and liver-related tumor lines is shown in figure 9.4. Activity was measured in cell lines that had been cultured for at least a year after transformation had occurred; the normal cell lines had all been in culture for the same length of time. It is apparent that the level of AHH activity cannot be used as a criterion for the transformed (or normal) status of cells in culture.

Additional studies aimed at biochemical characterization of these lines are in progress. Traces of rat serum albumin have been detected in one line tested. Base levels of tyrosine aminotransferase (Diamondstone, 1966; Thompson, et al., 1970) have been observed in two of the lines in preliminary studies of whole-cell homogenates—4.4 units (nmole *p*-hydroxyphenylpyruvic acid/mg protein/minute incubation at 37°C) in one line and higher levels in another. One line was found to catalyze N-acetylation (Lower and Bryan, 1973).

9.3 Conclusions

It is apparent that there are biochemical differences in cell lines of epithelial morphology derived from the same host tissue. However, it does not seem feasible to characterize the biochemical capabilities of cell lines frozen at early passage levels or to utilize them in studies of the carcinogens for which they are most suited.

9.3 AHH activity in TRL-8 line spontaneously transformed at passage 35. Although there was no morphological evidence for transformation, rats inoculated with cells from passage 35 developed nodules within 7 weeks after injection (see arrow).

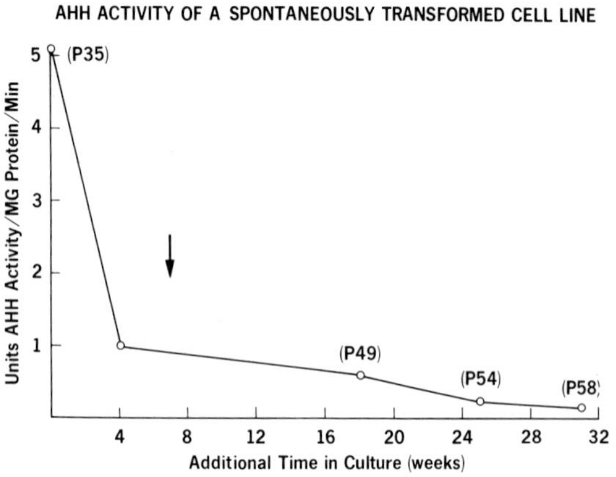

9.4 AHH activity in cell lines from (a) normal rat liver, (b) spontaneously transformed liver cells, and (c) liver cell tumor lines.

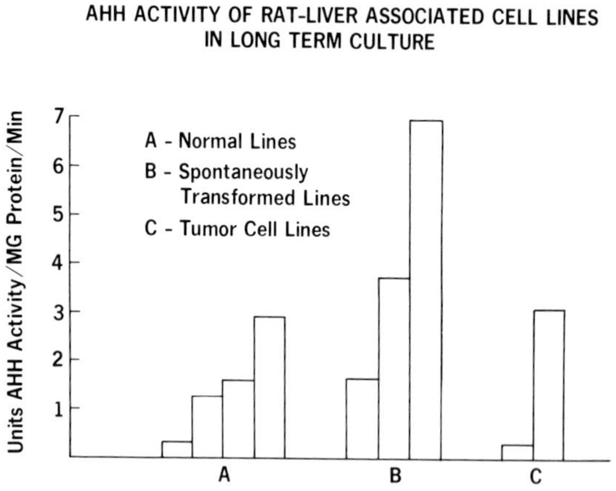

9.4 References

Diamondstone, T. I. 1966. Assay of tyrosine transaminase activity by conversion of *p*-hydroxyphenylpyruvate to *p*-hydroxybenzaldehyde. *Analyt. Biochem.* 16:395-401.

Gelboin, H. V. 1967. Carcinogens, enzyme induction, and gene action.*Adv. Cancer Res.* 10:1-81.

Lower, G. M. Jr., and Bryan, G. T. 1973. Enzymatic N-acetylation of carcinogenic aromatic amines by liver cytosol of species displaying different organ susceptibilities. *Biochem. Pharmacol.* 22:1581-1588.

MacPherson, I., and Montagnier, l. 1964. Agar suspension culture for the selective assay of cells transformed by polyoma virus. *Virology.* 23:291-294.

Nebert, D. W. and Gelboin, H. V. 1968. Substrate-inducible microsomal aryl hydroxylase in mammalian cell culture. I. Assay and properties of induced enzyme. *J. Biol. Chem.* 243:6242-6249.

Nebert, D. W., and Gelboin, H. V. 1969. The in vivo and in vitro induction of aryl hydrocarbon hydroxylase in mammalian cells of different species, tissues, strains and developmental and hormonal states. *Arch. Biochem.* 134:76-89.

Thompson, E. B., Granner, D. K., and Tomkins, C. M. 1970. Superinduction of tyrosine aminotransferase by actinomycin D in rat hepatoma (HTC) cells. *J. Mol. Biol.* 54:159-175.

Williams, G. M., Elliott, J. M., and Weisburger, J. H. 1973. Carcinoma after malignant conversion in vitro of epithelial-like cells from rat liver following exposure to chemical carcinogens. *Cancer Res.* 33:606-612.

Williams, G. M., Weisburger, E. K., and Weisburger, J. H. 1971. Isolation and long term cell culture of epithelial-like cells from rat liver. *Exp. Cell. Res.*69:106-112.

9.5 Discussion

Bannasch Do you know the morphology of your cells in culture? Have you any idea whether the alpha-fetoprotein (α_1 F) is produced by the membrane or the free ribosomes?

Weisburger We have not done any α_1 F studies with the cells of the culture.

Leffert We have tried to localize cells that produce α_1 F but without much success thus far. We have tried fluorescent dye stains but they are not sensitive enough. We have incubated cultures with ^{3}H-estradiol and find grains that are apparently perinuclear. We have studied cellular concentration of α_1 F both by direct radioassay and by incorporation of amino acids, and find that a correlation exists between the percentages of cells in the synthesis phase and the amount of nuclear protein. When we look at the released material, we find inverse correlation, suggesting that the cells that are synthesizing DNA do have α_1 F in them and are secreting it either during DNA synthesis or afterward. We would like very much to know which population in the culture produces the α_1 F.

Weisburger Insofar as function is concerned, if the binding data are upheld, we may be able to correlate α_1 F with lipoprotein metabolism, and you could argue that the function of this protein is to compete with other hepatocyte processes for the available estradiol. By secreting the estradiol via the protein, the liver effectively removes estradiol from the cells and in that way it may be related to DNA synthesis.

10

Immunology of Rat Hepatic Neoplasia

R. W. Baldwin

Cancer Research Campaign Laboratories
University of Nottingham
Nottingham, England

10.1 Introduction

Several distinct types of neoantigen may be expressed in carcinogen-induced tumors of the rat liver (Baldwin, 1973). These include tumor rejection antigens and cell surface antigens, both of which are characteristic for individual tumors as well as for cross-reacting fetal antigens. In addition, a variety of "abnormal" antigens may be demonstrated, sometimes transiently, during early stages of hepatocarcinogenesis; these are normal liver components modified by covalent interaction with carcinogen metabolites (Baldwin, 1962; Kitagawa et al., 1966). While it is unlikely that all of these neoantigens play a significant role in host immunosurveillance, they may be viewed as specific markers, characterizing transformed cells and useful in studying metabolic events during carcinogenesis.

10.2 Tumor-Associated Rejection Antigens

Tumor-associated rejection antigens are generally identified by their capacity to elicit immune rejection responses against tumor cells transplanted in syngeneic hosts. Transplantation may be achieved, for example, by surgical resection of a developing tumor graft, or by implantation of tumor cells attenuated by x or γ irradiation (Baldwin and Barker, 1967). Immunity may also be induced by implantation of viable tumor cells in admixture with immunological adjuvants such as bacillus Calmette Guérin (BCG antigen) (Baldwin, Cook, Hopper, and Pimm, 1974). The result is often rejection of the transplanted tumor and production of an enhanced immunological response.

Employing these procedures, hepatic tumors induced by feeding 4-dimethylaminoazobenzene (DAB) and 3′-methyl-DAB have been shown to express tumor-associated rejection antigens (Gordon, 1965; Baldwin and Barker, 1967). This phenomenon is exemplified by tests with a group of DAB-induced hepatic neoplasms in Wistar rats in which immunization, by either tumor excision or implantation of γ-irradiated

(15,000 R) tumor, consistently induced tumor immunity; treated rats rejected challenge with up to 5×10^5 viable cells of the immunizing tumor. So far there has been no attempt to characterize systematically the immunogenicities of hepatic neoplasms induced with other aminoazo dye carcinogens, although 5(*p*-dimethylaminophenylazo)quinoline-induced tumors also proved to be active (Baldwin, Harris, and Price, 1973).

Hepatic tumors induced in rats with diethylnitrosamine (DENA) are also immunogenic as defined by the capacity of γ-irradiated tumor cells to elicit rejection of transplanted tumor in syngeneic recipients (Baldwin and Embleton, 1971a). In contrast to these findings with aminoazo-dye- and DENA-induced hepatic tumors, tumors induced by 2-fluorenylacetamide (2-FAA) express little or no immunogenicity (Baldwin and Embleton, 1969). Thus, tumor rejection reactions could only be detected with three out of ten hepatic tumors, and the level of immunity (reflected by the maximum tumor challenge rejected by immunized rats) was low. It is notable that other tumors arising in FAA-treated rats, especially mammary carcinomas, also lack significant immunogenicity (Baldwin and Embleton, 1969, 1974). No satisfactory explanation for these observed differences in the immunogenicities of hepatic neoplasms induced by different chemical carcinogens has yet been provided (see Baldwin, 1973), but these observations indicate that the expression of tumor-associated rejection antigens is not a necessary concomitant of neoplastic transformation. This is further emphasized by studies showing variable immunogenicities in murine sarcomas induced by 3-methylcholanthrene (Prehn, 1970) and also in mouse prostate cells transformed in vitro with this carcinogen (Mondal et al., 1970; Embleton and Heidelberger, 1972; Heidelberger, 1973). Nevertheless, there is evidence suggesting that the frequency of expression of tumor-rejection antigens (fig. 10.1), as well as the immunogenic "strength" of transformed cells, depend upon some metabolic interaction with carcinogen metabolites, and the nature of this interaction deserves further investigation.

Undoubtedly the most significant feature of the tumor rejection antigens associated with chemically induced tumors is their great diversity; at the present time there are few exceptions to the general finding that each tumor has a characteristic antigen (Baldwin, 1973). This fact was established with DAB-induced rat hepatic tumors in tests showing that immunization elicited resistance to challenge only with cells of the immunizing tumor (Baldwin and Barker, 1967). Specificity is further illustrated by experiments showing that four distinct hepatic nodules arising in a rat treated with 3′methyl-DAB were antigenically distinct (Ishidate, 1970).

10.3 Tumor-Specific, Cell Surface Antigens

Neoantigens expressed at the cell surface of carcinogen-induced rat

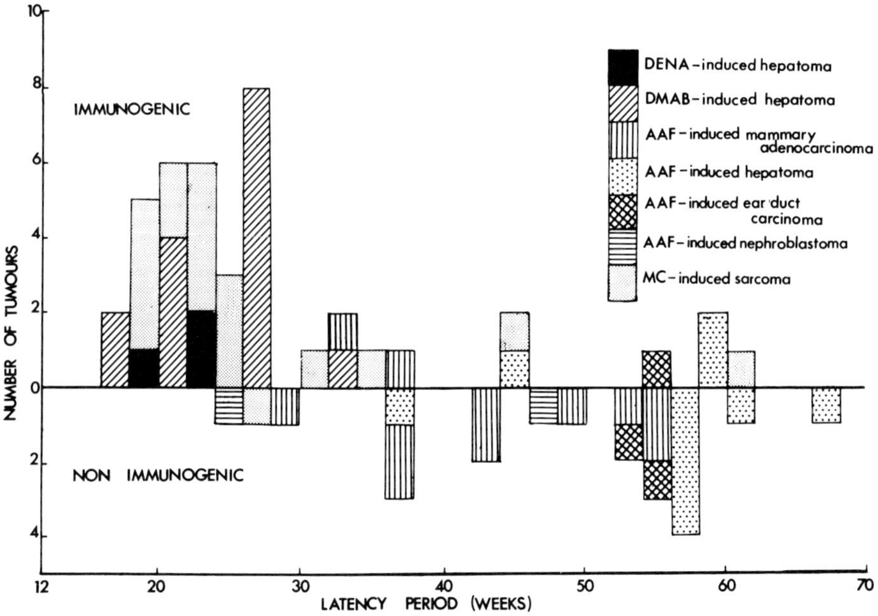

10.1 Correlation between immunogenicity and latent period of development in chemically induced rat tumors. DENA = diethylnitrosamine; DMAB = 4-dimethylaminoazobenzene; AAF = 2-acetylaminofluorene; MC = 3-methylcholanthrene.

hepatic neoplasms have been identified by analyses in vitro of cell mediated and humoral immunity elicited against tumors transplanted into syngeneic recipients (Baldwin, 1973). These reactions in vitro have closely correlated with the expression of tumor rejection antigens, and the cell surface neoantigens detected in this way are also distinctive for individual tumors. Even with those close similarities in specifities and the frequency of expression, it cannot be concluded that the in vitro assays *necessarily* detect tumor rejection antigens. Nevertheless, these cell surface antigens can be precisely defined and so are ideally suited for characterizing transformed cells.

Cell mediated immune reactions to neoantigens on DAB-induced rat hepatic neoplasms were initially demonstrated with the colony inhibition technique (Hellström, 1967), by showing that lymph node cells from rats immunized against syngeneic tumor transplants were specifically inhibitory for cells of the immunizing tumor (Baldwin and Embleton, 1971b). This technique has subsequently been modified as a microtototoxicity assay in which the survival of tumor cells plated in wells of microtest plates and exposed to sensitized lymphoid cells is compared with that of tumor cells exposed to normal control lymphoid cells (Hellström et al., 1971). Again, lymph node cells from rats immunized against individual DAB-induced hepatic neoplasms were cytotoxic for cells of the immunizing tumor. Also, the technique has been employed to detect specifically sensitized lymphoid cells in tumor-bearing hosts (Baldwin, Embleton, and Robins, 1973).

Antibody responses against neoantigens expressed at the cell surfaces of DAB-induced hepatic tumors have been detected by employing either the colony inhibition or microcytotoxicity assay to determine the complement-dependent cytotoxicity of serum from tumor-immune rats (Baldwin and Embleton, 1971b; Baldwin, Harris, and Price, 1973). In addition, tumor-specific antibody can be detected by the immunofluorescence staining of target tumor cells in suspension, and this assay has been extensively employed for characterizing neoantigens expressed at the cell surface of rat hepatic neoplasms (Baldwin, et al., 1971; Baldwin, 1973). All of the DAB-induced tumors examined elicited a specific antibody response when syngeneic rats were immunized against transplanted tumor cells. These responses were highly specific, so that the tumor-immune serum reacted with cells of the immunizing tumor but not with other DAB-induced hepatic tumors. Furthermore, because antibody in serum from rats immunized against individual hepatic tumors could be drawn out only by cells of the immunizing tumor, this assay has subsequently been employed to characterize neoantigens in subcellular fractions of tumor (Baldwin and Glaves, 1972; Baldwin, 1973).

Tumor-specific antibody has also been demonstrated by membrane immunofluorescence methods in serum of rats immunized against syngeneic transplants of FAA-induced hepatic neoplasms (Baldwin and

Embleton, 1971a). It should be noted that only those examples demonstrating tumor-specific rejection reactions showed significant antibody responses, and they were also directed specifically against cells of the immunizing tumor (table 10.1).

10.4 Tumor-Associated Embryonic Antigens

Reexpression of embryonic antigens on a variety of chemically induced tumors, including those arising in hepatic tissues, has been well-documented. A major component arising in the hepatic tumors of several species, including humans, mice, and rats, has been identified as an α-fetoprotein (α_1 F) (Abelev, 1974; Ruoslahti et al., 1974). This protein is found only in trace amounts in normal adult serum, but in high concentrations in sera of newborn or pregenant rats, as well as rats bearing hepatic neoplasms. Elevated serum levels of α_1 F have also been observed in rats during exposure to several hepatocarcinogens, often well before neoplastic change is evident. For example, Becker and Sell (1974) demonstrated elevated α_1 F levels in rats receiving approximately 1 percent of a carcinogenic dose of 2-FAA.

Hepatic neoplasm induced by carcinogens in the rat also express embryonic antigens, which, unlike α-fetoprotein, are immunogenic in the tumor-bearing host (Baldwin, Embleton, Price, and Vose, 1974). These embryonic antigens have been positively correlated with aminoazo-dye-induced rat hepatic tumors by reaction of tumor cells with either lymphoid cells or serum from multiparous female rats (Baldwin, Embleton, Price, and Vose, 1974). During pregnancy, female

Table 10.1 Specificity of tumor cell surface antigens on carcinogen-induced rat hepatic neoplasms.

Assay method[a]	Target cells[b] derived from	Reactions[c] with cells of hepatic tumors induced by DAB	FAA
Serum membrane immunofluorescence	Immunizing tumor	16/16	1/8
	Cross-test tumor	0/64	0/8
Complement-dependent serum cytotoxicity	Immunizing tumor	7/7	NT[4][d]
	Cross-test tumor	3/72	
Lymph node cell cytotoxicity	Immunizing tumor	7/7	NT
	Cross-text tumor	0/67	

[a]Membrane immunofluorescence assays were carried out employing viable tumor cells in suspension (Baldwin et al., 1971). Cytotoxicities of lymph node cells and serum were determined in vitro, employing colony inhibition or microcytotoxicity assays against cultured target cells (Baldwin and Embleton, 1974).
[b]Target cells were derived either from the tumor used for immunization of normal syngeneic rats or from another hepatic tumor induced by the same carcinogen as the immunizing tumor.
[c]Figures indicate the number of positive reactions compared with total number of combinations tested.
[d]Not tested.

rats are thought to become sensitized to a wide range of antigens that are expressed upon cells during specific stages of embryonal development but are not present on the cells after the rat has given birth (Baldwin and Vose, 1975). This change can be demonstrated, for example, in tests showing positive membrane immunofluorescence staining of multiparous rat sera with cells taken from 14- to 15-day-old fetuses but not with cells taken from older fetusess (Baldwin, Embleton, Price, and Vose, 1974; Baldwin and Vose, 1975). In a similar manner, embryonic antigens have been detected on aminoazo-dye-induced rat hepatic tumors by membrane immunofluorescence staining or complement-dependent cytotoxicity of multiparous rat serum for tumor cells. Lymph node cells from multiparous rats were also cytotoxic for the rat hepatic tumor cells (Baldwin, Embleton, Price, and Vose, 1974) (table 10.2).

Tumor-associated embryonic antigens have also been demonstrated on rat hepatic tumors induced by 2-FAA by reaction with multiparous rat serum, irrespective of whether tumor rejection antigens are expressed (Baldwin and Vose, 1974). Comparably cross-reacting embryonic antigens have been identified on a wide range of tumor types in the rat, including carcinogen-induced or spontaneous mammary carcinomas (Baldwin, Embleton, Price, and Vose, 1974; Baldwin and Embleton, 1974) and carcinogen-induced colon carcinomas (Steele and Sjörgren, 1974). These findings suggest that embryonic antigen expression may be a concomitant of neoplastic change, and so typing of

Table 10.2 Detection of embryonic antigens on aminoazo-dye-induced rat hepatic tumors by reaction with serum or lymph node cells from multiparous rats.[a]

Target cell	Multiparous rat serum tests: Membrane immunofluorescence indices	Multiparous rat serum tests: Complement-dependent cytotoxicity (%)	Cytotoxicity of lymph node cells from multiparous rats
Hepatic tumor			
D23	0.36–0.86	21–76	29–66
D30	0.30–0.65	23–70	21–56
D33	0.34–0.81	NT[2][b]	NT
Adult rat liver (freshly prepared)	0	NT	NT
Cultured rat lung fibroblasts	NT	0	NT
Cultured 13–16-day-old rat embryo cells	0.00–0.79	25–53	6–80

[a]Membrane immunofluorescence assays were carried out employing viable tumor cells in suspension (Baldwin et al., 1971). Cytotoxicities of lymph node cells and serum were determined in vitro by employing colony inhibition or microcytotoxicity assays against cultured target cells (Baldwin and Embleton, 1974).
[b]Not tested.

these antigens could provide a suitable preliminary screening assay for carcinogen-induced transformation. This possibility is exemplified by studies showing that mouse prostate cells transformed in vitro by chemical carcinogens generally express embryonic antigens.

The finding that rat hepatic tumors induced by aminoazo dyes express embryonic antigens at the cell surface suggested initially that these tumor components may be responsible for the immune rejection reactions elicited against these tumors. It also seemed conceivable that the tumor-specific cell surface antigens detected by reactions in vitro with serum or lymphoid cells from tumor-bearing or tumor-immune hosts were also the same antigens as those detected by reaction with the reagents from multiparous female rats. It has been possible, however, to distinguish between the tumor-associated embryonic antigens and tumor-specific cell surface antigens by analyzing their specifities (Baldwin, 1973; Baldwin, Embleton, Price, and Vose, 1974). Thus, the tumor-specific antigens are characteristic components of individual tumors. In comparison, the embryonic antigens are cross-reactive, as has been confirmed by absorption assays of the membrane immunofluorescence staining of serum from multiparous rats. These tests showed that antibody in multiparous rat serum that reacted with several tumor types could be totally absorbed with each of the reactive tumors. Consequently, reaction with multiparous serum reveals common embryonic antigens expressed upon several tumor types, including rat hepatic tumors and 3-methylcholanthrene-induced sarcomas (Baldwin, Embleton, Price, and Vose, 1974). Differentiation between tumor-specific and tumor-associated embryonic antigens expressed on DAB-induced hepatic tumors has also been done by comparing the capacity of sera from multiparous and tumor-immune rats to block tumor cells from attack by sensitized lymph node cells (Baldwin, Glaves, and Vose, 1974a). These tests showed that the in vitro cytotoxicity of multiparous rat lymph node cells for tumor or embryo cells could be blocked by pretreating target cells with multiparous rat serum. Lymph node cells from tumor-immune rats are also cytotoxic for both tumor and embryo cells, but only the reaction with embryo cells was blocked by multiparous rat serum.

Apart from these differences, the tumor-specific and embryonic antigens on DAB-induced rat hepatic tumors have markedly different biochemical properties (Baldwin, 1973; Baldwin, Embleton, Price and Vose, 1974). For example, tumor-specific antigen is intimately associated with the cell surface membrane and can be released only by degradative procedures such as digestion with papain (Baldwin and Glaves, 1972; Baldwin, Embleton, and Vose, 1973) or β-glucosidase (Baldwin, Bowen, and Price, 1974). In comparison the embryonic antigen, while expressed at the cell surface, is readily released following cell rupture, and localized in the cell cytosol fraction (Baldwin, Price, and Vose, 1974).

The role of tumor-associated embryonic antigens in tumor rejection is still not fully resolved, although it is unlikely that these components are responsible for immunity elicited against transplanted hepatic tumors. For example, most of the tumors examined express cross-reacting embryonic antigen, whereas tumor rejection responses are normally directed against the individually distinct components. Furthermore, attempts to induce immunity to transplanted hepatic tumors by immunization with embryo cells taken at a stage when the embryonic antigen is present have been generally unsuccessful (Baldwin, Glaves, and Vose, 1974b).

10.5 Conclusions

Carcinogen-induced rat hepatic neoplasms express a variety of neoantigens, although the most consistently demonstrable components are tumor-associated, embryonic antigens. These antigens have been demonstrated on all of the hepatic tumors so far analyzed, including both primary and transplanted tumors induced with aminoazo dyes and FAA-induced tumors. These observations, together with the now abundant literature on embryonic antigen expression on a range of tumor types with widely differing etiologies (Baldwin, 1973), suggests that typing of these antigens may provide suitable methods for characterizing transformed cells. In addition, studies showing the appearence of embryonic antigens on cultured cells transformed in vitro by chemical carcinogens suggest that antigen assay in these systems may provide relatively simple screening systems for preliminary evaluation of chemical carcinogens.

The most significant characteristic of carcinogen-induced tumors, including those arising in hepatic tissues, is the expression of the tumor-specific cell surface antigens that may function as tumor rejection antigens. This type of neoantigen is not always expressed on transformed cells, so it cannot be used as a criterion of malignant change. When the neoantigen is present, however, characterization of this specificity will be of considerable value in evaluating the role of carcinogenic agents in transformation.

10.6 Acknowledgments

These studies were supported by a grant from the Cancer Research Campaign.

10.7 References

Abelev, G. I. 1974. Alpha-fetoprotein as a marker of embryo-specific differentiations in normal and tumor tissues. *Transplant. Rev.* 20:3–37.

Baldwin, R. W. 1962. Studies on rat liver cell antigens during the early stages of azo dye carcinogenesis. *Brit. J. Cancer* 16:749–756.

Baldwin, R. W. 1973. Immunological aspects of chemical carcinogenesis. *Adv. Cancer Res.* 18:1–75.

Baldwin, R. W., and Barker, C. R. 1967. Tumor-specific antigenicity of aminoazo-dye-induced rat hepatomas. *Int. J. Cancer* 2:355–364.

Baldwin, R. W., and Embleton, M. J. 1969. 2-acetylaminofluorene-induced rat mammary adenocarcinomas. *Int. J. Cancer* 4:47–53.

Baldwin, R. W., and Embleton, M. J. 1971a. Tumor-specific antigens in 2-acetylaminofluorene-induced rat hepatomas and related tumors. *Isr. J. Med. Sci.* 7:144–153.

Baldwin, R. W., and Embleton, M. J. 1971b. Demonstration by colony inhibition methods of cellular and human immune reactions to tumor-specific antigens associated with aminoazo-dye-induced rat hepatomas. *Int. J. Cancer* 7:17–25.

Baldwin, R. W., and Embleton, M. J. 1974. Neoantigens on spontaneous and carcinogen-induced rat tumors defined by in vitro lymphocytotoxicity assays. *Int. J. Cancer* 13:433–443.

Baldwin, R. W., and Glaves, D. 1972. Solubilization of tumor-specific antigen from plasma membrane of an aminoazo-dye-induced rat hepatoma. *Clin. Exp. Immunol.* 11:51–56.

Baldwin, R. W., Barker, C. R., Embleton, M. J., Glaves, D., Moore, M., and Pimm, M. W. 1971. Demonstration of cell-surface antigens on chemically induced tumors. *Ann. N.Y. Acad. Sci.* 177:268–278.

Baldwin, R. W., Bowen, J. G., and Price, M. R. 1974. Solubilization of membrane-associated tumor specific antigens by beta-glucosidase. *Biochim. Biophys Acta* 367:47–58.

Baldwin, R. W., Cook A. J., Hopper, D. G., and Pimm, M. V. 1974. Radiation-killed BCG in the treatment of transplanted rat tumors. *Int. J. Cancer.* 13:743–750.

Baldwin, R. W., Embleton, M. J., Price, M. R., and Vose, B. M. 1974. Embryonic antigen expression on experimental rat tumors. *Transplant Rev.* 20:77–99.

Baldwin, R. W., Embleton, M. J., and Robins, R. A. 1973. Cellular and humoral immunity to rat hepatoma-specific antigens correlated in tumor status. *Int. J. Cancer* 11:1–10.

Baldwin, R. W., Glaves, D., and Vose, B. M. 1974a. Differentiation between the embryonic and tumor specific antigens on chemically induced rat tumors. *Br. J. Cancer*, 20:1–10.

Baldwin, R. W., Glaves, D., and Vose, B. M. 1974b. Immunogenicity of embryonic antigens associated with chemically induced rat tumors. *Int. J. Cancer*, 13:135–142.

Baldwin, R. W., Harris, J. R., and Price, M. R. 1973. Fractionation of plasma membrane-associated tumor-specific antigen from an aminoazo-dye-induced rat hepatoma. *Int. J. Cancer*, 11:385–397.

Baldwin, R. W., and Vose, B. M. 1974. Embryonic antigen expression on 2-acetylaminofluorene induced and spontaneously arriving rat tumors. *Brit. J. Cancer* 30:209–214.

Baldwin, R. W., and Vose, B. M. 1975. The expression of a phase-specific foetal antigen on rat embryo cells. *Transplantation*, 6:525–530.

Becker, F. F., and Sell, S. 1974. Early elevation of alpha-1-fetoprotein in N-2-fluornylacetamide hepatocarcinogenesis. *Cancer Res.* 34:2489–2494.

Embleton, M. J., and Heidelberger, C. 1972. Antigenicity of clones of mouse prostate cells transformed in vitro. *Int. J. Cancer* 9:8–18.

Gordon, J. 1965. Isoantigenicity of liver tumors induced by an azo dye. *Brit. J. Cancer* 19:387–391.

Heidelberger, C. 1973. Chemical oncogenesis in culture. *Adv. Cancer Res.* 18:317–366.

Hellström, I. 1967. A colony inhibition (C1) technique for demonstration of tumour cell destruction by lymphoid cells in vitro. *Int. J. Cancer* 2:65–68.

Hellström, I., Hellström, K. E., Sjögren, H. O., and Warner, G. A. 1971. Demonstration of cell-mediated immunity to human neoplasms of various histological types. *Int. J. Cancer* 7:1–16.

Ishidate, M. 1970. Antigenic specificity of hepatoma cell lines derived from a single rat. Abstract #365, *10th Int. Cancer. Congr., Houston*, p. 227. Houston Medical Arts Publishing Co.

Kitagawa, M., Vagi, V., Planinsek, J., and Pressman, D. 1966. In vivo localization of anticarcinogen antibody in organs of carcinogen-treated rats. *Cancer Res.* 26:221–227.

Mondal, S., Iype, P. T., Griesbach, L. M., and Heidelberger, C. 1970. Antigenicity of cells derived from mouse prostate cells after malignant transformation in vitro by carcinogenic hydrocarbons. *Cancer Res.* 30:1593–1597.

Prehn, R. T. 1970. Analysis of antigenic heterogeneity within individual 3-methylcholanthrene-induced mouse sarcomas. *J. Nat. Cancer Inst.* 45:1039–1045.

Ruoslahti, E., Pinko, H., and Seppälä, M. 1974. Alpha-fetoprotein immunochemical purification and chemical properties. Expression in normal state and in malignant and non-malignant liver disease. *Transplant. Rev.* 20:38–60.

Steele, G., and Sjögren, H. O. 1974. Embryonic antigens associated with chemically induced colon-carcinomas in rats. *Int. J. Cancer* 14:435–444.

10.8 Discussion

Leffert What evidence is there that the blocking material is an antibody, and what would you expect if you tried to produce, say, hepatomas in multiparous female rats? Or if you transfused multiparous serum into animals at different times after immunizing them with tumors. Also, have you shown directly that you can isolate IGG from one or more of those sera?

Baldwin I think the blocking factor in the tumor-bearing rat serum is antibody, and I was not sure I wanted to get into the blocking factor story. In the serum that contains antibody, there is convincing evidence that the blocking pattern is an antigen-antibody complex and not antibody. What we have done is a slightly different approach to prove that complexes will block. We have taken anti-tumor antibody which does not block and added antigen to make a complex to show that it blocks. So there is good evidence that immune complexes are at least more efficient blocking than antibody alone.

Leffert In those cases when you get rejection, do you have to give the blocking material before the tumor takes, or can you give it when the tumor has established itself?

Baldwin You cannot do any of these experiments when the tumor is established. I think there are simply too many cells.

Butler All your work is presumably done on carcinomas. Do you know anything about the earlier stages?

Baldwin Farber has worked on the liver nodules induced with FAA in which he demonstrates new antigens by producing antiserum to nodules. I question whether the use of antiserum will tell you really whether you have a new antigen or not. We tend to be rather prejudiced, and if you do an experiment where you take some tissue and put it into a rabbit and take serum and start absorbing to make it, shall we say, tumor-specific, I am not very convinced that will really tell you anything. You have got to really put your antiserum into a host to make sure that the host from which the tissue came can respond immunologically too.

When you immunize a foreign species and produce an antiserum and preabsorb, shall we say, with normal liver and then test it on preneoplastic liver, I am not sure that that is a suitable control. I think that the best way would be to put that serum back into a rat and then isolate the antibody again.

Becker There is one problem, again, with this, and that is that the antigen was found in all the nodules studied whether they were reversible nodules or not, so that there was no indication that they have anything to do with true premalignancy.

Bannasch Do you see any correlation between the different immunological behavior of the tumors and the morphology.

Baldwin No. The other point that I think is important is that we see no relationship between the immunology and the aggressiveness of the tumor. I think the trouble is that these interpretations are much too general.

The most important component of all of this is the release of antigen into the circulation which interferes specifically with cell-mediated immunity. The response, in terms of aggression, that you get with a tumor depends, first, on the tumor's capacity to shed tumor membrane into the circulation and, second, on the animal's capacity to produce an immune response to those antigenic shed membranes.

Squires Would you define your fluorescence index? And why did you pick 23 as the significant figure? How do you determine your viability for these cells?

Baldwin This is worked out on the system as we do it, and it can vary. The fluorescence index is the percentage of unstained cells in the controls. Essentially, it takes into account the fact of nonspecific staining from the control versus the specific staining in the test.

You can define what value you are going to take as a significant difference between the background versus the positive staining.

The viability is determined by dye exclusion. The cytotoxicity is done on the microplates and surviving cells are counted.

Becker I have a bit of a problem envisioning that system in which you use varied substances to remove what I assume was the embryonic-associated antigen or the tumor-specific antigen. You use papain and β-glycosidase. That would seem odd to me that papain would remove the substance identical to that removed by β-glycosidase.

Baldwin All of those are, in fact, the case. We think we have a common proteolytic activity and can use 3 M KCl as well. Cells can be used to immunize after enzyme treatment. Only a short period of enzyme treatment is used; it might be 30 minutes. You do not remove all of the cell surface antigen, and it regenerates very quickly anyway. Those cells are immunogenic. If the cells are killed, they are no longer immunogenic.

P. Newberne I wonder if you would say a few words about BCG.

Baldwin I did say that I was not sure that any human tumor was antigenic. One of the arguments that they are is because there is a certain amount of immunotherapy going on with BCG. We have been using BCG for immunotherapy. The models that we are working on are all based on clinical problems. For example, we have studied the possible treatment of pulmonary metastases. We inject tumor cells intravenously, and they lodge in the lung. This is then followed with BCG intravenously which goes to the lungs and stops tumor growth, and that effect can be obtained with both viable BCG and BCG which is dead after γ radiation. Developing from that model we now are collaborating with the Department of Animal Pathology, Cambridge University, where, in the dog, osteosarcoma of the leg is treated by amputation. The dog then gets BCG intravenously. The survival, thirty out of thirty-seven dogs on trial without BCG, was three to nine months. We treated ten dogs with BCG, and seven survived for more than twelve months. The other model that we are interested in is one for mesothelioma of human asbestos workers. The model there is intrapleural injection of tumor cells followed by BCG intrapleurally. For elimination of the tumor growth it is important that the tumor cells are in contact with the BCG. If the BCG is given in any other way, there is no effect.

P. Newberne Have you put BCG directly into a liver tumor or a colon tumor?

Baldwin We are stuck on that program. We have been trying to infuse BCG into the liver of animals fed azo dyes to try to look at primary hepatomas, and we have had very little success so far in getting the BCG to stay in the liver. We have taken animals fed DAB for five months. These animals are all going to die of primary hepatocarcinoma with pulmonary metastases. At three months, the rats were given 1 mg Glaxo BCG vaccine 15. Eventually, I think something like 90 percent of the controls got pulmonary metastases against about 30 percent BCG-treated animals.

11

Dietary Effects on Chemical Carcinogenesis in the Livers of Rats

Adrianne E. Rogers
Laboratory of Experimental Pathology
Department of Nutrition and Food Science
Massachusetts Institute of Technology
Cambridge, Massachusetts 02139

11.1 Introduction

In studies of chemical carcinogenesis in experimental animals, the composition of the diet is of major importance; diet not only affects the animals' general health but also influences the number and site of tumors induced. The amounts of calories and protein influence the development of both spontaneous and chemically induced tumors in rats (Ross and Bras, 1965 and 1973; Weisburger and Weisburger, 1958; Tannenbaum and Silverstone, 1953). Lipotropic factors, vitamin A, and antioxidants also influence tumor development in certain organs of rats and hamsters (Chu and Malmgren, 1965; Newberne et al., 1966; Newberne and Rogers, 1973; Rogers and Newberne, 1971a and 1973a; Rogers et al., 1974; Saffiotti et al., 1967; Ulland et al., 1973; Wattenberg, 1972a). In diets composed of unrefined natural ingredients, factors not yet positively identified affect the incidence of tumors induced in mice by irradiation and in rats by chemicals (Silverstone et al., 1952; Engel and Copeland, 1951; Ershoff et al., 1969; Commoner et al., 1970). Nutritional effects on carcinogenesis can be marked but more often are subtly expressed as changes in induction time or distribution of tumors. Marginal deficiencies or excesses of nutrients are probably more relevant to the epidemiology of human cancer than are severe dietary inadequacies.

To demonstrate nutritional effects on carcinogenesis and elucidate their mechanisms of action, diets of known, consistent composition must be fed. The natural-ingredient diets prepared for laboratory rodents satisfactorily support growth, development, and reproduction, but variations in the amount and quality of nutrients and the presence of contaminants may make them unsuitable for carefully controlled studies of nutritional effects (Newberne, 1975). Several purified diets (sometimes called semisynthetic diets) have been developed for laboratory rodents; they support normal growth, development and reproduction, are consistent in purity and composition, and can be modified to provide deficiency or excess of one or more nutrients (Newberne et al., 1973a,b; Rogers et al., 1973 and 1974; Rogers, 1975).

The effects of different types of diets and of specific nutrients or additives on development of chemically induced hepatic tumors in rats will be considered in this review.

11.2 Effects of Natural Ingredients or Purified Diets

Natural-ingredient diets give some degree of protection against hepatic carcinogenesis. Results of studies comparing tumor induction by N-2-fluorenylacetamide (FAA) in rats fed natural-ingredient or purified diets are summarized in table 11.1. In female rats fed a natural-ingredient diet, the incidence of hepatic and mammary tumors was decreased; the incidence of squamous carcinoma of Zymbal's gland was unchanged. In addition, the induction time of tumors was longer (Engel and Copeland, 1951). Partial modification or refinement of the diets by substitution of sucrose for whole or modified cereal grain and of casein for meat and bone scraps gave intermediate results. (Such modifications change the dietary content of many nutrients and cannot be used to induce specific nutrient effects.) The intake of diet and carcinogen was slightly decreased in rats fed purified diet, an observation that emphasizes their greater susceptibility to carcinogenesis. Commoner et al. (1970) reported similar findings in the incidence of liver tumors and tumors of Zymbal's gland in male rats fed FAA. In a study by Reuber (1965) in which two strains of rats were fed purified diet containing FAA and then were divided into groups fed either natural-ingredient or purified diet without carcinogen, tumor incidence was not affected.

The protective effect of natural-ingredient diets on chemical induction of tumors in other organs has been reported. Ershoff (1964) studied tumors of the ovary and uterus induced in rats by estrogens; the incidence of tumors was lower in animals treated with ethinylestradiol and fed a natural-ingredient diet than in rats fed a purified diet. No dietary effect was found on tumor induction by diethylstilbesterol. Rowe et al. (1970) studied salivary gland neoplasia induced by dimethylbenz(a)anthracene (DMBA) and found a higher incidence of tumors in rats fed purified diet than in those fed a natural-ingredient

Table 11.1 FAA carcinogenesis in rats fed natural-ingredient or purified diets.

Target organ	Percent rats with tumor fed: Natural-ingredient diet	Percent rats with tumor fed: Purified diet	Author
Liver	29	67	Engel and Copeland, 1951
Mammary gland	33	91	
Ear duct	46	57	
Liver	48	96	Commoner et al., 1970
Ear duct	73	53	

diet. Reddy et al. (1974) treated rats fed natural-ingredient or purified diets with 1,2-dimethylhydrazine dihydrochloride (DMH) and found a decreased incidence of colon tumors and a decreased number of tumors per colon in rats fed natural-ingredient diets.

In long-term studies, Ross and Bras (1965) fed male, Sprague-Dawley rats natural-ingredient or purified diets from weaning for periods up to three years. Rats fed the natural-ingredient diet had greater caloric intake, gained weight faster, and maintained a higher average weight than did rats fed purified diets. Their age-specific tumor incidence rate was greater and their life spans were shorter than rats fed purified diets. The distribution of tumor types was similar in both groups, except that lung tumors were found only in rats fed the purified diet. The differences may have been due to differences in caloric intake rather than in diet composition; hence there is a question about the nutritional adequacy of the purified diets. We have conducted long-term studies, using purified diets developed in our laboratory (table 11.2). Growth, reproduction, and survival of rats and mice fed Diet 1a (purified) were compared with those of animals receiving a natural-ingredient diet (Newberne et al., 1973a,b). Reproduction and growth rates of five generations of Sprague-Dawley rats were the same in the two dietary groups in all generations. Spontaneous tumors were found in 18 percent of rats fed the natural-ingredient diet and 23 percent of rats fed Diet 1a; distribution of tumors did not differ between the two groups. A simpler diet, Diet 1, has been used extensively in studies of carcinogenesis in our laboratory (Newberne and Wogan, 1968).

Table 11.2 Composition of experimental diets.[a]

	Percentage of components (by weight)		
Component	Diet 1a	Diet 1	Diet 2
Casein	20	22	3
Methanol-extracted peanut meal			12
Gelatin	4		6
Fibrin			1
Methionine	0.2		
Sucrose, dextrose, dextrin, cornstarch	60.8	56	36.3
Cellulose flour	5		2
Cottonseed oil	5	15	
Corn oil			2
Beef fat			30
Vitamin mix	1	2	2
Salts	4	5	5
Cystine			0.5
Choline chloride			0.2

[a] All three diets are adequate for growth and maintenance; diet 2 is marginally deficient in lipotropes.

The mechanism by which natural-ingredient diets modify the incidence of chemically induced tumors is not known. Components of foods induce microsomal oxidases in the liver, gastrointestinal tract, and possibly other organs (Marshall and McLean, 1969; Wattenberg, 1971 and 1972b,c). The role of these enzymes in activating and inactivating carcinogenic compounds has been extensively documented (see chapter 5). Microsomal oxidase induction in the liver protects against FAA-induced tumors, as the enzymes favor the production of inactive ring-hydroxylated products over the active N-hydroxylated (Miller and Miller, 1972). The metabolic pathways and active forms of other carcinogens discussed in this review are not well defined, but most are thought to require enzyme activation.

Antioxidants present in natural-ingredient diets also may provide protection against tumor induction. Cereal grains contain antioxidant tocopherols, and selenium, another antioxidant, occurs in variable levels in natural products. The antioxidants butylated hydroxyanisole (BHA) and butylated hydroxytoluene (BHT), are added to food components that are present in natural-ingredient diets (for instance, fats) and are also added to the diets themselves (Shamberger et al., 1972). BHT markedly lowered hepatic tumor incidence when it was fed to rats in combination with FAA or N-hydroxy-(N-OH-)FAA (table 11.3) (Ulland et al., 1973). BHA and/or BHT protected rats against induction of mammary tumors by DMBA (Wattenberg, 1972a) and N-OH-FAA, but not by FAA (Ulland et al., 1973). BHT failed to modify significantly the

Table 11.3 Effect of antioxidants on chemical carcinogenesis in rats.

Carcinogenic regimen[a]	Rats: Strain	Rats: Sex	Target organ
FAA, 0.022% diet, 24 wk.	S-D[c]	M	Liver
N-OH-FAA, 0.024% diet, 16 wk.	S-D	M	Liver
FAA, 0.015% diet, 16 wk.	Fischer	M	Liver
N-OH-FAA, 0.015% diet, 16 wk.	Fischer	M	Liver
DENA, 51 ppm, H_2O, 24 wk.	S-D	M	Liver
DENA, 51 ppm, H_2O, 24 wk.	S-D	M	Esophagus
DENA, 51 ppm, H_2O, 24 wk.	S-D	F	Liver
DENA, 51 ppm, H_2O, 24 wk.	S-D	F	Esophagus
FAA, 0.022% diet, 32 wk.	S-D	F	Mammary gland
N-OH-FAA, 0.024% diet, 32 wk.	S-D	F	Mammary gland
DMBA, 12 mg, ig	S-D	F	Mammary gland

[a]FAA, N-2-fluorenylacetamide; N-OH-FAA, N-hydroxy-FAA; DENA, N-nitrosodiethylamine; DMBA, dimethylbenz(a)anthracene.
[b]0.66% in diet, except in Wattenberg's study, in which a single dose of 200 mg/rat was given.
[c]Sprague-Dawley.

100 percent incidence of hepatic tumors induced by large doses of N-nitrosodiethylamine (DENA) (Ulland et al., 1973). It would be interesting to see results of similar studies using lower DENA dosage. Other antioxidants tested have proved ineffective.

BHA and BHT may also inhibit carcinogenesis through induction of microsomal oxidases (Ulland et al., 1973; Cumming and Walton, 1973).

11.3 Effects of Specific Nutrients

Early studies on the effects of calorie, fat, and protein content in the diet were reviewed by Tannenbaum and Silverstone (1953), and the role of riboflavin was reviewed by Rivlin (1970).

11.3.1 Protein Deficiencies

Madhavan and Gopalan (1968) studied the effects of dietary protein content on induction of hepatocarcinoma by aflatoxin B_1. Pair-fed rats were given a diet that was either adequate in protein (20 percent casein) or low in protein (5 percent casein) and identical doses of aflatoxin B_1. Fifty percent of the rats fed adequate protein had developed hepatocarcinoma at one year; there were no hepatocarcinomas in the low-protein group. Although caloric intake was controlled, weight gain in protein-deficient animals was less than in controls and the specificity of the protein effect could not be established.

Percent tumor incidence in rats given		
No additive	BHT[b]	Author
70	20	Ulland et al., 1973
60	15	
90	13	
64	50	
100	85	
50	55	
100	95	
10	20	
20	35	
80	40	
80	28	Wattenberg, 1972a

11.3.2 Lipotrope Deficiency

Deficiency of lipotropes (choline, methionine, folic acid, and vitamin B_{12}) has been related to enhanced hepatic carcinogenesis in experimental animals and may be associated with hepatic carcinogenesis in people. Lipotrope deficiency produces a fatty liver and cirrhosis in male rats similar to human alcoholic cirrhosis (Hartroft, 1969; Rogers and Newberne, 1973b). Hepatic and esophageal carcinoma occur with greater frequency in alcoholics, with or without cirrhosis, than in non-alcoholics (Steiner et al., 1959; Wynder and Bross, 1961; Martinez, 1969).

Lipotrope-deficient rats are more susceptible than normal rats to several chemical carcinogens (table 11.4). Rats fed a lipotrope-deficient diet in which peanut meal was the major protein source developed liver tumors, but no such tumors developed if choline was added to the diet

Table 11.4 Effects of lipotrope deficiency on chemical carcinogenesis in rats.

Carcinogenic regimen[a]	Rat Strain	Rat Sex	Target Organ
Aflatoxin-contaminated peanut meal	AES	M	Liver
DAB, diet	S-D	F	Liver
DAB, diet	Holtz-Man	F	Liver
Aflatoxin-contaminated peanut meal	S-D	M	Liver
AFB_1, 240μg ig	S-D	M	Liver
AFB_1, 350μg ig	Fischer	M	Liver
DENA, 40 ppm, diet, 18 wk.	S-D	M	Liver
DENA, 40 ppm, diet, 12 wk.	S-D	M	Liver
DBN, 3.7 mg/kg, sc	S-D	M	Liver[j]
DMN, 40 ppm, diet, 12 wk.	S-D	M	Liver
DMH, 30 mg/kg, ig	S-D	M	Colon[k]
DMH, 150 mg/kg, ig	S-D	M	Colon[k]
AFB_1, 375 μg ig	Fischer	M	Liver
DDCP, 195 mg/rat/ig	S-D	M	Liver
			Stomach
FANFT, 0.188%, diet, 15 wk.	S-D	M	Bladder
DMBA, 20 mg/kg, ig	S-D	F	Mammary gland

[a]DAB, *p*-dimethylaminoazobenzene; AFB_1, aflatoxin B_1; DENA, N-nitrosodiethylamine; DBN, N-nitrosodibutylamine; DMN, N-nitrosodimethylamine; DMH, 1,2-dimethylhydrazine; DDCP, 3,3-diphenyl, 3-dimethyl carbamoyl-1-propyne; FANFT, N-[4-(5-nitro-2-furyl)-2-thiazolyl] formamide; DMBA, dimethylbenz(a)anthracene.
[b]Deficient diet + choline, 0.2%.
[c]Deficient diet + B_{12}, 50μg/kg.
[d]Deficient diet + DL-methionine, 0.6%.
[e]Deficient diet + B_{12}, 50μg/kg and DL-methionine, 0.6%.
[f]Deficient diet + choline, 0.3% + vitamin B_{12}, 50μg/kg.

(Copeland and Salmon, 1946). Tumor induction was therefor attributed initially to choline deficiency. Later studies showed that the tumors resulted from aflatoxin contamination of the peanut meal (Salmon and Newberne, 1963; Newberne, 1965). Because peanut meal was used in both deficient and choline-supplemented diets, it appeared that addition of choline to the diet inhibited aflatoxin carcinogenesis.

Experiments in rats fed *p*-dimethylaminoazobenzene (DAB) in a severely lipotrope-deficient diet showed that supplementation with methionine, which greatly increased intake of food and carcinogen, decreased tumor incidence, but supplementation with vitamin B_{12} markedly increased hepatic tumor incidence. Again, supplementation resulted in a slight increase in food and carcinogen intake. Addition of both methionine and B_{12}, which further increased food intake, raised tumor incidence to a level probably consistent with the elevated intake

Percent tumor in rats fed		
Control diet	Lipotrope-deficient diet	Author
0[b]	30	Copeland and Salmon, 1946
78[c], 11[d], 33[e]	17	Day et al., 1950
78[c]	37	Miller et al., 1952
14[f]	33	Newberne et al., 1966
0[g]	29	Newberne et al., 1968
6[h]	22	Rogers and Newberne, 1971a
70[h]	88[i]	Rogers et al., 1974
24[h]	60[i]	Rogers, 1975
24[b,h]	64	Rogers et al., 1974
65[h]	46	Rogers et al., 1974
86[h]	100[i]	Rogers and Newberne, 1973a
56[h]	85[i]	Rogers and Newberne, 1973a
11[h]	87	Rogers, 1975
40[h]	67	Rogers, 1975
60[h]	33	Rogers, 1975
54[h]	61	Rogers, 1975
92	100	Rogers, unpublished

[g]Deficient diet + choline, 0.5%; DL-methionine, 0.5%, vitamin B_{12}, 50μg/kg.
[h]Control diet is adequate in all known respects for rats; deficient diet is marginally deficient in choline and methionine, contains no folate, and is adequate in vitamin B_{12} and high in fat.
[i]Induction time significantly decreased compared to control diet.
[j]Tumor incidence in other organs not affected: bladder, 80% of control rats and 72% of deficient rats; lung, 100% of control rats and 88% of deficient rats.
[k]Incidence of small bowel tumors not significantly affected.

of DAB (Day et al., 1950). The enhancement of tumor development by addition of vitamin B_{12} was confirmed in a subsequent study (Miller et al., 1952).

We have found enhancement of carcinogenesis by several chemicals in rats fed lipotrope-deficient diets (table 11.4). Induction of hepatocarcinoma by aflatoxin B_1 was increased by lipotrope deficiency severe enough to cause cirrhosis in some but not all experiments (Newberne et al., 1966; Rogers and Newberne, 1969). Marginal deficiency, which induces fatty liver, but not cirrhosis, and supports normal growth, consistently enhances hepatocarcinogenesis (Rogers and Newberne, 1971a; Rogers, 1975; Rogers et al., 1974). The marginal deficiency induced by Diet 2 (table 11.2) significantly increased the incidence of hepatic tumors in rats given DENA, N-nitrosodibutylamine, and 3,3-diphenyl, 3-dimethyl carbamoyl-1-propyne (Harris et al., 1971), but not by N-nitrosodimethylamine (DMN). Similar dietary effects on the induction of neoplasia in other organs have been reported (Rogers and Newberne, 1973a; Rogers et al., 1974; Rogers, 1975). Rats fed the marginally deficient Diet 2 ate, grew, and survived normally; thus, the differences in response did not result from nonspecific caloric effects. The marginally deficient rats were compared to rats fed Diet 1 (table 11.2) rather than a diet supplemented only with lipotropes. The two diets differ in fat and amino acid content, as well as in lipotrope content, but the significant deficiency appears to be that of lipotropes enhanced by high fat content. Studies in progress indicate that the increased dietary fat depresses induction of hepatic tumors and therefore is not responsible for the enhancement of hepatic carcinogenesis by Diet 2. However, the high fat content of Diet 2 may be responsible for its enhancement of colon tumor induction by DMH.

Miller and Miller (1972) studied the effect of increased dietary methionine on FAA carcinogenesis. Male rats fed 0.006 percent FAA in a natural-ingredient diet to which 2.5 percent methionine was added had a decreased incidence of hepatic tumors. Addition of cysteine or casein (which contains the sulfur-containing amino acids) produced a similar effect. Addition of methionine, choline, and betaine to an adequate diet inhibited, to varying degrees, the induction of liver tumors by ethionine (Farber and Ichinose, 1958).

It appears from these studies that lipotropic factors, with the possible exception of vitamin B_{12}, protect rats against induction of hepatic tumors by a number of chemicals. There is evidence of direct biochemical interaction between lipotropes and carcinogens. DAB administration protected lipotrope-deficient rats against development of renal necrosis, an acute effect of the deficiency, presumably by supplying methyl groups normally derived from methionine and choline (Jacobi and Baumann, 1942). Miller et al. (1952) demonstrated that transfer of methyl carbon from 3′-methyl-DAB to serine and choline was inhibited by a dietary deficiency of folate or vitamin B_{12}. The aminoazobenzenes have been reported to cause defects in transmethylation; studies of the effects of lipotropic agents on carcinogenesis by this group of compounds have been contradictory (Haven and Bloor, 1956).

DENA appears to enhance lipotrope deficiency. Poirier and Whitehead (1973) observed that rats fed 0.01 percent DENA excreted levels of urinary formiminoglutamic acid, although dietary folate was adequate. Increased formiminoglutamic acid excretion is evidence of metabolic folate deficiency. Formiminoglutamic acid excretion was blocked by feeding 1.5 percent methionine, 1 percent choline, or 1.1 percent betaine but not by addition of 50 μg vitamin B_{12} per kilogram of diet. Decreased hepatic folate stores in DENA-fed rats were restored by feeding methionine but not by feeding folic acid. We found that marginally lipotrope-deficient rats fed 40 ppm DENA had increased hepatic lipid, compared with deficient rats not given DENA, and that DENA feeding induced renal necrosis, which is associated with severe but not marginal lipotrope deficiency. Methylation of macromolecules by nitrosamines and other carcinogens may influence or be influenced by lipotrope content of tissues.

Haven and Bloor (1956) reviewed other dietary effects on carcinogenesis that may be related to lipotrope and lipid metabolism. An increase in dietary lipids often increases tumor incidence. Studies on aminoazobenzene carcinogenesis demonstrated that certain lipids, including corn oil, enhanced carcinogenesis by reducing hepatic riboflavin. Riboflavin is needed for hepatic metabolism of these carcinogens to noncarcinogenic derivatives (Miller and Miller, 1953). Furthermore, corn oil, which enhanced tumor induction in comparison to hydrogenated coconut oil, intensifies dietary lipotrope deficiency, whereas coconut oil ameliorates the deficiency (Patek et al., 1966; Zaki et al., 1966).

The various lipotropic factors interact in the 1-carbon cycle. Methylation of homocysteine to methionine required folate and vitamin B_{12} as cofactors and may use methyl groups from choline (Finkelstein and Kyle, 1968). Increased dietary fat increases the requirement for these nutrients (Zaki et al., 1963). There are complex interactions between lipotropic nutrients in both humans and rats. The interactions between choline and methionine have been recognized since the early investigations of these agents (Mulford and Griffith, 1942; Best et al., 1950). Dietary deficiency of vitamin B_{12} produces a metabolic deficiency of folate, even if folate is in adequate supply in the diet. This deficiency can be corrected by supplementation with either B_{12} or methionine (Vitale and Hegsted, 1967; Herbert, 1973; Thenen and Stokstad, 1973). In humans, deficiency in vitamin B_{12} and/or folate occurs in alcoholism, Addisonian pernicious anemia, sprue, and a variety of other disorders (Herbert, 1973). All these conditions may be associated with deficiencies of methionine and choline resulting from decreased intake and absorption (Halsted et al., 1973; Leevy et al., 1970). Two of these conditions are associated with an increased risk of cancer: alcoholism with cancer of the liver and esophagus and pernicious anemia with gastric cancer (Lowenfels, 1973).

Further points of interest in the relationship between lipotrope deficiency and carcinogenesis concern antioxidants and orotic acid. The antioxidants BHA and BHT, which reduce induction of tumors under certain circumstances, have a lipotropic effect (Newberne et al., 1969; Wilson et al, 1973). The mechanism of this action is not known, but it raises the question of whether BHA and BHT reduce carcinogenesis by means of a lipotrope-sparing action. Orotic acid decreases tumor incidence and protects against lipotrope deficiency under certain circumstances. Sidransky and Verney (1970) found that 1 percent dietary orotic acid protected male rats against hepatocarcinoma induction by either ethionine or FAA but not by 3′-methyl-DAB. Simon et al (1969) reported that orotic acid protected against lipotrope deficiency.

Lipotrope deficiency may affect chemical carcinogenesis by (1) depressing microsomal drug metabolism in the liver and other organs (Rogers and Newberne, 1971a,b); (2) decreasing the concentration of nucleophiles in the tissues (Miller and Miller, 1972); (3) decreasing cell-mediated immunity and therefore the effectiveness of immune surveillance for tumors (Williams et al., 1975). The relationship between microsomal drug metabolism, levels of hepatic mixed-function oxidases, and tumor induction is complex (Miller and Miller, 1972; Wattenberg, 1972c). Studies of carcinogen metabolism in vitro cannot be used at present to predict in vivo tumor induction. In preliminary studies we have found that in vitro activation of aflatoxin B_1 to a bacterial mutagen is decreased if livers from lipotrope-deficient rats are used. This result correlates with the observed decrease in microsomal oxidases but not with our observation of increased carcinogenesis. The expectation was that livers from deficient rats would be associated with increased mutagenesis.

Lipotrope deficiency depresses the tissue levels of individual nucleophiles, in particular methionine, but whether the total content of nucleophiles is decreased is not known. It should be borne in mind that many lipotrope-deficient diets, including our marginally deficient Diet 2 (table 11.2), contain the nucleophile, cystine. Cystine has been shown to decrease the incidence of tumors induced by FAA (Miller and Miller, 1972).

Cell-mediated immunity is depressed by lipotrope deficiency in rats and by folate deficiency in humans (Gebhardt and Newberne, 1974; Gross et al., 1975; Williams et al., 1975). The role of cell-mediated immunity in inhibition of tumor growth is uncertain, but increased tumor incidence in immunosuppressed patients is evidence that favors its significance.

11.3.3 Vitamin A Deficiency

Vitamin A will be considered only briefly because the majority of published studies have concerned organs other than the liver and have used hamsters (Chu and Malmgren, 1965; Saffiotti et al., 1967; Smith, Rogers, Herndon, and Newberne, 1975; Smith, Rogers, and Newberne, 1975). Administration of aflatoxin B_1 to rats lowered the hepatic content of vitamin A, but the feeding of increased levels of vitamin A did not affect the incidence of hepatocarcinoma. Vitamin A deficiency, however, increased the incidence of colon tumors and decreased the incidence of hepatocarcinoma (Newberne and Rogers, 1973).

11.4 Effects of Nutrients on Toxicity of Carcinogens

In our studies of rats fed purified diets, we have often found it impossible to give the high doses of carcinogens reported in studies using natural-ingredient diets because of the toxicity of the compounds, especially with FAA and N-[4-(5-nitro-2-furyl)-2-thiazolyl]formamide (Rogers, 1975).

Toxicity of aflatoxin B_1 is affected by deficiency of either lipotropes or protein, but the direction of the effect depends on the number of doses given. Lipotrope-deficient rats were resistant to the toxicity of a single dose of aflatoxin B_1, given either intragastrically or intraperitoneally. However, their sensitivity to toxicity of repeated small doses increased (tables 11.5, 11.6) (Rogers and Newberne, 1971a; Butler and Neal, 1973). Lipotrope-deficient rats that survived administration of repeated small (carcinogenic doses of aflatoxin B_1 developed foci of hyperplastic hepatocytes with abnormal histochemical reactions earlier and in greater number than did rats fed an adequate diet (figs. 11.1 and 11.2). Deficient rats were also more sensitive to the toxicity of repeated doses of aflatoxin G_1 and had marked proliferation of bile ducts and loss of hepatocytes, with persistence of foci of abnormal hyperplastic cells, during administration of aflaxtoxin G_1 at a time at which rats fed

Table 11.5 Toxicity of a single dose of aflatoxin B_1 in male lipotrope-deficient rats.

Diet	Stain	Route of administration	Toxicity (mg/kg)	Author
1	Sprague-Dawley	1g	LD_{50} = 7, LD_{80} = 9	Rogers and Newberne, 1971a
		1p	LD_{100} = 7	
2	Sprague-Dawley	1g, 1p	No deaths at 7,9	
1	Fischer	1g	LD_{100} = 7	
2	Fischer	1g	No deaths at 7	
Natural ingredients	Fischer	1p	LD_{75} = 4, LD_{90} = 8	Butler and Neal, 1973
2	Fischer	1p	LD_{10} = 8	

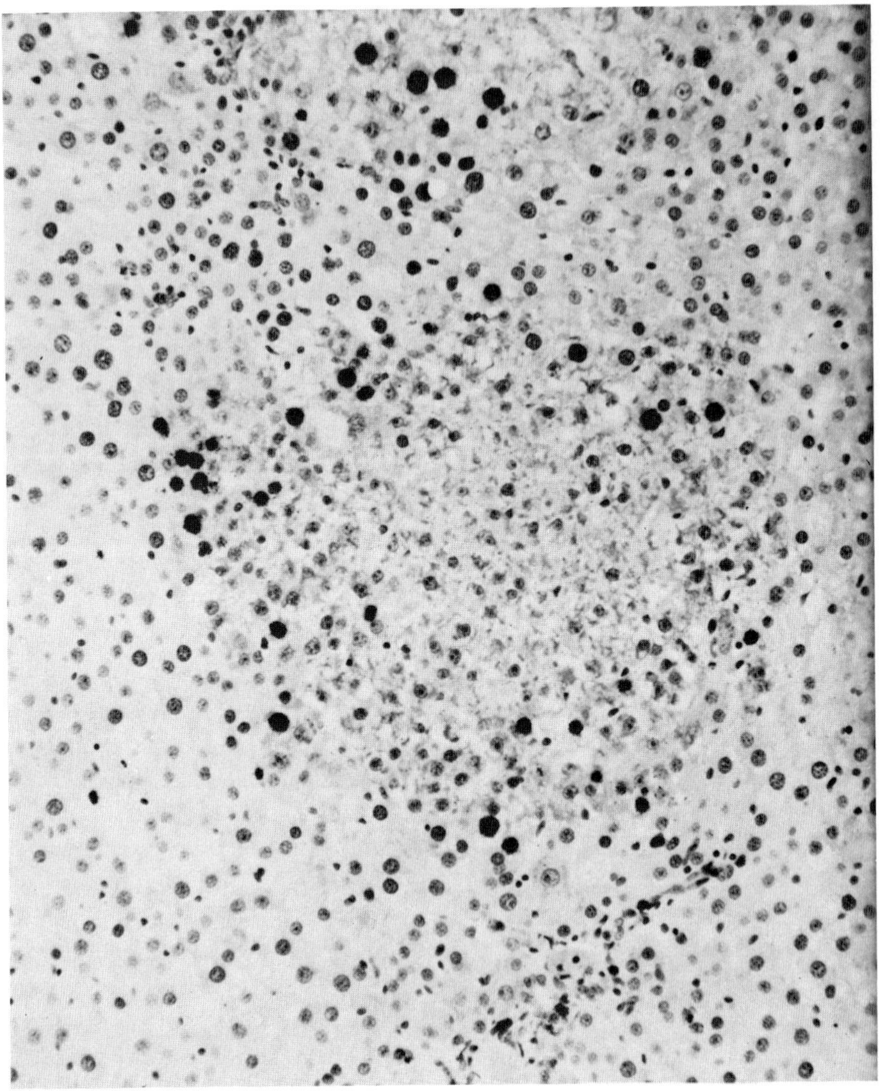

11.1 Focus of hyperplastic hepatocytes in liver of rat fed Diet 2, given 350 μg aflatoxin B_1 given ^{3}H-thymidine and killed 24 hours after the last dose. There are many ^{3}H-labeled nuclei in the basophilic focus in the center. (Autoradiograph, H and E, × 180.)

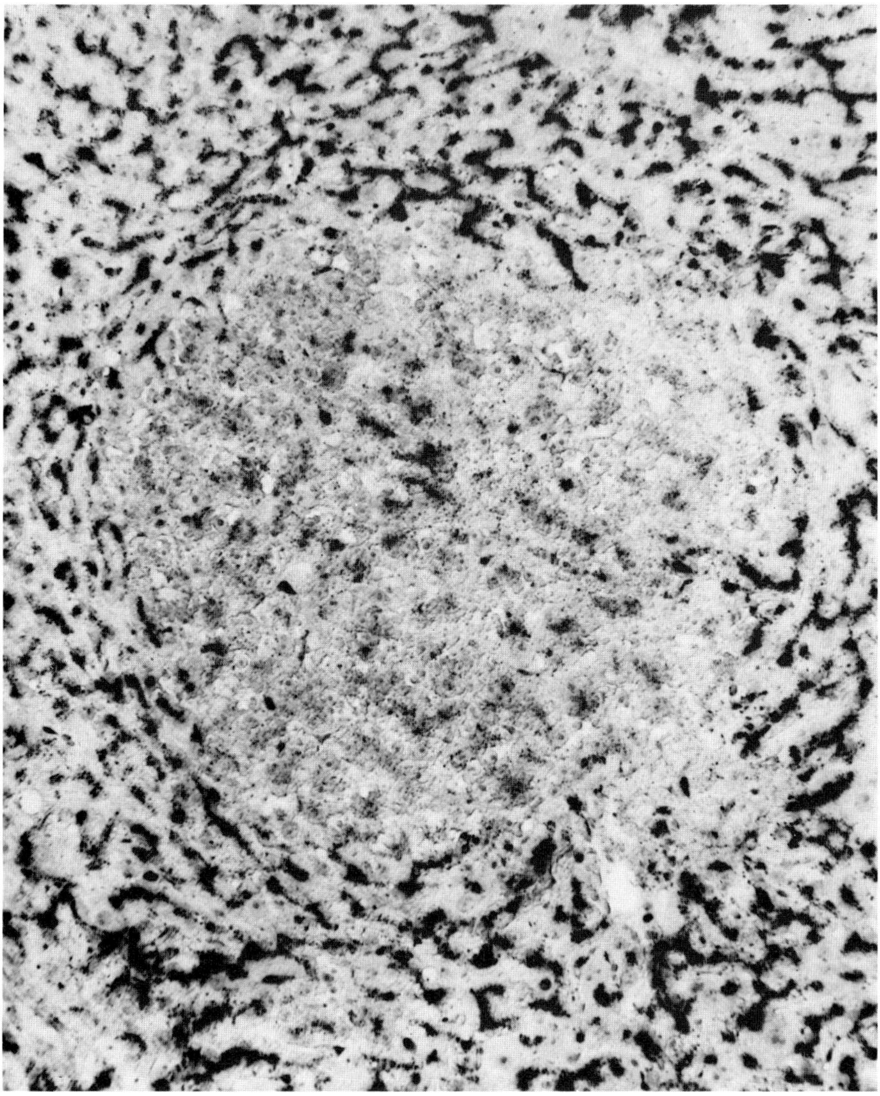

11.2 Focus of hyperplastic hepatocytes with decreased and diffuse acid phosphatase reaction in rat treated as above. (Acid phosphatase (Gomori), × 190.)

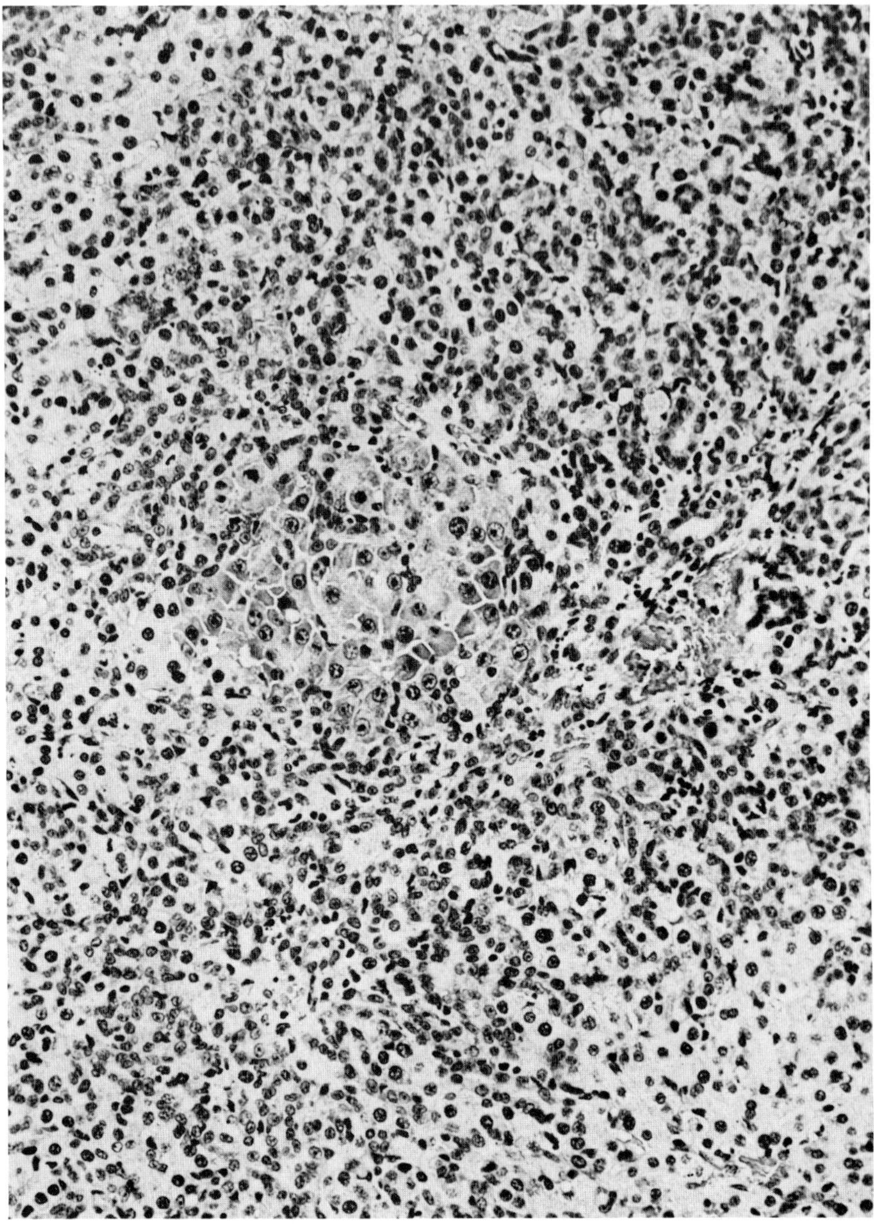

11.3 Liver of rat fed Diet 2 and given aflatoxin G_1, 50 $\mu g \times 9$ and killed 24 hours after the last dose. There is a focus of abnormal hepatocytes in the center of the field surrounded by bile duct cells and degenerating hepatocytes. (H and E, × 140.)

Table 11.6 Toxicity of repeated doses of aflatoxins in Fischer male lipotrope-deficient rats.

Aflatoxin	Diet	Intragastric dose (μg/day × no. of days)	Mortality (% during treatment)	Author
AFB_1	1	35 × 10	0	Rogers and Newberne, 1971a
	2	35 × 10	50	
AFG_1	1	50 × 9	0	Rogers, 1975
	2	50 × 9	80	

an adequate diet showed no abnormality of the liver (fig. 11.3). The toxicity of a single dose of the pyrrolizidine alkaloids lasiocarpine and monocrotaline was decreased by lipotrope deficiency (Rogers and Newberne, 1971b; Newberne et al., 1971). Protein deficiency did not alter toxicity of a third alkaloid, retrorsine, but it decreased toxicity of DMN (McLean, 1970). Lipotrope-deficient female rats were more susceptible to the acute toxicity of DMBA but less susceptible to its carcinogenic effects (Tanaka and Dao, 1965; Rogers, unpublished observations). Vitamin-A-deficient rats had increased susceptibility to the acute toxic effects of aflatoxin (Reddy et al., 1973).

11.5 Conclusions

Chemical induction of tumors in rat livers can be modified by changes in composition of the diet. With certain exceptions, diets composed of natural, unrefined ingredients tend to lower the incidence of tumors in rats compared with rats fed purified diets. Rats fed diets deficient in lipotropic factors are more susceptible to several hepatocarcinogens. These dietary effects may be mediated through changes in the mixed-function oxidases. Other mechanisms, however, such as alteration in tissue electrophiles and immunological processes, must be considered. In some instances the food antioxidants BHA and BHT protect against chemical carcinogenesis in liver and other organs. They may protect by means of their antioxidant properties or by altering the activity of mixed-function oxidases. Vitamin A protects against the induction of gastrointestinal neoplasia by aflatoxin B_1 and other carcinogens but does not consistently affect the induction of hepatic tumors by aflatoxin. Dietary effects on the toxicity of carcinogens may or may not correlate with their effects on carcinogenicity. Dietary manipulation offers a valuable tool for the exploration of the mechanisms of chemical carcinogens.

11.6 References

Best, C. H., Lucas, C. C., Ridout, J. H., and Patterson, J. M. 1950. Dose-response curves in the estimation of potency of lipotropic agents. *J. Biol. Chem.* 186:317-329.

Butler, W. H., and Neal, G. E. 1973. The effect of aflatoxin B_1 on the hepatic structure and RNA synthesis in rats fed a diet marginally deficient in choline. *Cancer Res.* 33:2878-2885.

Chu, E. W., and Malmgren, R. A. 1965. An inhibitory effect of vitamin A on the induction of tumors of forestomach and cervix in the Syrian hamster by carcinogenic polycylic hydrocarbons. *Cancer Res.* 25:884-895.

Commoner, B., Woolum, J. C., Senturia, B. H., and Ternberg, J. L. 1970. The effects of 2-acetylaminofluorene and nitrite on free radicals and carcinogenesis in rat liver. *Cancer Res.* 30:2091-2097.

Copeland, D. H., and Salmon, W. D. 1946. The occurrence of neoplasms in the liver, lungs and other tissues of rats as a result of prolonged choline deficiency. *Amer. J. Path.* 22:1059-1067.

Cumming, R. B., and Walton, M. F. 1973. Modification of the acute toxicity of mutagenic and carcinogenic chemicals in the mouse by prefeeding with antioxidants. *Fd. Cosmet. Toxicol.* 11:547-553.

Day, P. L., Payne, L. D., and Dinning, J. S. 1950. Procarcinogenic effect of vitamin B_{12} on *p*-dimethylaminoazobenzene-fed rats. *Proc. Soc. Exp. Biol. Med.* 74:854-855.

Engel, R. W., and Copeland, D. H. 1951. Protective action of stock diets against the cancer-inducing action of 2-acetylaminofluorene in rats. *Cancer Res.* 11:211-215.

Ershoff, B. H. 1964. Effects of diet on pituitary tumor induction by estrogens. *Exp. Med. Surg.* 22:28-32.

Ershoff, B. H., Bajwa, G. S., Field, J. B., and Bavetta, L. A. 1969. Comparative effects of purified diets and a natural food stock ration on the tumor incidence of mice exposed to multiple sublethal doses of total-body X-irradiation. *Cancer Res.* 29:780-788.

Farber, E., and Ichinose, H. 1958. The prevention of ethionine-induced carcinoma of the liver in rats by methionine. *Cancer Res.* 18: 1209-1213.

Finkelstein, J. D., and Kylc, W. E. 1968. Ethanol effects on methionine metabolism in rats liver. *Proc. Soc. Exp. Biol. Med.* 129:497-501.

Gebhardt, B., and Newberne, P. M. 1974. Nutrition and immunological responsiveness. T-cell function in the offspring of lipotrope- and protein-deficient rats. *Immunology* 26:489-495.

Gross, R. L., Reid, J. V. O., Newberne, P. M., Burgess, B., Marston, R., and Hift, W. 1975. Depressed cell-mediated immunity in megaloblastic anemia due to folic acid deficiency. *Amer. J. Clin. Nutr.* 28:225-232.

Halsted, C. H., Robles, E. A., and Mezey, E. 1973. Intestinal malabsorption in folate-deficient alcoholics. *Gastroenterology* 64:526-532.

Harris, P. N., Gibson, W. R., and Dillard, R. D. 1971. The oncogenicity of 6 analogs of 1,1-diphenyl-2-propynyl N-cyclohexyl-carbarnate. *Proc. Amer. Assoc. Cancer. Res.* 12:26.

Hartroft, W. S. 1969. Experimental hepatic injury. In *Diseases of the Liver,* ed. L. Schiff, 3rd edition. Philadelphia: J. B. Lippincott.

Haven, F. L. and Bloor, W. R. 1956. Lipids in cancer. *Advan. Cancer Res.* 4:258-275.

Herbert, V. 1973. The five possible causes of all nutrient deficiency: illustrated by deficiencies of vitamin B_{12} and folic acid. *Amer. J. Clin. Nutr.* 26:77-88.

Jacobi, H. P., and Baumann, C. A. 1942. Choline in tumor-bearing animals and choline-like effect of butter yellow. *Cancer Res.* 2:175-180.

Leevy, C. M., Tamburro, C., and Smith, F. 1970. Alcoholism, drug addiction, and nutrition. *Med. Clin. N. Amer.* 54:1567-1575.

Lowenfels, A. B. 1973. Etiological aspects of cancer of the gastrointestinal tract. *Surg. Gyn. Obstet.* 137:291-299.

McLean, A. E. M. 1970. The effect of protein deficiency and microsomal enzyme induction by DDT and phenobarbitone on the acute toxicity of chloroform and a pyrrolizidine aklaloid, retrorsine. *Brit. J. Exp. Path.* 51:317-321.

Madhaven, T. V., and Gopalan, C. 1968. The effect of dietary protein on carcinogenesis of aflatoxin. *Arch. Path.* 85:133-137.

Marshall, W. J., and McLean, A. E. M. 1969. The effect of nutrition and hormonal status on cytochrome P-450 and its induction. *Proc. Biochem. Soc.* 115:27-28.

Martinez, I. 1969. Factors associated with cancer of the esophagus, mouth and pharynx in Puerto Rico. *J. Nat. Cancer Inst.* 42:1069-1094.

Miller, E. C., and Miller, J. A. 1972, Approaches to the mechanisms and control of chemical carcinogenesis. Bertner Foundation Award Lecture in *Environment and Cancer, 24th Annual Symposium.* pp. 5-39. Baltimore, Md.: Williams and Wilkins.

Miller, E. C., Plescia, A. M., Miller, J. A., and Heidelberger, C. 1952. The metabolism of methylated aminoazo dyes. 1. The demethylation of 3-methyl-4-dimethyl-C^{14}-aminoazobenzene in vivo. *J. Biol. Chem.* 196:863-874.

Miller, J. A., and Miller, E. C. 1953. The carcinogenic aminoazo dyes. *Advan. Cancer Res.* 1:346-351.

Mulford, D. J., and Griffith, W. H. 1942. Choline metabolism. VIII. The relation of cystine and of methionine to the requirement of choline in young rats. *J. Nutr.* 23:91-100.

Newberne, P. M. 1965. Carcinogenicity of aflatoxin-contaminated peanut meals. in *Mycotoxins in Foodstuffs*, ed. G. N. Wogan. Cambridge, Mass.: MIT Press.

Newberne, P. M. 1975. Influence on pharmacological experiments of chemicals and other factors in diets of laboratory animals. *Fed. Proc.* 34:209-218.

Newberne, P. M., and Rogers, A. E. 1973. Rat colon carcinomas associated with aflatoxin and marginal vitamin A. *J. Nat. Cancer Inst.* 50:439-448.

Newberne, P. M., and Wogan, G. N. 1968. Sequential morphologic changes in aflatoxin B_1 carcinogenesis in the rat. *Cancer Res.* 28:720-731.

Newberne, P. M., Bresnahan, M. R., and Kula, N. 1969. Effects of two synthetic antioxidants, vitamin E and ascorbic acid on the choline deficient rat. *J. Nutr.* 97:219-231.

Newberne, P. M., Glaser, O., and Friedman, L. 1973a. Safety evaluation of fish protein concentrate over five generations of rats. *Toxicol. Appl. Pharmacol.* 24:133-141.

Newberne, P. M., Glaser, O., and Friedman, L. 1973b. Biologic adequacy of fish protein concentrate in five generations of mice. *Nutr. Reports Int.* 7:181-192.

Newberne, P. M., Harrington, D. H., and Wogan, G. N. 1966. Effects of cirrhosis and other liver insults on induction of liver tumors by aflatoxin in rats. *Lab. Invest.* 15:962-969.

Newberne, P. M., Rogers, A. E., and Wogan, G. N. 1968. Hepatorenal lesions in rats fed a low lipotrope diet and exposed to aflatoxin. *J. Nutr.* 94:331–343.

Newberne, P. M., Wilson, R. B., and Rogers, A. E. 1971. Effect of a low lipotrope diet on response of young male rats to the pyrrolizidine alkaloid, monocrotaline. *Toxicol. Appl. Pharmacol.* 18:387.

Patek, A. J., Jr., Kendall, F. E., de Fritsch, N. M., and Hirsch, R. L. 1966. Cirrhosis-enhancing effect of corn oil. *Arch. Path.* 82:596–601.

Poirier, L. A., and Whitehead, V. M. 1973. Folate deficiency and formiminoglutamic acid excretion during chronic diethylnitrosamine administration to rats. *Cancer Res.* 33:383–388.

Reddy, B. S., Weisburger, J. H., and Wynder, E. L. 1974. Effects of dietary fat level and dimethylhydrazine on fecal acid and neutral sterol excretion and colon carcinogenesis in rats. *J. Nat. Cancer Inst.* 52:507–511.

Reddy, G. S., Tilak, T. B. G., and Krishnamurthi, D. 1973. Susceptibility of vitamin A-deficient rats to aflatoxin. *Fd. Cosmet. Toxicol.* 11:467–470.

Reuber, M. D. 1965. Development of preneoplastic and neoplastic lesions of the liver in male rats given 0.025 percent *N*-2-fluorenyldiacetamide. *J. Nat. Cancer Inst.* 34:697–709.

Rivlin, R. S. 1970. Riboflavin metabolism. *New Eng. J. Med.* 283:463–472.

Rogers, A. E. 1975. Variable effects of a lipotrope-deficient, high-fat diet on chemical carcinogenesis in the rat. *Cancer Res.* 35:2469–2474.

Rogers, A. E., Herndon, B. J., and Newberne, P. M. 1973. Induction by dimethylhydrazine of intestinal carcinoma in normal rats and rats fed high or low levels of vitamin A. *Cancer Res.* 33:1003–1009.

Rogers, A. E., and Newberne, P. M. 1969. Aflatoxin B_1 carcinogenesis in lipotrope-deficient rats. *Cancer Res.* 29:1965–1972.

Rogers, A. E., and Newberne, P. M. 1971a. Diet and aflatoxin B_1 toxicity. *Toxicol. Appl. Pharmacol.* 20:112–121.

Rogers, A. E., and Newberne, P. M. 1971b. Lasiocarpine: factors influencing its toxicity and effects on liver cell division. *Toxicol. Appl. Pharmacol.* 18:356–365.

Rogers, A. E., and Newberne, P. M. 1973a. Dietary enhancement of intestinal carcinogenesis by dimethylhydrazine in rats. *Nature* 246:491–492.

Rogers, A. E., and Newberne, P. M. 1973b. Fatty liver and cirrhosis induced by lipotrope deficiency: an animal model of human disease. *Amer. J. Path.* 73:817–820.

Rogers, A. E., Sanchez, O., Feinsod, F. M., and Newberne, P. M. 1974. Dietary enhancement of nitrosamine carcinogenesis. *Cancer Res.* 34:96–99.

Ross, M. H., and Bras, G. 1965. Tumor incidence patterns and nutrition in the rat. *J. Nutr.* 87:245.

Ross, M. H., and Bras, G. 1973. Influence of protein under- and overnutrition on spontaneous tumor prevalence in the rat. *J. Nutr.* 103:944–963.

Rowe, N. H., Grammer, F. C., Watson, F. R., and Nickerson, N. H. 1970. A study of environmental influence upon salivary gland neoplasia in rats. *Cancer* 26:436–444.

Saffiotti, U., Montesano, R., Sellakumar, A. R., and Borg, S.,A. 1967. Experimental cancer of the lung. Inhibition by vitamin A of the induction of tracheobronchial squamous metaplasia and squamous cell tumors. *Cancer* 20:857–864.

Salmon, W. D., and Newberne, P. M. 1963. Occurrence of hepatomas in rats fed diets containing peanut meal as a major source of protein. *Cancer Res.* 23:571.

Shamberger, R. J., Tytko, S., and Willis, C. E. 1972. Antioxidants in cereals and in food preservatives and declining gastric cancer mortality. *Cleveland Clinic Quart.* 39:119–124.

Sidransky, H., and Verney, E. 1970. Influence of orotic acid on liver tumorigenesis in rats ingesting ethionine, N-2-fluorenylacetamide, and 3′-methyl-dimethylamino-azobenzene. *J. Nat. Cancer Inst.* 44:1201–1215.

Silverstone, H., Solomon, R. D., and Tannenbaum, A. 1952. Relative influences of natural and semipurified diets on tumor formation in mice. *Cancer Res.* 12: 750–755.

Simon, J. B., Scheig, R., and Klatskin, G. 1969. Hepatic ATP and triglyceride levels in choline-deficient rats with and without dietary orotic acid supplementation. *J. Nutr.* 98:188–192.

Smith, D. M., Rogers, A. E., Herndon, B. J., and Newberne, P. M. 1975. Vitamin A (retinyl acetate) and benzo(a)pyrene-induced respiratory tract carcinogenesis in hamsters fed a commercial diet. *Cancer Res.* 35:11–16.

Smith, D. M., Rogers, A. E., and Newberne, P. M. 1975. Vitamin A and benzo(a)pyrene carcinogenesis in the respiratory tract of hamsters fed a semisynthetic diet. *Cancer Res.* 35:1485–1488.

Steiner, P. E., Camain, R., and Netik, J. 1959. Observations on cirrhosis and liver cancer at Dakar, French West Africa. *Cancer Res.* 19:567–580.

Tanaka, Y., and Dao, T. L. 1965. Effect of hepatic injury on induction of adrenal necrosis and mammary cancer by 7,12-dimethyl-benz(a)anthracene in rats. *J. Nat. Cancer Inst.* 35:631–640.

Tannenbaum, A., and Silverstone, H. 1953. Nutrition in relation to cancer. *Advan. Cancer Res.* 1:451.

Thenen, S. W., and Stokstad, E. L. R. 1973. Effect of methionine on specific folate co-enzyme pools in vitamin B_{12} deficient and supplemented rats. *J. Nutr.* 103: 363–370.

Ulland, B. M., Weisburger, J. H., Yamamoto, R. S., and Weisburger, E. K. 1973. Antioxidants and carcinogenesis: butylated hydroxytoluene, but not diphenyl-*p*-phenyenediamine, inhibits cancer induction by N-2-fluorenylacetamide and by N-hydroxy-N-2-fluorenylacetamide in rats. *Fd. Cosmet. Toxicol.* 11:199–207.

Vitale, J. J., and Hegsted, D. M. 1967. Vitamin B_{12} deficiency in the rat: effect on serum folate and liver formiminotransferase activity. *Amer. J. Clin. Nutr.* 20: 311–316.

Wattenberg, L. W. 1971. Studies of polycyclic hydrocarbon hydroxylases of the intestine possibly related to cancer. Effect of diet on benzpyrene hydroxylase activity. *Cancer* 28:99–102.

Wattenberg, L. W. 1972a. Inhibition of carcinogenic and toxic effects of polycyclic hydrocarbons by phenolic antioxidants and ethoxyquin. *J. Nat. Cancer Inst.* 48: 1425–1450.

Wattenberg, L. W. 1972b. Dietary modification of intestinal and pulmonary aryl hydrocarbon hydroxylase activity. *Toxicol. Appl. Pharmacol.* 23:741–748.

Wattenberg, L. W. 1972c. Enzymatic reactions and carcinogenesis. in *Environment and Cancer, 24th Annual Symposium of the M.D. Anderson Hospital.* pp. 241–254. Baltimore, Md.: Williams and Wilkins.

Weisburger, E. K., and Weisburger, J. H. 1958. Chemistry, carcinogenicity and metabolism of 2-fluorenamine and related compounds. *Advan. Cancer Res.* 5:333–341.

Williams, E. A. J., Gross, R. L. and Newberne, P. M. 1975. Effects of folate deficiency on the cell-mediated immune response of rats. *Nutr. Reports Int.* 12:137–148.

Wilson, R. B., Kula, N. S., Newberne, P. M., and Conner, M. W. 1973. Vascular damage and lipid peroxidation in choline-deficient rats. *Exp. Mol. Path.* 18: 357–368.

Wynder, E. L., and Bross, I. J. 1961. Study of etiological factors in cancer of the esophagus. *Cancer* 14:389–413.

Zaki, F. G., Bandt, C., and Hoffbauer, F. W. 1963. Fatty cirrhosis in the rat. IV. The influence of different levels of dietary fat. *Arch. Path.* 75:654–660.

Zaki, F. G., Hoffbauer, F. W., and Gramde, F. 1966. Prevention of renal necrosis by coconut oil in choline-deficient rats. *Arch. Path.* 81:85–89.

11.7 Discussion

Butler We looked at acute toxicity of aflatoxin in choline deficiency (*Cancer Res.* 33:2878–2885, 1973). The animals survived, and we did not see hepatic necrosis. At the electron microscope level, soon after treatment there was the same degree of cytoplasmic disruption in the periportal zone in choline-deficient and normal animals. The choline-deficient animals rapidly recovered. Nucleolar segregation appears similar in the two groups, although on the whole the choline-deficient animals recovered more quickly. We only went on for forty-eight hours and the segregation was still present. RNA polymerase was initially suppressed regardless of diet, but in the choline-deficient animals this was restored to normal while in animals on control diet it was not. We have done no carcinogenicity trials with this dietary regime.

Rogers It would be interesting to follow by electron microscopy the acute effects of giving repeated doses, since there we get reversal of toxicity.

Butler That is a mammoth task. What we have done with Dr. Neil is to look at DNA synthesis by incorporation of thymidine. The results are a little puzzling, because if you give a single dose, which we know will produce carcinoma, there is an inhibition of thymidine incorporation. But if you feed aflatoxin B_1 to the animals in the diet at a level to give the same total dose you get an increase in DNA synthesis (*Cancer Res.*, in press).

J. Newberne Maybe I missed the implication of the quick recovery in the choline-deficient animal. Would you elaborate a little more on this? What is the recovery mechanism, or do you know?

Butler We do not know—I assume that it's just another manifestation of the fact that the compound is not as toxic. The thing that interested me was the mechanism of acute injury regardless of carcinogenesis. At two hours after aflatoxin treatment the livers were indistinguishable at the electron microscopic level, and yet the control diet animals go on to produce periportal necrosis; the choline-deficient animals do not. The electron microscope changes reverse.

Rogers But I think you found also that in the deficient animals the changes never progressed down into the pericentral cells; they were not making whatever the toxic compound was.

Butler Yes, that is right because normally in a control diet one sees a graded effect through the lobule. In the centrilobular cells which will not necrose, cytoplasmic change is seen which reverses, but we never saw that in the choline-deficient animals. Changes stopped about halfway down the lobule.

J. Newberne That tends to support Dr. Gillette's suggestion this morning of accumulation or perhaps of a toxic effect gradient.

Butler This, I think, is quite a common finding; but what you cannot predict, as he was saying earlier, is whether it is going to be centrilobular or periportal in distribution.

Weisburger I would like to make one comment. We did a study with BHT to find out its mechanism of action, and though this may not explain the entire protective effect, BHT does seem to enhance the metabolic detoxification and excretion to some extent, so it may have some enzyme-inducing activity like some of the other compounds Dr. Gillette mentioned.

Rogers Yes. BHA and BHT differ from α-tocopherol in that respect; that is why I think they are effective and α-tocopherol is not.

Butler As a matter of interest, do food companies put these compounds into diet as a matter of course?

Rogers They are in all commercial diets for laboratory animals. If you look carefully at the label you will see it someplace.

Butler We have run into problems using our stock diet. The LD_{50} of aflatoxin will be 1 mg per kilo and then three weeks later it will be 15 mg per kilo.

Rogers It may depend on diet; it may depend on which particular vegetables are in there and whether they are the ones that induce the enzymes or not. If you are making semisynthetic diets and using solid fats, as we are using beef fat, they also have BHT in them. The vegetable oils generally do not. It is on the label if they do. The Mazola oil that we use does not, nor the Wesson oil.

Gillette I know many times we have seen, in comparing the saline controls with our corn-oil controls, that we find induction by corn oil itself.

P. Newberne Well, indirectly let me speak to that point with another compound—monocrotaline. In looking at the dietary effects and examining the polysome profiles in the livers of marginally deficient and normally fed groups, we found that if rats were fed marginally deficient diet and given monocrotaline, the polysomes did not break down, but polysomes of adequately fed animals did. Looking further at what we believed to be a metabolite of this compound—a pyrrole—if you put that directly into the portal vein, then the polysomes did break down. Apparently the liver was not converting monocrotaline to the active metabolite under conditions of marginal deficiency. In any case, it looks as though the diet has some direct effect on the way the compounds are metabolized, and it may well be the same with aflatoxin. From what little we know about it now, the urinary metabolites are different when you feed deficient diet.

Rogers We can also say that phenobarbital will induce hepatic microsomal oxidases in deficient rats; aflatoxin did not.

Leffert A question about the composition of the oils. Corn oil is mainly unsaturated fatty acids, and coconut oil is saturated. Am I correct?

Rogers We fed cottonseed oil, which is unsaturated. Coconut oil is a medium-chain tryglyceride rather than long-chain and is more highly saturated. Are you thinking of the old effect on DAB carcinogenesis?

Leffert In the current growth control studies, certain receptor changes and membrane fluidity changes are dependent on the composition of the fatty acids of the membrane; unsaturated fatty acids would tend to make the membranes more fluid, as would decreased cholesterol-phospholipid ratios. In line with some of the hormone work, some of these early changes could be related to alterations of the fatty acid membrane composition, and that could in fact change polysome profiles.

We have just started, with Dan Steinberg's group, some work in which we have been trying to look at effect of cholesterol on DNA synthesis. We cannot suppress the HMG reductase activity with cholesterol, but in our first experiments, we could with a metabolite: cholesterone. So I would certainly be interested to know what happens to the cholesterol metabolism of the liver under these deficient-diet conditions and subsequent refeeding, in line with the idea that you are altering the composition of the membrane. That would be the rationale.

P. Newberne I think one of the problems is that in this dietary treatment the animal does not synthesize lipoproteins at a normal rate, which would relate to membrane renewal.

Gillette I would like to raise a point there. The enzyme that forms the 7-α-hydroxylated cholesterol is distinctly a different enzyme based on a variety of measurements. The fact that cholestyramine can enhance the activity of the 7-α-hydroxylase without altering the activity of any of the enzyme systems and, if we take into account our measurement as one indication of this, the fact that the alterations have a difference in turnover time—that is, in the turnover level of the enzyme itself—clearly indicate a difference.

Bannasch Have you tested the reversibility of the nodules which developed after choline deficiency without contamination with aflatoxins?

Rogers Nodules do not develop just with marginal choline deficiency alone, but with severe deficiency they are a function, I think, of the regeneration going on and the active growth of connective and vascular tissue. Cirrhosis is clearly an abnormality, which we do not understand, of the connective tissue elements as well as the hepatocytes. Once one gets a cirrhosis with choline deficiency it is not reversed if one puts animals back on an adequate diet, though the increased turnover of liver cells will revert to normal. The liver cells lose their fat, but the nodules persist.

Tate Do you have any laboratories that are working with your semisynthetic diets?

P. Newberne We have more people at MIT not looking so much at deficiencies but using a semisynthetic diet. There are still some problems with them, particularly related to breeding. For practical purposes there is really no difference, and why some are having trouble with breeding I do not understand, because we do not. We have a very good breeding and longevity record. Diet is one additional factor we can control, and we must consider it in crucial experiments. I am not sure you want to switch totally to them, because of considerations of expense and time.

12 Summary

Editors' Note: These comments were taken from the edited tapes of the meeting.

Grice In the final analysis, our animal studies lead us to the point at which we have to consider the tools that are available for extrapolation and application to man. Certainly two of the major tools are the animal studies and the epidemiological information that is available in man. I think that one of the more important aspects we have been talking about here, and what we are coming to now, is a need for standardization of our protocols. When we speak of standardization we have to consider the type of cancer study that we are doing and what the objectives of the study are. Perhaps we can categorize cancer studies into three or four particular groups. In one fairly straightforward type of experiment, we are trying to find out whether or not a compound is carcinogenic, and I think many studies supported by the National Cancer Institute are like that. I think we could get some standardization in this type of study.

If we are looking at drugs or food additives, there are other factors that we have to consider:

1. Is the compound likely to get into the fetus?
2. Is it likely to get in through the milk?
3. Is it going to be used in the young individual or is it going to be used in the geriatric individual?

In these instances we are going to set up our cancer study perhaps to answer some of these specific questions, and I do not think that we are going to be able to use a standard protocol.

Going beyond that to the type of work that Dr. Rogers and Dr. Newberne are doing, looking more at pathogenesis or mechanisms of cancer than at carcinogenic agents, standardization of protocol is going to be pretty difficult.

We have talked of the strain of animals that we use; there could be some standardization there. There perhaps should be some standardization in diets and certainly some standardization of histological diagnosis of tumors—that is what part of this workshop is about. There perhaps could be standardization of our toxicological procedures. There could be standardization of the number of animals on test and the number of animals that are available for study at the end of the test.

As far as the epidemiological aspects are concerned, there is need, I think, for more interagency and international collaboration on defining spontaneous tumor types in different species of animals and establishing additional registries for assessing them.

J. Newberne Would you comment on the issue of the "positive control" as a simultaneous exercise with the study?

Grice I have never thought that positive controls serve a useful purpose for this reason: if we are dealing with an unknown, we never know what positive control to use. If we do not know where the compound is going to cause cancer, I can see little value in using a positive control. I suppose if you knew the mechanism of action of the carcinogen you were working with and had positive control with the same mechanism, then it might serve some purpose.

J. Newberne Is there one possible utility: perhaps in a very useful compound that might be used in some special lifesaving situation, a positive control could be used to rank the relative potency of the unknown compound.

Grice The relative potency of the chemical, yes. I think that is true. If more of our studies were directed toward categorizing compounds according to their relative potency, I think we could list several categories as being highly potent descending to those compounds that are of low carcinogenic potency. Then in our studies, particularly of unknowns, we would relate these to the different categories. That is about all we can do at this point with current knowledge. The type of thing Dr. Gillette is talking about, where we go down to such very low concentrations, may not be of any real concern because there are so many insults that can come in to alter the results.

Goyings Dr. Newberne, you referred to the differences between the mouse and the rat with respect to metabolism and tumor induction. Is there evidence for differences in metabolism between the mouse and rat which might account for the differences in susceptibility to chemical carcinogens, and could such differences prove useful in attempts to extrapolate rodent results to man? Is there any literature evidence, or have you done the work yourself or has anybody else here?

P. Newberne Most of the studies have attempted to associate a difference in metabolism or metabolites with cancer induction or toxic responses. Most of the reports are rather brief and do not give you much information. They tell you their findings and sometimes are able to indicate a rational basis for metabolites to be considered as carcinogenic or innocuous, but this area, as Jim Gillette has so elegantly described, holds much promise.

J. Newberne The chemicals described by the National Institute of Occupational Safety and Health as potential carcinogens for man and therefore representing public health hazards were really selected rather empirically, as far as I know. The commission went through a long list of compounds and decided on the basis of total use and manufacturing practices the likelihood of exposure. They reviewed compounds that had already been banned on epidemiologic grounds, like benzidine and some others. After all the work we have done this week, we are ending up with the dull job of trying to apply some of the data to real-life problems. I think that just for the purposes of discussion and guiding my thoughts a little, I might list a few things that I think are worthy of consideration, most of which we have discussed already in one way or another.

It is rather a popular sport in some of the committees that I have worked with in the government and industry to identify the structure-activity relationship and say, "Ah-hah, you know, this is a carcinogen." Soon, however, when more data are in, aside from some very well-defined potent carcinogens, this exercise proves to be not too helpful in the kinds of drugs and other compounds that we work with. Now, obviously the chemical groups are important, and we should consider them; but if two compounds are chemically related, I guess essentially all compounds in the world could be considered chemically related in one way or another in a very superficial relationship: arrangements of atoms, distribution of electrons, chemical reactivity, and a host of other factors.

Further, we have whatever information we can gather on biological activity, and here I have seen a serious pitfall in situations where the pharmacological activity has been used to indict compounds. The best example of this is the case of the β-adrenergic blocking agents, which to me was very sad; in the United States, all research was brought to a halt on β-blocking agents for at least two and a half years because of a few nodules which appeared in mouse livers and which could not be duplicated in other laboratories. Finally, we are back now working again in the field. I do not know how many people really benefited from this. I am reasonably sure that some have suffered. But the study of biological activity can be helpful, and we know that all biological activities, carcinogenic or otherwise, are a reflection of some event at the molecular level. So we put together as much of this as we can and try to use it. We can get help from our profile of safety tests, which is standard to most of you and involves the acute toxicity tests (the various short-term tests at different dose levels, establishing dose ranges and looking for target organ insults). Metabolic disposition is indeed important. Dr. Gillette did an excellent job elucidating that this morning. We did not talk a great deal about absorption, total body distribution, bile transformation, and enzyme induction. We did discuss protein binding, displacement from binding sites, variable excretion patterns and so on.

I was delighted to hear Dr. Gillette's comment with respect to selecting the model animal species, because I have felt for a long time that there is no perfect model species. In fact, the most variable animal in this regard is probably man himself. We have also discussed a host of environmental factors including diet, housing, handling of animals, when you start your experiments, when you stop them, everything that relates to the general housekeeping, and what the influences of these might be. I will not repeat these, but they are worthy considerations.

The discussion about the mouse, earlier, is quite pertinent to the consideration of species, because if the bad time we gave the mouse during the workshop in England last spring did result in its being dropped from favor as a test animal in some quarters, then we really are faced with the problem of trying to come up with a different species. In the meantime, I think we might consider the intraspecies variation as a possible alternate. In some of my tests I have found enormous differences between two rat strains given the same compound that is carcinogenic in two different organ systems. Some of the data Dr. Rogers pointed out this morning helps support this contention. The rat came out in this workshop rated a little better than the mouse in regard to understanding and using it as a test animal. I still think we have to ask the fundamental question, though, with respect to the utility of the rat. In using all the techniques at our disposal and with all of those we have discussed here, have we developed the objective criteria for distinguishing benign and malignant liver nodules in the rat? I get the impression that there may be a little difference of opinion here yet, a little nagging worry that maybe those basophilic foci may be precancerous and so on. But I think we are in better shape with the rat than with the mouse.

Regardless of our systems and recognizing how imperfect they are, somewhere, sometime, somebody has to use the data to make a judgment and this becomes a pretty crushing burden on a number of people in and out of government. I think we have to start thinking about better ways to establish risk-to-benefit ratios or risk factors, if you will, and devote some effort in that area. Even in the face of a carcinogenic response in an animal, the judgment has to be made: Is the risk factor appropriate with respect to its intended use? If we do this we have to accept the concept of dose relationship, and I would like to hear some more discussion on that point. As in the field of teratology, this has been largely overlooked and it is an important point. I think that the all-or-none attitude that we have been faced with violates all the pharmacological criteria we have been using all these years which have worked out pretty well so far in toxicology.

Harris I have been concerned from time to time with knowing just what to do when we come up with a question in which it is important to try to decide whether our data are really applicable to man, and as I say, we really do not have the answer. If you have a compound that is carcinogenic for a rat, for example, in a low incidence, you have to consider the nature of the compound, how it is to be used, and over what period of time before deciding that you should not even submit it to the Food and Drug Administration for consideration. If it is a potentially lifesaving compound and produces tumors in a low incidence, then of course it is worth following up. If it is not very useful, then of course you abandon it: and if you should have something that is highly carcinogenic then you would use it only in cases of malignancy in which there is no other satisfactory treatment. But how you are going to apply your animal data to man? That is one of the great unsolved problems.

P. Newberne Thus far, we have brought up for discussion the biological behavior of tumors. The histological evidence we have seen on the screen and generally agreed with, however, is not always enough. It is what the tumor does in the animal or person that really counts. I do not know that we are ready to address that at this workshop. We do not have the time, but it is lurking in the background of everything that we have said here.

Butler Recently, I think, there are two instances where there was evidence of compounds being carcinogenic in the rat which since then have been shown to be carcinogenic in man or produced neoplasm in man. One is the reported incidence of benign liver adenomas in young women who have received contraceptive agents for a number of years. It was known in about 1960 that these compounds were carcinogenic for the rat, and in fact the British Committee on the Safety of Medicine agreed that there were carcinogenic for rats and mice; despite this they claimed that this did not represent a risk for women and the use has continued. The other one is vinyl chloride. Maltoni, I think demonstrated angiosarcomas in the rat before the cases were picked up in man, as far as I know. And now, the permitted level of 50 ppm of vinyl chloride is being reduced to 1 ppm for an eight-hour period. It was known as a carcinogen in rats before it was found to be carcinogenic in man. So these are two possible examples of where time is evidence that what was found in the rat is applicable to man.

Unfortunately, these are somewhat rare, and in most cases it's the other way around; carcinogens have been discovered in man first. One other comment about aflatoxin, which is that what Dr. Newberne says is absolutely true. The World Health Organization basically went on the evidence from MIT and the MRC Toxicology Unit for the recommended levels in supplemented food as 30 ppb. This is a known carcinogenic level for the rat, and it is necessary to balance the risk of malnutrition against the possibility of cancer.

Goyings You showed a slide of some of these mouse liver lesions which you had sent around to several people for classification and you received many different responses. I gather from what we have seen here today that you generally have a sufficient number of malignant changes in a rat liver to make them more convincing than changes in mouse liver; there does not appear to be as much of a problem in the rat liver as there currently is in the mouse. Is this correct?

P. Newberne I think that is generally true. Certainly it is with me and most of the people I deal with. The rat liver is easier to assess, not necessarily to assess on purely histological grounds but to interpret what it means.

Goyings To carry this a little further, Tomatis published an article about the good correlation between the mouse liver hepatomas, as he called them, and ability to produce neoplasia in other species, I think the rat and hamster. If you go down that list, it is striking that some of the compounds that were strong enzyme-inducers produced lesions in the mouse liver but were negative in the rat and often in the hamster.

P. Newberne The literature review was exhaustive, but more than half of those chemicals were already known to be tumorigenic in the mouse, so predictive test was not actually developed, but a comparison of what he could find in the literature was made with something that was already known. Another thing about the report that disturbs me and is an unknown quantity is that you do not have any idea of what the term "tumor" actually means in the context of that publication. Nodules, tumors, and cancers may not be that well defined in many of the papers from which the report was drawn, and there is no way of sorting it out. I think it is a valuable review of the literature in terms of how the livers of these three species respond, but I do not see that it helps us a great deal in answering the questions we are trying to get at here.

To get back to your original question, did we look at it from the point of view of whether or not many of the chemicals induce enzymes or centrilobular changes in the liver? A lot of them do, and my impression is that the centrilobular hypertrophy that we see with many of the insecticides can be associated with nodule formation. This is particularly true with the mouse, less so with the rat. I believe that is pretty clear if you look at a broad spectrum of chemicals which induce enzymes. You have cell death, proliferation, and nodule formation. The difficulty arises in attempting to evaluate whether or not the nodule is a true cancer or simple hyperplasia, and in the mouse liver it is very difficult. It is difficult enough in the rat liver with these compounds, because they do not produce the same histological change, in my opinion, as many of our well-known chemical carcinogens.

Squires Will you elaborate on your statement that the significance of the mouse lesion is unclear? From what you just said it sounds like the status is the same as it may be in the rat.

P. Newberne I am referring to the lesion caused by enzyme-inducing chemicals. The lesion in the mouse liver is not convincing; I am not convinced with many of them that we are dealing with a real carcinogen.

We look at mouse liver with well-known carcinogens, and they generally produce a pretty clear-cut liver cell carcinoma that I do not think most people would have difficulty agreeing to. They are quite different from insecticide-induced liver lesions; they are fast growing, usually, and trabecular or anaplastic types—histological changes which I find relatively easy to diagnose whether in the rat liver or in the mouse liver. The insecticides have not done that. There are very mild changes. They appear to continue to harass and injure the liver slowly but surely, and there is no question that there is proliferating liver. But I am yet to be convinced that most are real liver cell carcinoma. If you look at the whole spectrum of mouse strains, you find what people refer to as nodules, tumors, or cancer, anywhere from 3 to 5 percent on up to in some of the substrains, where there is as much as 100 percent. If there is an average of 10 percent of animals with spontaneous liver tumors, and you are imposing an agent on this and getting 80 percent tumors, I do not know whether you are dealing with a carcinogen and developing 70 percent tumors on top of 10 percent that were there already or whether you are just enhancing whatever effect is there already from some unknown agent of mechanism.

Squires Aside from clear-cut carcinoma, why are the equivocal lesions in the rat less confusing than those in the mouse?

P. Newberne The rat generally does not have equivocal lesions, and most strains develop nodules only if you induce them; and usually when you do induce them you can determine, on histological grounds only in most cases, if it is a hyperplastic nodule or if it is a cancer. It is not that easy in the mouse; he has all kinds of histologic patterns, many of which I cannot truthfully evaluate.

Squires So it is based on the spontaneous background of hepatoma in the mouse, in contrast to the rat, at least in part, and also on their behavior?

P. Newberne Yes. The biological behavior in the mouse is an important aspect, and we do not always know that.

Goyings I have problems separating the lumps in rat liver and trying to decide whether or not they are actual tumors. Would you mind putting your criteria for separating those hyperplastic lumps, or whatever you want to call them, from actual carcinomas?

P. Newberne I have criteria, but I really do not think I want to attempt to lay them out for you here. I have no difficulty with the rat, as far as most cases of tumors are concerned, but with the mouse I do encounter difficulty. Rather than trying lay down strict criteria for histolocical diagnosis of liver nodules, we have tried to describe what is known about the biology of the rat liver, including morphology, and its behavior when exposed to chemicals.

Goyings To indicate how difficult it is to diagnose nodules, we had a lesion produced in a mouse which we had seven or eight pathologists look at. Some of these people have had quite a bit of experience in both the mouse and the rat, and they would readily admit that in the mouse they had no good criteria to separate lesions, unless it was obviously invasive carcinoma. They really were at a loss as to how it should be classified.

P. Newberne I do not think you have to have evidence always for invasion or metastasis to classify something in a mouse liver as a carcinogen. If I have an unknown compound, however, and I am looking at it in mice and rats, I want to see metastasis in some animals in that group. If you are going to continue to work with that same compound you do not have to find lung metastasis or invasion in every case or even in the next group that you look at if you are studying the same compound, but if it is an unknown compound, I want to see good, clear-cut evidence.

J. Newberne I want to mention again that we are looking at two different problems here, and certainly each one deserves its own kind of attention. One is the routine screening of compounds for carcinogenic activity and the other is experimental carcinogenesis, which is a totally different issue, and in that case I think the mouse has some utility. The liver is exquisitely sensitive. It responds to all kinds of things. I have no faith in the mouse whatever.

Squires Is it not a problem partially because of our lack of information about the biological behavior in the mouse? Considering the cost of substituting any other species in screening, it might be more fruitful to investigate the biological nature of the mouse lesion in much the same way that we have heard about the rat lesion, using every methodology available and determining whether it is a suitable model. It may be if you knew what those lesions were.

J. Newberne I agree on that point, but I disagree on the cost. If you make one bad mistake on an important compound, you could never count the cost. I would much rather spend more as overhead and feel safer about it than to gamble on the mouse.

P. Newberne We have worked with mouse tumors that looked very benign yet transplant very easily and metastasize. If you looked at it as a histological slide you would say, "Oh well, it is just a nodule and will probably never amount to much," and yet when you look carefully you can find metastatic lesions. On the other hand, in the mouse you can also find lesions which look highly anaplastic, but you cannot find metastases. Thus, we are faced with difficult decisions if an important chemical is in question.

Squires Do they die earlier?

Butler No. It is interesting in some of the DDT studies in the mouse where they had a very high incidence of liver nodules, they lived longer, except in the highest-dose group.

P. Newberne It may well be that there will be something like covalent binding or something new that will come up that will be a good predictor of a potential carcinogen. I agree that we ought to get on with trying to learn what the mouse liver lesion means. That is the crux of the whole matter because you are not going to get rid of the mouse; he will outlive us all and perhaps he should. People are going to use the mouse for one reason or another—for testing, for basic carcinogenesis—and we must expand our knowledge about spontaneous lesions.

Laqueur I must say that this has been a most enjoyable meeting for me because, for one thing, I have learned a lot. And what I have learned is not so much in the area of microscopy, my field of experience, but in other areas, such as biochemistry, pharmacology, immunology, and so one. I want to make a suggestion: Would it be possible for some of our speakers who have given us such very interesting papers, in one or two sentences, to state their ideas about what, at the present state of our knowledge, they think is the most important consideration for future studies? I have nothing to add on how to interpret experimental data in terms of human disease. I always have been struck that the three compounds which occur in nature, and to which man is exposed, have not been shown in one instance to produce cancer in man. I am talking about aflatoxin, bracken fern, and cycasin. This may be happening without being detected, but we really have not heard about it. Working with a natural carcinogen, which people eat, and not being able to document its effect on man, it becomes much more difficult for me to talk about a chemical which is not widely used and discover it has carcinogenic activity. My own feeling is that much more work needs to be done along the lines which have been suggested. This is basic research and is, I think, much, much more important to emphasize than the attempt to extrapolate what happens in a rat to man. But I am very much encouraged about all the information we have had and I think we are very much better off talking about the rat than we were talking about the mouse and lumps.

Baldwin I have not really done much work with the mouse system. For me the situation is really rather clear at a research level. I think I am willing to say that I know what would be the immunological characteristics of chemically induced tumors in the rat, and I know that because I know the biology of the tumor; I know I can induce the tumor and I can transplant it, and I can then characterize antigens on transformed cells. And I think I know—this is not quite certain yet—that I can transform rat cells in vitro with chemicals and show similar antigens on cell surface. Therefore, as a first approximation in that system, I would say that I had a change which one could loosely call neoplastic change, and one then would have to be careful about the use of that compound. Comparing that to the mouse, the problem seems to be that there has not been anywhere near enough work done on the biology of the mouse.

Dr. Newberne mentioned some of these tumors that will transplant very easily, but we have really moved past that stage. If I had that system, I would like to know why it transplanted easily; I would want to look at the immunology of those transplantable mouse tumors. The story I get from colleagues is exactly the opposite, that they have mouse tumors that they cannot transplant, tumors that they have induced with carcinogens. They want to study the immunology of these tumors, but they cannot transplant them, so there seems to be a glaring defect in the immunology of the mouse. I would say that from the immunological angle, we now know what sort of biological research needs to be done. Then if you ask me, what would you suggest that one do in terms of screening programs, I would say that we now know enough about the new antigens on transformed cells to be able to set up in vitro culture systems using pure cell lines. I think I would feel confident to be able to type antigens on those cells and get a first approximation as to whether they were something produced by the agent that would predict for me that you were going to get a malignant change.

P. Newberne I did not mean to imply that all mouse tumors transplant easily. We have had some that would not transplant at all.

Bannasch I am afraid I will not have very much to say expect for a very subjective statement from my point of view that we have to look a bit deeper in this interesting phenomenon of hepatocellular glycogenesis and its possible relation to the development of liver cancer. I think that it would be very helpful to ask if we could come together one day with our slides so that we could follow up the experiments with different substances which we cannot all repeat.

Leffert When I began this work, hormones were things that I had learned about in medical school, and I began to work with Dr. Gordon Sato. He had been interested in trying to develop cloned cell lines that would require single hormones for growth. He began with transplantable tumors from Dr. Furth's animals; these tumors remain differentiated. I carried on these studies with Bob Holley at the Salk Institute, and for the first year or two the feelings were that hormones would not be involved, that new factors would be involved, some unknown substance to be purified from serum. What it has all come to is that everybody's factors, including hormones, seem to be important, at least in in vitro studies. We have found in primary cultures of fetal liver cells that some of the hormones that are now known to stimulate growth in rats are also involved in vitro. What is of interest is that these are concerned in evolution as well. For example, insulin and growth hormone and cortisone are involved in the regeneration of the tail in amphibians, and there are at least four hormones that are active in the in vitro system with fetal rat hepatocytes.

So where does this leave carcinogenesis? For someone who studied growth control in the cell cultures, a major philosophical question is: Why does the nucleus replicate its DNA, and what are the internal signals inside the cell, that is, that stimulate DNA replication? You could assume that whatever the agents are that stimulate growth, whether they are in well-coordinated cells that require many hormones or in cells altered by carcinogens and have lost their requirements for control, it seems to me that the final common signals are the same. So that in a regenerating liver, which is normal, and in neonatal and fetal livers, which are fast-growing, the replication signals, it seems to me, will be proved to be the same. Now if that is the case, then it seems to be that what carcinogens are doing is gradually altering the signal responsivity of cells to factors which are normally regulating growth. Cells may be waiting for increased internal levels of nutrients that limit their growth. In simple words, the proper food supply regulates the population density or the population adjusts to the food supply.

There are many mechanisms by which that could happen. It could be because of transport or disposition of these nutrients by intracellular pathways. But it is a simple idea that could resolve a lot of the complexity in carcinogenesis, that is to say, if a hepatic cell merely needs to have internal levels of all of the amino acids or sugars or who knows what, then with time, as mutations take place in systems that permit those levels of all of the amino acids or sugars or who knows what, then with time, as mutations take place in systems that permit those levels to increase inside the cell, one would ultimately get changes that would lead to DNA synthesis. You could say that the cell has to adjust to the protein-synthesizing machinery available in the sense that if transcriptional processes become overburdened by the demands, then the cell either has to make more messengers to take care of that, for example, if the cell is flooded with amino acids, or the cells has to make a new set of genes in order to make a new set of messengers. Now I do not know if anybody has put the problem into those terms, but there has to be a physical reason for replicating DNA and I think ultimately we will find that carcinogenesis in liver is related to the regeneration of liver after partial hepatectomy.

P. Newberne In a sense what you have said has been borne out in part by studies that showed if you reduce the caloric level, you reduce the incidence of tumors, which might be part of the mechanism. You reduce the levels of substrates in the cell.

Jones When we started in this work in carcinogenesis I read the book by Foulds called *The Neoplastic Development I.* He put forth a very simple scheme and I found it useful to keep it in mind when we have been doing some of our experiments. It is not a very high-powered science. He applied it to all tissues, but one can apply it to the liver quite well. Whenever you give a chemical and it affects an organ, it does a variety of things. Most of them will produce some sort of toxic effect and you can also produce a whole range of biochemical effects, and we can identify and categorize these and study them. When you come to the liver, they frequently produce nodular lesions, and they, of course, may also produce carcinoma.

We can identify all these processes; we can look at them morphologically, immunologically, and in all ways, but the thing we then have to do is to try and decide whether A goes to B or is at all related to our final carcinoma, whether this progresses through the biochemical state and whether these early lesions have any relationship to the nodules and whether these are, in fact, at all related to the final lesion or whether we are looking at a whole series of changes that just occur in parallel. I think this is the biggest difficulty we have, in trying to connect all these things we have heard about this week. It is always worth bearing in mind that all these changes we see in the liver might not have any bearing on the production of the tumors.

Rogers I think we have brought together metabolic, tumor-inducing and growth control effects, and so on, and found that we do know quite a bit. We are at a stage now at which we can look much more carefully at the progression of these lesions, at the reversible and irreversible lesions, and decide which ones are important and what they mean. Then, at the same time, we should pursue all the metabolic differences that have been brought out, keeping in mind particularly the environmental and dietary effects on all these changes.

Knook There is great complexity of all the lesions we have seen, all produced in the same animal—the rat—but different strains, different diets, different environmental conditions. There are very great differences in the various strains of rats and this may influence a lot of the results shown.

The only suggestion I can easily make is to publish explicit data on the conditions of experiments: the strain of the rats, the age of the rats, the life span of the strain. We found that there are many age-dependent lesions.

Gillette There are certain things that have occurred to me. Over the years I guess I have really become less and less of a generalist and some of the things that have been stated here should be clarified. For example, a chemical structure, what do we mean by it, whether or not there is a relationship between the chemical structure and the way in which a compound is metabolized. In the last year we came up with a rather startling view point in one particular area, in the metabolism of nitrosamine. Although we have shown we can form nitrosamines with certain, rather complicated or not too complicated antihistamines, for example, we find that actually during the formation of the nitrosamine we can get increased covalent bonding. The bonding is by no means what one would ordinarily expect. When we consider that under ordinary circumstances one can make nitrosamines from secondary amines and then get bonding, one might suggest that a demethylation reaction is followed by rearrangement to form an active metabolite. But it is not that way in this particular compound. It actually is formed through an epoxide, and there is no evidence at all that there is any demethylation.

Another possibility that I think should be put in perspective is absorption of a compound. Is the rate of absorption an important aspect in toxicity when you are talking about the formation of reactive metabolites? Certainly, on the basis of just simply the pharmacokinetics of the reaction, there is no reason at all to believe that the rate of absorption has anything to do with the formation of a tissue lesion. In fact, the term for absorption never occurs in the equation. There are certain instances where this is going to be important, but I am not sure it is going to be too important, especially in the liver. Certainly, in painting a compound on the skin, we are talking about relative rates of formation of reactive metabolites and rates of diffusion through the skin and reaching the blood vessels. These relative rates are going to be very important in determining how much of that compound is going to be covalently bonded to the skin. Since the rate of absorption through the skin is relatively slow, the rate of metabolism can be relatively slow under those conditions.

When one talks about going through the rather thin membranes of the lung, you may have an entirely different story. The possibility that a compound must be inhaled in order to show lung damage is not necessarily true. You may find just as much lung damage after administration intraperitoneally or intravenously as you do by inhalation experiments.

The third thing that I would like to bring up is transplacental carcinogenesis. Most people say that the rates of absorption are important in transplacental toxicity, and yet they give a single dose and they will measure it at various periods afterward. But people do not give drugs by a single dose, usually. They do it by multiple doses, and they keep the plasma levels within relatively narrow ranges. The concentration of the drug in fetus should reach just as high as the concentration in the mother. So this concept that there is some barrier mechanism to the fetus may apply when you use one dose but not when you use multiple doses. There is no mechanism that I know of by which most foreign compounds can be removed from the fetus actively. You have the small amount of enzyme systems within the liver; even though they be very low, they can be quite important as far as carcinogenesis is concerned under those particular conditions. On the other hand, you could also argue when the mother's blood first contains the compound you are interested in just before birth, then much of the carcinogenic activity is not related to the prebirth metabolism but to the amount of metabolism that occurs after birth, during the development of these enzyme systems that I described this morning. So I think we have to be a little careful in interpreting the results and not jump to the conclusion. When one does not know very much about the kinetics, then the pragmatic approach of applying a compound by the route by which it is going to be used in man is probably a pretty good one. But in research, I think, one has to consider all of these possibilities.

Butler At the end of the mouse meeting, those of you who were there will remember that my comment was that what impressed me is how ignorant we were of the mouse. But in this one I think we are a shade better off. What we demonstrated to you in the light microscopy of the carcinomas represents to us most of the range of lesions which we consider to be carcinoma. Electron microscopy does not tell you a cell is a malignant cell, but this does not say that you should discard it, because I think it is most important that you understand as much as you can and be able to recognize and characterize as far as you can, all lesions. I think it is important in doing this or indeed any form of morphology to make a clear distinction between testing situations and studying mechanisms. If you are concerned with testing you have to devise a series of criteria which satisfies you and falls reasonably well within the context of normal pathology. But when one is studying mechanism you are much more able to say you do not know.

P. Newberne I think we are a little closer to a better understanding of rat liver than we were. I am surely much more pleased with this workshop than I was with the one on the mouse and not through any fault of the participants. But I think we all have to keep in mind that if you have got cancer here, that is the end of the journey. I hope we are not going to continue to pile another million nodules on top of the billion we have already and try to make sense out of them. I hope we can do as suggested and get at some of the mechanisms and meanings of what is in those nodules. Just beyond our current wisdom's reach lie tomorrow's answers; let us reach out for them.

Index